Data Analysis by Resampling

Concepts and Applications

www.duxbury.com

Valuable resources @ no additional charge

Data Analysis by Resampling

Concepts and Applications

Clifford E. Lunneborg

University of Washington

Australia • Canada • Mexico • Singapore • Spain • United Kingdom • United States

Sponsoring Editor: Carolyn Crockett
Marketing Team: Tom Ziolkowski, Beth Kroenke,
    Laura Hubrich
Editorial Assistant: Ann Day
Production Coordinator: Kelsey McGee

Production Service: Robin Gold/Forbes Mill Press
Permissions Editor: Mary Kay Hancharick
Cover Design: Christine Garrigan
Cover Printing, Printing, and Binding: R. R. Donnelley
    and Sons/Crawfordsville

Printed in United States of America

10 9 8 7 6 5 4 3 2

Library of Congress Cataloging-in-Publication Data

Lunneborg, Clifford E.
    Data analysis by resampling : concepts and applications / Clifford
E. Lunneborg
        p.   cm.
    Includes bibliographic references and index.
    ISBN 0 534-22110-6 (case)
    1. Resampling (Statistics) I. Title.
QA278.8.L86 2000   99-40284
519.5--dc21

Table of Contents

Part II: Resampling Applications

Preface

Data Analysis by Resampling

This book discusses, from the perspective of modern statistical computing, those questions empirical researchers most commonly ask of their case-based studies. Increases in computer power and decreases in computer costs have revolutionized the role of computing in statistics. One important aspect to this is the growing importance of data analyses based on recycling the scores constituting a data set, a collection of techniques broadly known as resampling. These computer-intensive techniques repeat a data analysis many times on replicate data sets (the resamples), all based on an observed set of data.

Why are resampling techniques important? Until inexpensive computing power made replicate data analysis practical, the drawing of statistical inferences from a set of data almost always required that we accept an idealized model for the origin of those data. Such models can be either inappropriate or inadequate for the data in our study. Resampling techniques allow us to base the analysis of a study solely on the design of that study, rather than on a poorly-fitting model.

How have models dictated data analysis? Here is a prime example. To compare the responses of cases (e.g., patients, students, or animals) receiving different treatments (e.g., alternative anti-inflammatory drugs, reading instruction, or fat-controlled diets), researchers commonly have adopted an analysis of variance (ANOVA) model for the data. Briefly, ANOVA models the responses to treatment as those of cases that were: (a) randomly sampled from a population or populations in which the distribution of responses to each treatment would (b) have the same or homogeneous variance and (c) follow that of a normal random variable

The ANOVA model leads to simple data analysis computations.

More often than not, at least one of the assumptions of the ANOVA model will be palpably or, at least, arguably untrue. To take an example, we assess the reading abilities of students participating in our study comparing methods of reading instruction with a test yielding integer scores between 0 and 50. How might the ANOVA model be inappropriate to these data?

1. The population distribution of such scores could not be normal, that is, it could not include all magnitudes between negative and positive infinity.

2. Although the population distributions might have a common variance, we could never know this from our sampled data. All we can expect to do is to identify some data sets for which an underlying population homogeneity of variance is highly unlikely.

3. Finally, of course, our cases might not have been sampled randomly from any student population, rather they might have been all the first-graders enrolled last autumn in Shady Brook primary school.

The resampling inference techniques featured in this book either eliminate or restrict to a minimum any unwarranted or unverifiable assumptions about our data. These techniques depend almost exclusively on the data and on our knowledge of how the data were collected—our knowledge of the design of the study. Our resampling data analyses, then, can be described as design-based rather than model-based.

Our resampling techniques let us perform data analyses when we believe one or more of the assumptions of a classical, model-based technique are inappropriate.

The range of classical statistical models is limited. The ANOVA model, to pursue this important example, allows the researcher to estimate confidence intervals for contrasts among population distribution means. A mean contrast, though, might not be the researcher's choice for a treatment comparison. A researcher might prefer a confidence interval for a population effect size measure, such as d which scales the difference in means for a control and treatment population by the standard deviation of the control population. But there is no classical model for estimating a confidence interval for this effect size measure. Our resampling methods adapt to the questions that researchers regard as important, rather than requiring that researchers ask only questions for which there are model-based answers.

We characterized the resampling approaches to statistical inference as design-based. In this book, we develop and illustrate three distinct resampling approaches, targeting the three major and distinct data collection designs employed in case-based studies:

1. Bootstrap Population Inference for studies in which cases (patients, animals, customers, student subjects, classrooms, clinics, agricultural or forest tracts, and so on) have been randomly sampled from one or more populations.

2. Rerandomization Causal Inference for studies in which cases, though not randomly chosen, have been randomly allocated among two or more treatments.

3. Subsample Descriptive Inference for studies in which cases have been neither randomly sampled nor randomly allocated.

This book has two audiences. The first comprises graduate and advanced undergraduate students interested in the design and analysis of empirical research, perhaps their own. Students should have had an introduction to statistical ideas at the level of standard errors, confidence intervals, and tests of statistical hypotheses.

The second audience consists of active researchers, trained in classical, model-based, statistical methodology but who are now interested in learning how to use the more realistic data analyses that modern statistical computing has made available.

Goals

The major goals of the book are the following:

1. Establish the importance of matching the data analysis for a study to

    a. the sampling design of the study

    b. the amount of data collected

    c. the precision of the data collected

    d. what is known about any populations sampled

    e. what the researcher wants to learn from the study

2. Introduce three resampling approaches to data analysis that satisfy these important criteria.

3. Show how each of these resampling approaches can be applied across the range of questions commonly asked of case-based empirical data.

4. Provide explicit algorithms for a wide variety of bootstrap, rerandomization, and subsampling applications.

5. Illustrate the use in resampling analyses of three widely available statistical packages.

A critical and unique feature of this book is the firm linkage developed between study design and analysis. A number of otherwise very good resampling books have appeared in recent years (e.g., Davison & Hinkley,

1997, Efron & Tibshirani, 1993, Good, 1994, Manly, 1997, and Sprent, 1998) that unfortunately blur the distinctions that should be drawn among the principal case-selection designs—random sampling, randomization, and nonrandom.

In addition, the range of applications treated here, extending from one- and two-group designs through more complex multiple treatment designs to linear and logistic regression models, is well suited to the needs of researchers conducting case-based studies.

Organization

The book is organized into two parts.

Part I – Resampling Concepts with Applications

Concepts 1 through 16 begin with a review of basic ideas in statistical inference and then introduce the three approaches to resampling inference. These Concepts are short, each the basis for a single lecture, and are accompanied by Applications that provide examples and exercises to illustrate the Concepts.

Part II – Resampling Applications by Study Design

Applications 17 through 23 develop resampling applications for the more common case-based study designs. These applications begin with the single data set, move through analyses of multiple independent data sets or treatments and repeated measures, and finish with techniques for linear models with measured or categorical response data. At each stage, we outline the limitations of classical inference and develop applications of each of the three resampling approaches.

A final Postscript summarizes resampling approaches to statistical inference.

Selection of Topics

Instructors might want to choose among the Concepts and Applications, or to give greater emphasis to some than to others, depending on the amount of instructional time available, the background of students, and the goals of the course.

- Students having completed a previous course at the level of Moore and McCabe's *Introduction to the Practice of Statistics* will be able to review quickly the material in Concepts and Applications 1 through 5 and have time in a one-semester course to survey, at least, the full range of Applications.

• For students less-well grounded in classical statistical inference, instructors can give more time to the initial Concepts and, where time is short, limit the course to introducing resampling ideas, extending the course perhaps no further than Concepts and Applications 16.

• Where more time is available, some instructors might choose to emphasize Applications 19 through 21, if the goal is to develop a nonparametric alternative to the analysis of variance, whereas others might feature Applications 22 and 23, if linear models are the central concern.

In my experience, resampling course instruction is divided profitably and fairly equally between two activities, one-half of each week's meetings given to lecture and discussion of reading, the other half to directed (exemplar) and free (assigned homework) computer lab work. It is important to provide, early on, ample time for students to acquaint themselves with any software that is new to them.

The bootstrap, rerandomization, and subsampling all use random sequences of resamples. As a result, students working on the same problem will obtain slightly different answers. A computational exercise that I have found valuable early in the course demonstrates how the variability in student answers decreases as the number of resamples is increased. Students can learn in this way how to choose an appropriate number of resamples.

Computational Support

Resampling is computer-intensive. Applications of the techniques presented in this book require some programming in a statistical computing package. Some statistical packages make resampling easier than others. This text uses three statistical computing packages that have features well adapted to resampling computations, S-Plus (MathSoft, 1997), SC (Dusoir, 1997) and Resampling Stats (Bruce, Simon, & Oswald, 1995). These are not the only possibilities, and students, their instructors, or their institutions might have other well-established statistical computing preferences.

Where possible, adopting one of the three featured packages to support a course taught from this text seems natural. But another package certainly can be used. In Part I, logical, step-by-step computational algorithms are presented for each resampling application in turn. These are then illustrated in enough variety in Part II that students should be able to program any application in another package with which they are proficient.

The text illustrates the use of three statistical packages. We don't have enough room, however, to provide a primer or introduction to the use of those packages. If students are not already familiar with a package, instructors should provide some additional support in getting students started. Each of the three featured packages has excellent documentation, and S-Plus has a number of short third party getting-started texts.

Accessibility

The mathematical presentation is at the level of college algebra. This reflects the fact that the resampling approaches to data analysis do not depend on deep mathematical reasoning. In particular, no knowledge of calculus is required. There are a few brief algebraic developments, associated with the definition and estimation of confidence intervals and with hypothesis testing. In Applications 22, the Q-statistic, a nonparametric version of the F-test, is presented using the notation of linear algebra. Because the necessary computational details are easily encapsulated, readers unfamiliar with matrix operations should not be disadvantaged in applying the approach. The linear log-odds models described in Applications 23 will be understood more easily if the student is familiar with logarithmic and antilogarithmic transformations. All the student needs by way of mathematical preparation, beyond college algebra, is an understanding of probability as taught in the typical introductory applied statistics course. The functional notation for cumulative distribution functions (cumulative proportions or probabilities) used to unify the resampling approaches might be new to some readers and therefore is introduced with some care.

Examples

The text includes a fairly large number of real-data examples and exercises. These were chosen to illustrate: (a) the three basic case-selection types—random samples, random allocation, and nonrandom; (b) a variety of experimental designs; and (c) a range of substantive content.

One invaluable source of data has been the publication, *A Handbook of Small Data Sets* (Hand, D. J. et al., 1994). Instructors searching for additional examples might find what they are looking for in that excellent collection.

Acknowledgments

My understanding of and approach to resampling and statistical inference have been shaped over the years by contact and correspondence with a number of statisticians actively working in the area. These include Gordon Bear, Peter Bruce, Steve Buckland, Brian Cade, David Draper, Tony Dusoir, Eugene Edgington, Bradley Efron, David Hand, Andrew Hayes, Tim Hesterberg, Chris Jones, Cyrus Mehta, Patrick Onghena, Normand Péladeau, Galen Shorack, Bruce Thompson, Robert Tibshirani, Jon Wellner, and Bruno Zumbo. They are not responsible, singly or collectively, for the choices I made and the resulting form of the book.

More recently, I am greatly indebted to the reviewers of earlier drafts of this book: Peter Bruce of Resampling Stats, Inc., Christine Anderson-Cook of Virginia Polytechnic Institute and State University, Bruce Thompson of Texas A&M University and Baylor College of Medicine, Rodney Wolff of Queensland University of Technology, and Arthur Yeh of Bowling Green State University. Their thoughtful critiques have been vital to shaping the final content and form of this book. Finally, three cheers for my editors at Duxbury-Brooks/Cole—Alex Kugushev, Carolyn Crockett, and Kelsey McGee—and for Robin Gold at Forbes Mill Press for turning draft into reality.

PART I

Resampling Concepts

Before we can use resampling methods to carry out data analyses, we must do two things. Of course, we will have to learn what those methods are. But, equally important, we will need some ground rules or principles that guide us in their application. Under what circumstances do we use one approach rather than another? The goal of Part I of this book is to describe the methods and establish the contexts in which they are useful.

We begin with a review of the principles of classical, parametric statistical inference as taught in a typical first-level applied statistics course.

Concepts 1, Terms and Notation, establishes the terms and notation that will be used throughout. For the most part, the terms are common to statistical analysis, but you should pay particular attention to how we use the cumulative distribution function and its close relative the inverse cumulative distribution function to describe a collection or distribution of scores on an attribute. These paired functions are at the heart of statistical inference.

Concepts 2, Populations and Random Samples, defines and illustrates the key concepts of case populations and random samples from case populations. We pay particular attention to how a simple random sample of cases can be drawn.

Concepts 3, Statistics and Sampling Distributions, introduces the concept of a sampling distribution for a statistic computed from a sample score distribution. From the sampling distribution we develop the ideas of the bias of an estimator, the standard error of a sample statistic, quantiles of a distribution, and the confidence interval for a population parameter. If you are comfortable already with these very important ideas, you can review this unit quickly.

Concepts 4, Testing Population Hypotheses, provides a brief review of some terms and concepts associated with testing population hypotheses. The null and alternative statistical hypotheses are defined, as is the extension of the sampling distribution to a null sampling distribution for a test statistic. The uses of this null distribution in fixed α-level hypothesis testing and in defining a p-value for a test statistic are outlined.

Each of the Concepts is accompanied by an Applications unit, which gives you an opportunity to apply these ideas as they are developed. Concepts 1 through 4 might be familiar to you from an earlier statistics course. Concepts 5 and 6 provide a bridge between classical and resampling inference.

Concepts 5, Parametrics, Pivotals, and Asymptotics, establishes that sampling distributions, null and ordinary, are unrealizable idealizations. To make them useful to us, mathematical statisticians have been ingenious at deriving asymptotic (large sample) results, defining pivotal forms for test statistics, and developing parametric requirements for standard statistical inference problems.

Concepts 6, Limitations of Parametric Inference, reminds us that we do not always have (a) random samples of cases, (b) population score distributions with known parametric forms, (c) large numbers of cases, and (d) access to statistics with known pivotal forms. To overcome these limitations of classic, parametric inference, we propose three approaches to resampling inference. Figure 6.1 links bootstrap, rerandomization, and subsample inference to crucial random elements of empirical study design and provides a key to the rest of the book.

Inference based on bootstrap resamples from an estimated population distribution is taken up in Concepts 7 through 11 as the first resampling approach.

Concepts 7, The Real and Bootstrap Worlds, introduces a bootstrap world parallel to the real world of population inference. First, we learn how to create a numeric estimate of the unobservable (real-world) population distribution. This nonparametric estimate requires only the score distribution for our one (real-world) random sample to provide a nonparametric estimate. Or, we might take account of additional knowledge we have of the population sampled to create a parametric, smoothed, or model-based estimate.

Concepts 8, The Bootstrap Sampling Distribution, describes the formation of a bootstrap sampling distribution as the result of drawing repeated random samples from our estimated population distribution. We use the bootstrap sampling distribution to compute bootstrap estimates of the bias and standard error of our sample estimate. The unit finishes with the description of some simple bootstrap estimates of confidence intervals (CIs).

Concepts 9, Better Bootstrap CIs: The Bootstrap-*t*, develops the first of two improved bootstrap CI estimates. The bootstrap-*t* is recommended when the parameter estimated is a location parameter and the estimator used has a known standard error estimate. This approach to CI estimation is fully documented, keeping in mind the needs of readers who will want to program the estimation in a new statistical package. If you use a preprogrammed bootstrap-*t* command (as in the package SC), you need not read the algorithm as carefully.

Concepts 10, Better Bootstrap CIs: BCA Intervals, covers the second improved and general purpose bootstrap CI estimate, the bias corrected and accelerated or BCA estimate. This is available as a command in both SC and S-Plus, so the full description of its derivation and solution need be read carefully only if you need to implement it in a new computing environment.

Concepts 11, Bootstrap Hypothesis Testing, deals with using bootstrap resampling for testing population hypotheses. We develop two approaches—one based on the bootstrap-*t* sampling distribution, the other on BCA confidence intervals. The latter permits you to test hypotheses (or, estimate p-values) without having to make unwarranted assumptions about a null population distribution.

Bootstrap inferences are population inferences and require random samples of cases. More often, our cases are not randomly sampled from a population but are instead a set of available cases. If those available cases are randomly allocated among alternate treatments, we can use that randomization to drive a second kind of statistical inference, local causal inference. Concepts 12 through 15 develop this through a second kind of resampling, rerandomization.

Concepts 12, Randomized Treatment Assignment, describes both the experimental and the statistical functions of case randomization. Based on a null treatment-effect hypothesis, we can rerandomize cases and their treatment response scores to produce a null reference distribution for our treatment comparison statistic. We use that distribution to test the null hypothesis against an alternative or to assign a p-value to our treatment comparison.

Concepts 13, Strategies for Randomizing Cases, applies local causal inference, via rerandomization, to a range of common randomization designs. These include the independent randomization of cases, the simultaneous or complete randomization of all available cases, the independent randomization of several blocks of cases, and restricted randomization, appropriate when some treatments are not available to all cases. A key concept is that rerandomization must be faithful to (a) the randomization design and (b) the null treatment-effect hypothesis.

Concepts 14, Random Treatment Sequences, moves us from between-cases to within-cases designs. Available cases are randomized not to alternative treatments but to alternative treatment sequences, each case receiving two or more treatments. We present the role of treatment position in developing null treatment-effect hypotheses and guiding the rerandomization of cases. Alternative strategies for randomly allocating treatment sequences among cases are described.

Concepts 15, Between- and Within-Cases Designs, provides an orientation to the resampling analysis of between/within designs, studies in which some treatment factors are to be evaluated between cases and others within cases. We develop the idea of an interaction between factors and the related notions of main and simple effects of a treatment factor.

Our third resampling approach becomes important when the cases in a study are neither sampled randomly nor assigned randomly. In the absence of either source of randomization, neither population nor causal inference is possible. What we can do is describe the results of the study. And we want that description to be a good description.

Concepts 16, Subsamples: Stability of Description, develops the use of subsampling from our available cases to establish the stability of a description. Briefly, a stable description is one that is not materially influenced by including (or, excluding) specific cases in the analysis. Criteria are developed for using a series of subsamples to evaluate stability.

Each of Concepts 7 through 16 is accompanied by an Applications unit that elaborates and illustrates resampling topics and shows the use of three statistical packages, Resampling Stats, SC, and S-Plus. Some of these Applications employ statistics such as the trimmed mean or the odds ratio that might be new to you. One advantage of resampling is that we can use a broader range of statistics for inference than has been encompassed by classical parametric inference.

Concepts 1

Terms and Notation

Before we can describe resampling ideas and how they are applied in data analysis, we need a vocabulary. Obviously, we can communicate better with one another if we agree about what certain key terms mean. Defining terms will also clarify what kinds of studies our resampling analyses are intended for.

Cases, Attributes, Scores, and Treatments

The data analyses of interest to us involve the observation of the scores on one or more attributes for one or more cases. Actually, except in one very special situation, all our analyses are based on multiple cases.

Case

Case is our generic term for what you might recognize as the experimental unit or the unit of observation. The case can be an individual animal or human, a classroom, a plot of agricultural land, a run of an industrial process, an organization, or a political subdivision. The case receives treatment in an experimental study and is observed in an observational study.

Attribute

Actually, what we observe is not the case itself but the score for that case on one or more attributes. An attribute can be some characteristic of the case, such as its age or sex, or a response of the case to some treatment, such as weight loss while following a diet. Attributes include response and explanatory variables (or, dependent and independent variables) and auxiliary variables used to form blocks of cases or as covariates in analyses such as the analysis of covariance.

Score

A case's score on an attribute often is numeric, representing the amount or strength of that attribute. On occasion, however, the score will be a label, identifying the category to which a case belongs. Thus, the score of the Boeing Company on the attribute Type of Business might be Manufacturing.

Treatment

In experimental studies, one important attribute of a case is the treatment received by that case. Treatment is another generic term. It has no clinical significance. Most experiments are comparative; that is, they involve more than one treatment. We refer to the different treatments as different levels of treatment, even when the distinction among them is qualitative rather than quantitative.

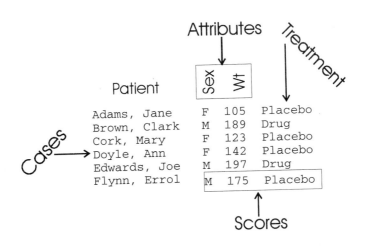

Figure 1.1:   *Cases, Attributes, Treatments, and Scores*

Figure 1.1 sketches the relations among cases, attributes, treatments, and scores.

Sequence of Observations

In many experimental studies, each case receives a single level of treatment, but, in other studies, each case might receive several levels. In the latter instance, we speak of the position of a particular treatment level in the sequence of treatments. The sequence can be temporal, as in crossover designs, in which patients cross over from one drug treatment to another to compare the effectiveness of the two drugs. Or, the sequence can be spatial as in agricultural field studies in which, for instance, different subplots in a main plot are seeded at different densities. In observational studies, as well, some attribute of the same case can be scored at different points in time, such as in a longitudinal study of the forgetting of learned material.

Experimental and Observational Studies

A simple distinction was drawn between experimental and observational studies. In an experiment, each case's score on at least one attribute is determined by the experimenter. That attribute is the treatment received by the case, with the score being the level of treatment. Although the experimental-observational distinction appears in examples and exercises throughout this text, we pay particular attention to randomized experiments. In the randomized experiment, the experimenter employs a random mechanism to assign cases to treatment levels. The random mechanism plays an important role in our analysis and interpretation of the outcome of the randomized experiment.

Data Sets, Samples, and Populations

The scores on all attributes for all cases in a study compose the full data set for that study. For some analyses, we can ignore parts of the full data set and work with a reduced data set. For instance, we could carry out separate analyses for male and female cases.

Data set is both the generic and the preferred term. Only in very well defined circumstances will we refer to a data set as a sample or, more accurately, as belonging to a sample. Whenever we refer to a sample, in any subsequent Concept or Application, we mean it to be a simple random sample.

The simple random sample is a set of cases selected from a well-defined population of cases. The simple random sample is selected from the population by a process that ensures that every sample containing the same number of cases has the same chance of being the one selected. We shall discuss random selection processes more later. For now, note the following two requirements for a sample to be described as random:

1. A well-defined population of cases from which to sample

2. A well-defined random process for selecting the sample

We cannot have a random sample without knowing the size and boundaries of the population sampled. We cannot have a random sample without specifying the random process by which the sample was drawn.

We regard the use of the term sample in any other than the random context as misleading in two respects. First, we cannot specify with any accuracy the population from which a so-called sample of convenience was drawn. Are the students in your data analysis class a sample from the population of students in your major? From the population of students at your class level? From the population of students taking statistics courses? From the population of students enrolled in your university? Or, from some population larger than any of these? You might answer that they are a sample from each of those populations! This brings us to our second concern: How can we decide which of these populations the sample is intended to represent? With explicit random sampling, we will have no uncertainty about the population.

Populations, then, are populations of cases. Samples are random selections of cases. When our cases are not a random sample, we'll refer to them as a set of available cases.

As a first bit of notation, we refer to the number of cases making up a population as N. We refer to the number of cases contributing to our full data set as n. Often we need to refer to splits among these cases. If, for example, our n cases were divided among three treatment levels, A, B, or C, we refer to the sizes of the different treatment groups in this notation:

$$n = n_A + n_B + n_C$$

Parameters, Statistics, and Distributions

When we collect the scores on an attribute for a number of cases, we refer to the collection as a score distribution. If we have the attribute scores for all N cases that compose a population, we call the score distribution a population score distribution and often denote it as X or Y. The score distribution is a sample score distribution if we have scores for a random sample of cases. And it is simply a score distribution if the contributing cases form neither a population nor a sample. To distinguish the n-element sample or data set score distribution from a population one we'll use the lower case x or y as its name.

One important data analysis task is to answer questions about characteristics of population score distributions. Is the mean IQ score for Oregon sixth-graders who were Head Start participants greater than 110? What proportion of the eligible voters in King County plans to vote in the September primary election? We refer to characteristics such as these as parameters of a population score distribution. The value of the parameter could be calculated if we had the entire population score distribution.

In developing some concepts we'll need to refer to a generic parameter. We'll use θ to refer to a nonspecific population score distribution parameter.

We refer to the mean of a population score distribution as a parameter. If we had the mean of a data set score distribution instead, it would be called a statistic. More generally, a statistic is a quantity computed from a data set. If our data refer to a random sample of cases, we can calculate a statistic from the sample score distribution that provides an estimate of a parameter of the score distribution for the population that was sampled.

We use s to refer to a nonspecific statistic. We use t to refer to a statistic that is used to estimate θ.

Distribution Functions

The attribute scores for a population, random sample, or set of available cases form a score distribution. Where the scores are numeric, they often can be sensibly ordered, from smallest to largest. The score distribution then is summarized conveniently by a pair of distribution functions, the cumulative distribution function and the inverse cumulative distribution function.

The cumulative distribution function is written as $F(q)$ and defined as

$$F(q) = \frac{\text{Number of scores in distribution that are } \leq q}{\text{Total number of scores in distribution}}$$

$F(q)$ is the proportion of the score distribution that consists of scores smaller than or equal to q. $[1 - F(q)]$ is the proportion of the score distribution that consists of scores larger than q. The function $F(q)$ returns a proportion, a value between 0 and 1. If q is smaller than the smallest value in the score distribution, then $F(q) = 0$. If q is larger than the largest value in the score distribution (or, if q is equal to that largest value), then $F(q) = 1$. For q intermediate in value, $0 < F(q) < 1$.

The cumulative distribution function (cdf) takes as its argument, q, a potential value in the distribution, and returns a number between 0.0 and 1.0, a proportion. Paired with the cdf is a second function, the inverse cumulative distribution function (icdf) or quantile function that takes as its argument a proportion and returns a distribution value.

The icdf is written $F^{-1}(\alpha)$, where α is a number greater than 0 but less than 1. The value returned is the α-quantile of the distribution. By definition, the α-quantile is the result of solving $F(q) = \alpha$ to obtain a unique q, a value exceeded in the distribution by exactly the proportion $(1 - \alpha)$.

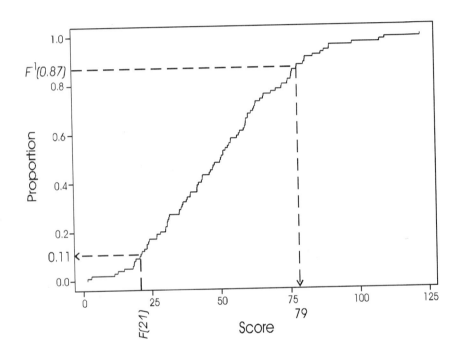

Figure 1.2:   *A Cumulative Distribution Function or cdf*

Figure 1.2 illustrates both $F(q)$ and $F^{-1}(\alpha)$. The figure plots the cumulative proportions of 100 numeric distribution elements, along the y axis, as the value of the elements increase, along the x axis. The range of values in the distribution is from 0 to 125. The cdf can be evaluated using the figure by moving from a value on the x axis, Score, upward to the curve and then leftward to the y axis, Proportion. Thus, $F(21)$ evaluates in the figure as 0.11. That is, 11% of the distribution consists of values of 21 or smaller. Similarly, the icdf can be evaluated using the figure by moving from a value on the y axis, Proportion, rightward to the curve and then downward to the x axis. Here, a cumulative proportion of 0.87 evaluates to a distribution value of 79; 13% of the distribution consists of values no smaller than 79.

In practice the limited size or finiteness of the distribution as well as the presence of tied values or of gaps between adjacent values all militate against finding a unique value for $F^{-1}(\alpha)$.

To avoid indeterminacy but, more importantly, to anticipate the use we shall make of the α-quantile of a distribution, we modify the definition of the quantile function.

We need a bit of notation for the definition. We let n be the number of elements making up the distribution. We sort these from smallest to largest, denoting the smallest as $x_{[1]}$ and the largest as $x_{[n]}$. Extending this notation, $x_{[j]}$ is the j-th smallest of the n elements, $j = 2, \ldots, (n-1)$. A second notational convention is the use of int$\{q\}$ to denote the integer part of q, the result of rounding q down to an integer.

Now, we can define our quantile function:

1. If α is 0.50 or smaller, then

$$F^{-1}(\alpha) = x_{[\text{int}\{\alpha(n-1)\}]}$$

2. If α is greater than 0.50, then

$$F^{-1}(\alpha) = x_{[(n+1)-\text{int}\{(1-\alpha)(n+1)\}]}$$

For example, if $n = 1000$ and $\alpha = 0.05$,

$$F^{-1}(0.05) = x_{[\text{int}(0.05 \times 1001)]} = x_{[50]}$$

and, if $n = 1000$ and $\alpha = 0.95$,

$$F^{-1}(0.95) = x_{[1001-\text{int}(0.05 \times 1001)]} = x_{[951]}$$

The smallest 5% of the distribution consists of elements $x_{[1]}$ through $x_{[50]}$, and the largest 5% consists of elements $x_{[951]}$ through $x_{[1000]}$.

Our new quantile function is symmetric and non-interpolating. It is symmetric in that we count up from the lower end of the distribution to define quantiles less than or equal to the 50th and down from the upper end to define higher quantiles. The function is non-interpolating in that it always returns the value of an element in the distribution, rather than returning a number intermediate in value between two adjacent distribution elements. Our icdf allows us to define quantiles of a score distribution. In particular, we say that $F^{-1}(\alpha)$ defines the α quantile of the distribution.

Although we use a mathematical notation for the cumulative and inverse cumulative distribution functions we have no cause to be interested in the shape or form of either function. In general, $F(q)$ and $F^{-1}(\alpha)$ serve only as names for specific proportions and quantiles of a distribution. We leave to the computer any counting of the elements in a distribution needed to find values corresponding to those names.

Applications 1

Cases, Attributes, and Distributions

Concepts 1 introduced some terms and notation we use in developing and applying resampling statistical techniques to the analysis of case-based studies. Here, those ideas are applied to a series of studies.

Attributes, Scores, Groups, and Treatments

The data from a case-based study consist of scores on one or more attributes. Attributes include treatment (Diet) and response to treatment (Weight Gain) as well as pretreatment characteristics (Initial Weight) and group membership (Sex) of cases. Scores on these attributes can be either numeric or categories.

Table 1.1 is taken from Agresti (1990) and summarizes the outcome of a study comparing radiation therapy to surgery in the treatment of cancer of the larynx. The cases are 41 patients randomly divided between the two treatments. Each case has scores on two attributes, treatment and outcome. Both scores are categorical rather than numeric.

Table 1.1:   *Outcomes of a Randomized Treatment Study*

Treatment	Cancer	
	Controlled	Not Controlled
Surgery	21	2
Radiation	15	3

Distributions of Scores and Statistics

A collection of n scores is said to form a distribution, x. In the distribution we aggregate scores that are identical and separate those that differ. If the scores are numeric, the distribution is defined by ordering the scores from smallest to largest. The distribution of ordered scores is described by a pair of functions, the cumulative distribution function (cdf) and the quantile function or inverse cumulative distribution function (icdf). These can be defined in slightly different ways, and we use the following.

The cdf is written $F(q)$, takes as its argument a score value, q, for some attribute, and returns the proportion of the n scores at or below q.

The quantile function (icdf) is written $F^{-1}(\alpha)$, takes as its argument a proportion, α, and returns:

1. A score value $x_{[\text{int}\{\alpha(n+1)\}]}$, if α does not exceed 0.50

2. A score value $x_{[(n+1)-\text{int}\{(1-\alpha)(n+1)\}]}$, if α is greater than 0.50

A collection of M values of a statistic, computed from a sequence of M data sets, also forms a distribution that can be described by distribution and quantile functions.

Exercises

1.   The two attributes in the Table 1.1 study could have been scored numerically as follows: For Treatment, surgery $= 0$, radiation $= 1$. For Outcome, cancer controlled $= 1$, cancer not controlled $= 0$. How would you interpret a strong positive correlation between treatment and outcome? A strong negative correlation? A near zero correlation? What is the magnitude of the correlation for the results in Table 1.1?

2.   Table 1.2 gives the results of an experiment on the water uptake of amphibia reported in Mead, Curnow, and Hasted (1993). Eight toads and eight frogs were used in the study. Each group of animals was randomly divided into four treatment groups of two animals each; these treatment groups differed in the conditions under which the animals were kept (wet or dry) and in whether a water balance hormone was injected immediately before the experiment or not. The experiment proper consisted of submerging each animal in water for two hours and measuring the percent increase in body weight. There were 16 cases in this study, each with scores on four attributes.

Discriminate among the attributes of the amphibia of Table 1.2. Which atttibutes are responses, treatments, and pretreatment characteristics? What design characteristic distinguishes amphibian type from hormone status?

Table 1.2:   *Weight Gains Among Submerged Amphibians*

Condition	Percent Body Weight Increase	
Toads, Wet, Control	$+2.31$	-1.59
Toads, Dry, Control	$+17.68$	$+25.23$
Toads, Wet, Hormone	$+28.37$	$+14.16$
Toads, Dry, Hormone	$+28.39$	$+27.94$
Frogs, Wet, Control	$+0.85$	$+2.90$
Frogs, Dry, Control	$+2.47$	$+17.72$
Frogs, Wet, Hormone	$+3.82$	$+2.86$
Frogs, Dry, Hormone	$+13.71$	$+7.38$

Table 1.3:   *Serum Cholesterol Levels for 25 Males*

| 106, 138, 146, 153, 153, 158, 160, 174, 180, 180, 190, 191, 198 |
| 203, 207, 215, 217, 223, 241, 242, 251, 260, 265, 280, 305 |

3.   Table 1.3 gives the serum cholesterol levels (mg per 100 ml of blood) for 25 males and is taken from Koopmans (1987). The scores have been arranged from smallest to largest. This facilitates the computation of

$$F(153) = 5/25 = 0.20 \, , \ F(215) = 16/25 = 0.64$$

$$F^{-1}(0.35) = x_{[\text{int}(0.35\times26)]} = x_{[\text{int}(9.1)]} = x_{[9]} = 180$$

$$F^{-1}(0.80) = x_{[26-\text{int}(0.20\times26)]} = x_{[26-\text{int}(5.2)]} = x_{[21]} = 251$$

For the distribution of Table 1.3 find the following:

a. The median, Mdn

b. $F(170)$

c. $F(225)$

d. $F^{-1}(0.25)$

    e. $F^{-1}(0.75)$

    f. The interquartile range (IQR), $F^{-1}(0.75) - F^{-1}(0.25)$

    g. The interquartile ratio, IQR/Mdn

The interquartile range, rather like the standard deviation, assesses the spread or scale of a distribution in the metric in which the scores are measured, here, in mg per 100 ml. The interquartile ratio, on the other hand is dimensionless; the mg per 100 ml appears in both numerator and denominator and cancels out. The ratio assesses spread (IQR) relative to location (Mdn).

4.   Table 1.4 is adopted from Efron and Tibshirani (1993) and gives the paired scores on two attributes, the average Law School Admission Test score (LSAT) and average undergraduate grade point average (GPA) for the entering class of 1973, for a population of U.S. law schools, $N = 82$. The population distribution here is bivariate and might be characterized by five parameters: μ_{LSAT}, μ_{GPA}, σ^2_{LSAT}, σ^2_{GPA}, and ρ. The first pair of parameters are the attribute means, the second pair are the attribute variances, and the final parameter is the correlation between the two. This last is computed over the population in the same way we would compute a Pearson product moment correlation coefficient for a sample:

$$\rho = \frac{(1/N)\sum_{i=1}^{N}[(\text{LSAT}_i - \mu_{\text{LSAT}})(\text{GPA}_i - \mu_{\text{GPA}})]}{\sigma_{\text{LSAT}}\, \sigma_{\text{GPA}}}$$

Draw 50 samples from the population distribution described in Table 1.4. You can choose the size of the sample and how the cases are sampled. Note, however, that when you sample a school you must bring along with it both LSAT and GPA scores. For each of your 50 samples, compute a product moment correlation, r. For the resulting distribution of correlations:

    a. Find and describe $F(0)$, $F(0.2)$, $F(0.4)$

    b. Find and describe $F^{-1}(0.1)$, $F^{-1}(0.9)$

    c. Compare the mean of your 50 r values with ρ, the population correlation

Table 1.4:   *Population Distribution of LSAT and GPA Scores*

LSAT	GPA	LSAT	GPA	LSAT	GPA	LSAT	GPA
622	3.23	546	2.99	575	2.92	637	3.33
542	2.83	614	3.19	573	2.85	572	3.08
579	3.24	628	3.03	644	3.38	610	3.13
653	3.12	575	3.01	545	2.76	562	3.01
606	3.09	662	3.39	645	3.27	635	3.30
576	3.39	627	3.41	651	3.36	614	3.15
620	3.10	608	3.04	562	3.19	546	2.82
615	3.40	632	3.29	609	3.17	598	3.20
553	2.97	587	3.16	555	3.00	666	3.44
607	2.91	581	3.17	586	3.11	570	3.01
558	3.11	605	3.13	580	3.07	570	2.92
596	3.24	704	3.36	594	2.96	605	3.45
635	3.30	477	2.57	594	3.05	565	3.15
581	3.22	591	3.02	560	2.93	686	3.50
661	3.43	578	3.03	641	3.28	608	3.16
547	2.91	572	2.88	512	3.01	595	3.19
599	3.23	615	3.37	631	3.21	590	3.15
646	3.47	606	3.20	597	3.32	558	2.81
622	3.15	603	3.23	621	3.24	611	3.16
611	3.33	535	2.98	617	3.03	564	3.02
		595	3.11			575	2.74

Concepts 2

Populations and Random Samples

In Concepts 1, we noted that an important goal of data analysis is to learn about the characteristics of some population of cases by observing the scores for a limited sample of cases from that population. This process of drawing population inferences from a sample is facilitated if the sample is a random one. In fact, we use the terms sample, random sample, and simple random sample interchangeably in the development of resampling techniques. Because of the importance of random sampling to population inference we revisit the topic here.

Varieties of Populations

Populations are collections of cases. Those that concern us have three characteristics:

1. They consist of cases that have or could have scores on an attribute that we believe important, for example, response time to an auditory signal.

2. We have an interest in learning about some characteristic of the distribution of scores for this attribute on the entire population, for example, the 0.25 and 0.75 quantiles of the distribution of ages for all United Airlines pilots.

3. The population is well enough defined that we can draw from it a random sample of cases, for example, we can place the names of all United Airlines Pilots in a figurative hat in which the names can be thoroughly shuffled and from which we can draw a sample of those names.

This third characteristic is important. Unless we can randomly sample a population, we have no basis for drawing any inference about a distribution of scores on that population.

United Airlines pilots form a natural population. We distinguish natural populations from those that are either prospective or constructed.

Natural Populations

A natural population has a degree of permanence to it. It is defined for purposes other than our response time study and has an existence both before and after that study. This permanence makes the natural population an appealing target of study. An important consideration, however, is that while the population persists, its composition might not be static. United Airlines, to pursue our example, regularly hires and retires pilots. Does the changing composition of a population invalidate the knowledge we gain of a score distribution on that population as it was constituted at one point in time? It might, so the possibility should not be ignored.

Some examples of other natural populations: U.S. cities with populations in excess of 500,000, public high schools in the State of Texas, New England manufacturing firms employing 50 or more workers, and University of Washington undergraduates registering for 15 or more credits in autumn quarter 1998.

Prospective Populations

Prospective populations are linked with experiments. The definition of a natural population changes with the delivery of a treatment to a sample from that population. Consider this example, typical of a randomized experiment with two levels of treatment.

1. We begin with a well-defined population, those 432 children aged 5–12 years who were seen at Mercy Hospital on at least one occasion between January 1 and December 31, 1998, for an asthma-related consultation and have agreed to continuing treatment through the hospital's Asthma Clinic.

2. We draw a random sample of 24 children from this population.

3. We divide the sample randomly into two groups of 12 children, Group A and Group B.

4. Physicians for Group A children prescribe Drug A for the relief of asthma symptoms whereas physicians for Group B children prescribe Drug B.

The step 3 random division of the random sample into two groups produces two random samples. Every possible set of 12 children from among the 432 has the same chance of forming Group A and every possible set of 12 children has the same chance of forming Group B.

The step 4 prescription of Drug A to the Group A random sample creates a random sample from a prospective population. Group A is now a random sample from the population of 432 asthmatic children that would exist if all were prescribed Drug A. Similarly, the prescription of Drug B to Group B creates a random sample from a second prospective population, the population of 432 asthmatic children, all prescribed Drug B.

We have illustrated an important principle of experimental design. If we draw a random sample from a natural population, randomly divide that sample into two or more treatment groups, and provide those groups with different levels of treatment, then those treatment groups are random samples from different prospective populations. Each prospective population contains exactly those cases in the natural population, all having received a certain level of treatment.

The advantage of having random samples from two or more populations, as we shall see, is that we can compare score distributions for the populations, not just for the groups. Would the average time taken to resume normal breathing for the 432 children be faster if all had been prescribed Drug A, or if all had been prescribed Drug B?

Figure 2.1 describes a three-step process for obtaining random samples from prospective populations:

1. Obtain a random sample from some natural population

2. Divide the sampled cases at random into groups

3. Apply a different level of treatment to each of the several groups

The populations sampled are prospective because the entire collection of cases making up the natural population has not received treatment. In our example, all 432 asthmatic children have not yet been treated either with Drug A or with Drug B. When we compare the characteristics of score distributions for these populations, we are comparing prospective score distributions, the score distributions that would result if all cases received treatment.

This sounds more complicated than it is. We want to know what would happen if all the members of a population were to receive some treatment. We can't treat the whole population, but we can a random sample. And we have a procedure for getting just such a random sample.

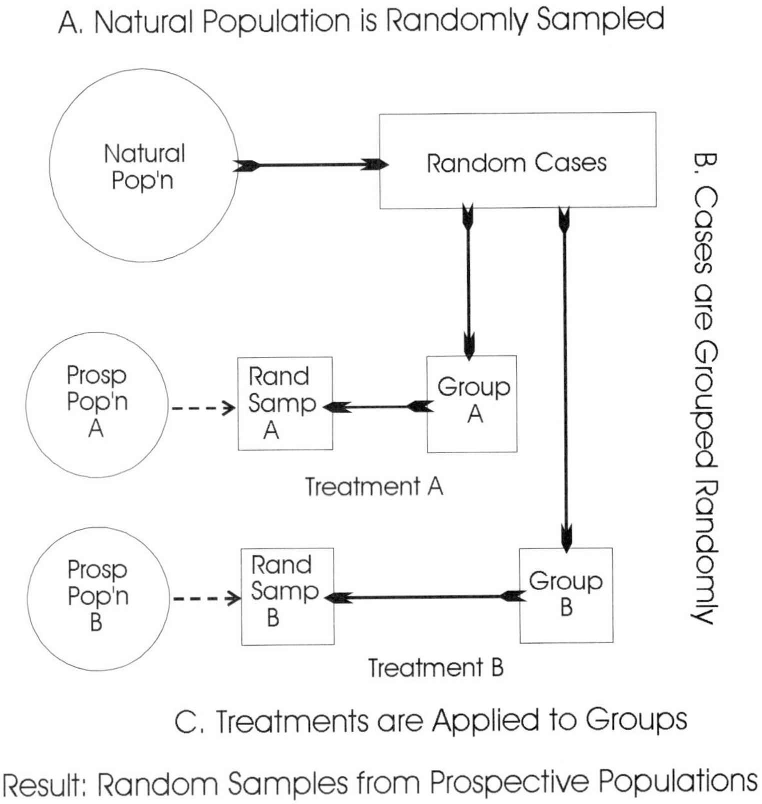

A. Natural Population is Randomly Sampled

Figure 2.1: *Obtaining Random Samples from Prospective Populations*

Constructed Populations

Our third variety of population, the constructed population, is as fully defined as the natural population but lacks its permanence. Indeed, some of these populations are constructed solely to provide random samples. Consider an example:

You are asked, as a course assignment, to administer an opinion survey to a random sample of 75 students. As noted earlier, students form many populations. Which population should you sample from? There might be an ideal one, say, all university students in the United States. But obtaining a random sample from that population is beyond your means. It might be extraordinarily difficult or costly to get a random sample even of the students on your own campus. But you might be able to construct a population similar to this second one, a population from which it will be easy to sample. Here is one way of doing that:

1.   Define your population first. You might let it be the first 1,000 students to enter the student union through a main door next Thursday beginning at 1:00 P.M. That's not your entire student body, but you might be very persuasive in arguing that those entering the student union at that time are very representative of the entire student body.

2.   To form a random sample of these 1,000 students start by putting a distinctively marked ticket for each of them in a box. You don't know their names, of course, so they can't be inscribed on the tickets. In fact you don't know anything at all distinctive about these students, except, in prospect, the order in which they will pass through the entry. One of them will be first, another second, and so on. So, you can inscribe the tickets 1 through 1,000 corresponding to the first through the last (1,000th) member of your population to enter the building.

3.   Shuffle the tickets thoroughly and draw out 75 of them. They identify the cases making up your random sample. To make it easier to recognize your sample, arrange the selected tickets in order from smallest to largest. This might give you a sequence beginning 23, 37, 66, and finishing up 823, 901, 920. Now, all you have to do is to count off students as they enter the building and collect a survey from each of those who make up your sample. You even can go for an afternoon coffee after the 920th student has entered and responded to your survey!

The constructed population, then, is one that easily can be sampled. It should be constructed, of course, to resemble as much as possible the natural population you would like to sample, if you only had the resources. It is important to distinguish the random sample drawn from a constructed population, on the one hand, from a sample of convenience on the other hand. The so-called sample of convenience is not a random sample from any population and, hence, does not support any of the statistical inference techniques appropriate to random samples. The random sample from the constructed population does support such inference, and you can generalize from the sample to that population. This ability to generalize, as we see in later Concepts, is a great advantage.

What are some other examples of constructed populations? In many departments of psychology, the undergraduates—often for additional course credit—form a pool of volunteer subjects, available to be recruited by researchers requiring human cases. Typically, they are allocated in a nonrandom fashion, diminishing the strength of statistical inferences that can be drawn from the studies in which they participate. This limitation would be reduced if researchers were to randomly sample their subjects from this local population.

Similarly, the stock of laboratory animals, for example, mice, available to researchers in a certain laboratory could be considered a population to be randomly sampled whenever cases are required for a new study. This, too, is seldom done and places limitations on the generalizations that can be drawn from many animal studies.

Random samples tend to be underused in experimental studies. One major reason is the difficulty of generating random samples from what are considered ideal populations. Greater attention should be given to constructing local populations that have some of the properties of these ideal populations but can be sampled, thus facilitating statistical generalization of results.

Random Samples

Once a population has been identified, how should we select a random sample of cases? We talked earlier of placing an inscribed ticket for each population member in a box, thoroughly shuffling the tickets and then drawing out the requisite number of tickets to form the sample. This description is somewhat a metaphor, but only somewhat. When we use our computer to help us sample, we don't need a physical box to hold the tickets while they are shuffled, but we do need something close to tickets and shuffling still can be very much the appropriate image. This shuffling takes advantage of our computer's ability via program or command to do two things:

1. To generate random numbers. Actually, because the computer is a nonrandom device we can program it only to generate pseudo-random numbers. These numbers, though, are as close as computer scientists can make them to being truly random—and we shall treat them as such.

2. To sort a list of numbers, random or otherwise, into an increasing or decreasing order.

Here is an example of how we might use our computer to select a random sample of n tickets from among those identifying all N members of a population.

1. Provide the computer with distinctive tickets for each of the N cases making up the population. These will be in the form of a column or vector containing N entries, each tied to a particular case. These entries can be case names or their scores on some attribute. More often, perhaps, the entries simply will be the integers from 1 through N, identifying the population members by number.

2. Attach a randomly generated number to each of these tickets. A very common computer routine generates uniform random numbers, each a number between 0 and 1. These are generated, briefly, so that all numbers in that interval appear equally often and so that the sequence of numbers generated is random.

3. Sort the tickets into order by these attached random numbers. At this stage we have achieved a random shuffling of the order of the cases making up the population.

4. Choose as the n members of the random sample those whose tickets are at the top of this randomly shuffled order. Each possible set of n tickets has the same chance of being the first n tickets and, hence, being selected.

Steps 2 and 3 are combined when we issue a command to the computer, as we can when using many statistical computing packages, to shuffle the N tickets. Put another way, our computer uses steps 2 and 3 to perform a shuffle.

Critical to the selection of a random sample, then, is the identification (and tagging) of each case in the population. Only then can the cases be shuffled.

The random sample obtained by the process we have described can be characterized further as a nonreplacement sample. That is, as a case is selected, at random, from the population it is not replaced in the population before the next case is sampled. Typically, this is how samples are formed, and whenever we refer to a sample or random sample it will be a nonreplacement sample.

Applications 2

Random Sampling

Concepts 2 offers descriptions of three varieties of populations from which researchers draw cases. Natural populations have a continuing existence whereas constructed populations can be formed specifically to be sampled on a single occasion. Prospective populations are created from natural or, on occasion, constructed populations by providing differential treatment to two or more random samples from the initial population. We present examples now of constructing populations and drawing simple random samples from populations.

Simple Random Samples

To draw a simple random sample of n cases from a population of N cases we need a procedure that ensures that each of the possible samples of size n have the same chance of being chosen. The number of these possible samples is given by

$$M = \frac{N!}{n!(N-n)!}$$

where $n!$ is read n factorial and denotes the product of the positive integers from 1 through n. Thus, the number of distinct samples of size $n = 5$ that can be obtained from a population of $N = 10$ cases is

$$M = \frac{10 \times 9 \times 8 \times 7 \times 6 \times 5 \times 4 \times 3 \times 2 \times 1}{(5 \times 4 \times 3 \times 2 \times 1) \times (5 \times 4 \times 3 \times 2 \times 1)} =$$

$$\frac{10 \times 9 \times 8 \times 7 \times 6}{5 \times 4 \times 3 \times 2} = 2 \times 9 \times 2 \times 7 = 252$$

A random sample of size 5 from a population of size 10, then, is one chosen in such a way that each of 252 samples has the same chance of being selected.

Shuffling the Population

You can draw a simple random sample of size n from a population of size N in a number of different ways. Some are more efficient than others, and your statistical software might implement a different one in different circumstances. For consistency, however, we shall use the image of first thoroughly shuffling the order of the N cases and then taking the n that are at the top of the shuffled arrangement. The shuffling mechanism, of course, must be random, one that ensures that all possible orderings have the same chance of occurring at each shuffle.

In the following example we use a computing package, Resampling Stats, to shuffle our population and calculate a statistic of interest from each of a series of random samples. The example serves two purposes:

1. To describe the process of drawing random samples

2. To illustrate the variability in a statistic computed over a series of random samples from the same population

Table 2.1:   *Voter Participation by State in the 1844 Presidential Election*

67.5	76.1	80.3	89.6	84.9
65.6	73.6	54.5	44.7	76.3
65.7	81.6	79.1	82.7	74.7
59.3	75.5	94.0	89.7	68.8
39.8	85.0	80.3	83.6	79.3

Table 2.1 is taken from Noreen (1989) and shows the percent of eligible voters from each of the then 25 United States who participated in the 1844 presidential election. These states constitute a population ($N = 25$), and the participation scores make up a population distribution.

A characteristic or parameter of this population distribution is its 20% trimmed mean, the mean of the scores that remain after the smallest 20% and largest 20% have been trimmed from the distribution. Figure 2.2 shows a set of Resampling Stats commands that computes this parameter. The computing script then draws a random series of 100 random samples ($n = 10$) from the population of states and computes for each a plug-in estimate of the parameter, the 20% trimmed mean of the 10 scores composing the sample distribution.

We'll look at this computing script in some detail. Lines 1 through 5 create the vector of scores making up the population distribution, PART. Lines 6 through 8 compute the population 20% trimmed mean, TRMEAN.

```
 1: copy (67.5 65.6 65.7 59.3 39.8) PART
 2: concat PART (76.1 73.6 81.6 75.5 85.0) PART
 3: concat PART (80.3 54.5 79.1 94.0 80.3) PART
 4: concat PART (89.6 44.7 82.7 89.7 83.6) PART
 5: concat PART (84.9 76.3 74.7 68.8 79.3) PART
 '
 6: sort PART HILO              'Sorted scores
 7: take HILO 6,20 TRIM         'Middle 60%
 8: mean TRIM TRMEAN            '20% Trimmed Mean
 '
 9: repeat 100
10:    shuffle PART PART
11:    take PART 1,10 SAMP
12:    sort SAMP HILO
13:    take HILO 3,8 TRIM
14:    mean TRIM STAT
15:    score STAT DIST
16: end
17: mean DIST XBAR
18: stdev divn DIST SDEV
19: print TRMEAN XBAR SDEV
```

Figure 2.2:   *Trimmed Means for a Series of Random Samples (Resampling Stats)*

The set of instructions included between lines 9 and 16 is to be repeated 100 times. Lines 10 and 11 draw a random sample of 10 cases, placing their participation scores in SAMP. Lines 12 through 14 compute a 20% trimmed sample mean, STAT. The trimmed sample means are saved, for the 100 random samples, as the vector DIST.

Lines 17 and 18 compute the mean and standard deviation of the 100 sample trimmed means. These summary statistics are printed out, together with the population parameter, by the last command.

The results of one run of the Figure 2.2 program are these: The population 20% trimmed mean is 76.34% while the mean and standard deviation of the 20% trimmed means for a random sequence of 100 random samples of 10 states are 76.401% and 2.8263%. If the program were run again, the 76.34% would not change, but the latter two values would. Why?

Exercises

1. a. Describe how you would construct a population from which you then obtain a random sample of 50 newborn infants. Describe the random sampling process as well.

b. Describe how you would construct a population of laboratory animals from which you then sample, at random, 16 animals.

c. Describe how you would construct a population of volunteer subjects for psychology studies from which you then sample 25 cases for a particular study.

2. The States and Territories of Australia make up a small population. They are eight in number: Queensland, New South Wales, Victoria, Tasmania, South Australia, Western Australia, and the Northern and Australian Capitol Territories. How many samples of size two are there? Enumerate them.

3. Write the names of each of the Australian States and Territories on eight cards. Thoroughly shuffle the cards. The names appearing on the top two cards constitute a random sample of two cases. Repeat this exercise 40 times, recording the outcomes. Each of the eight States and Territories should have been selected about the same number of times. What were your results?

For this next exercise you may want to use the shuffling capacity of your statistics package. If you do it by hand, you will want to inscribe 82 cards with the average LSAT scores for the law schools.

4. Select a series of 100 random samples of size 15 from the 82 law schools contributing data to Table 1.4 and compute the interquartile range of the

average LSAT scores for each sample. How is that done? What is the average of the 100 sample interquartile ranges? How does it compare with the interquartile range of the population distribution of 82 LSAT averages?

Concepts 3

Statistics and Sampling Distributions

At the heart of statistical inference based on random samples is the concept of a sampling distribution. By providing an answer to the question "What might have happened if I had drawn a different random sample?," the sampling distribution allows us to assess the accuracy of estimates of population distribution parameters and to formulate an approach to population hypothesis testing.

Statistics and Estimators

We noted in Concepts 1 that a statistic is some quantity computed from a data set. The statistic may be based on the scores on one attribute, as the median age of the cases, or on two or more attributes, as the correlation between seasonal rainfall and corn yields, or the predicted salary of a university graduate based on sex, academic major, and the selectivity of the undergraduate institution attended. The statistic might be based on all cases or on a portion of them, for example, as separate mean response times for left- and right-handed subjects.

Statistics often are classified as to their use. Descriptive statistics provide a summary of some aspect to a data set. Estimators are statistics computed from a random sample and used to estimate parameters of population distributions. Predictions are estimates of the magnitude of some as yet unobserved scores for additional cases from the same population we already have sampled. Inferential statistics include estimators and predictions (inferences of the size of population distribution parameters or of future observations) as well as statistics used in hypothesis tests, as F-ratios and chi-squared statistics. The same statistic can play different roles; the data set mean of an attribute can be used to describe the typical size of scores on that attribute in the data set or, if the data set is a random sample, to estimate the mean of the distribution of scores for the attribute over the entire population.

Here we focus on two varieties of inferential statistics, first estimators and then hypothesis-test statistics.

Accuracy of Estimation

When we estimate a parameter of some population score distribution from the scores for a random sample from that population, an obvious question is "How accurate is our estimate?" The population distribution parameter takes a value, θ. We estimate it to be a value t. How close are θ and t?

We can't answer the question directly as θ is unknown; that's why we are interested in estimating its value. We can assess the accuracy of our t indirectly, however, by considering other values that our estimator might have taken. How might the value of t differ from what we computed it to be, if we had a second random sample from this same population and computed t there? What might be true for t computed from a third or fourth random sample of the same size, n, as our first sample? Would all the values of t be about the same size? How closely would they be clustered about θ? We aren't able to determine the closeness of our particular t to θ, but we might be able to say something about the closeness of t, in general, to θ.

The Sampling Distribution

The vehicle we use for our accuracy estimation is the sampling distribution for our statistic, t. The sampling distribution is based on a fairly simple idea.

1. We start by letting X stand for the population distribution of scores on some attribute. The population distribution contains the scores for all N cases and is characterized by a parameter θ, whose value we want to estimate.

2. Next let $x_{(1)}$ be the sample distribution of scores on the attribute for one random sample of n cases from that population. We compute from this sample the statistic $t_{(1)}$ to be used to estimate θ.

3. Similarly, we let $x_{(2)}$ be the sample score distribution for a second random sample of n cases from the same population and $t_{(2)}$ the estimate of θ computed from $x_{(2)}$.

4. If we can consider a second random sample, we can consider other, additional random samples, all of size n, from the population, giving a sequence of sample score distributions, $x_{(3)}, x_{(4)}, \dots, x_{(50)}, \dots$ and corresponding estimates of θ, $t_{(3)}, t_{(4)}, \dots, t_{(50)}, \dots$.

5. We let this sequence expand to include all M possible random samples of size n that can be formed by random sampling (sampling without replacement) from the N cases making up the population:

$$M = \frac{N!}{n!(N-n)!}$$

where, as earlier, $n!$ is read "n factorial" and stands for the product of the positive integers from 1 through n, e.g., $3! = 3 \times 2 \times 1 = 6$.

6. The collection of estimates computed for all possible samples of size n from our population of N cases, $t_{(1)}, t_{(2)}, \ldots, t_{(50)}, \ldots, t_{(M)}$ makes up the sampling distribution for the estimator, t.

The formation of a sampling distribution is illustrated in Figure 3.1.

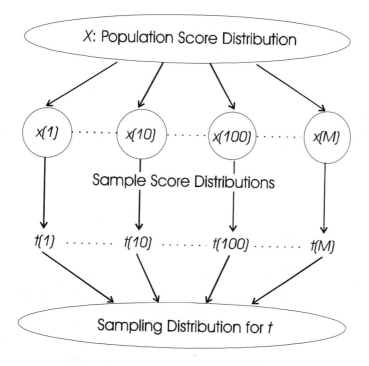

Figure 3.1: *Formation of a Sampling Distribution*

Two points about the sampling distribution should be noted. First, we typically have a single random sample and, hence, only $t_{(1)}$. Second, the $t_{(1)}$ that we compute from our one (random) sample score distribution is just one of M equally likely $t_{(i)}$s. That is, any one of the other $(M-1)$ members of the sampling distribution is just as likely as the one we computed.

The sampling distribution of t, as defined here, is specific to X, N, and n. That is, the same estimator—say, a 20% trimmed mean—will have a different sampling distribution depending on the size and contents of the population score distribution and the size of the random sample drawn from the population.

Here are two examples of the computation of M in step 5, above. For a very small population, $N = 10$, the number of distinct samples of size $n = 2$ is

$$\frac{10!}{2!(10-2)!} = \frac{10 \times 9 \times 8 \times 7 \times 6 \times 5 \times 4 \times 3 \times 2 \times 1}{(2 \times 1)(8 \times 7 \times 6 \times 5 \times 4 \times 3 \times 2 \times 1)} = \frac{10 \times 9}{2} = 45$$

More realistically, the number of distinct samples of size $n = 10$ from a population containing $N = 100$ cases,

$$M = \frac{100!}{10! \ 90!}$$

is reported by one statistical computing package as 1.7310309e+013. Moving the decimal point thirteen places to the right, we see that the number is something in excess of seventeen trillion: 17,310,309,000,000. So, sampling distributions can contain quite a number of elements!

We don't, of course, actually construct sampling distributions. Aside from the very large number of samples that might be involved, the cost of creating even a small number of additional sample score distributions can be quite prohibitive. Each might require rerunning a time-consuming or expensive experiment to obtain the needed attribute scores, e.g., the number of days of work lost to illness during the year following treatment for back pain.

Although we cannot construct the sampling distribution, we can estimate it, often by making assumptions about the properties of X or θ or t. How we do that is the province of later Concepts. For now, let's look at some properties of the sampling distribution and what we could learn from it if we had it available to us.

Bias of an Estimator

An obviously important property of the sampling distribution of the estimator t (conditional upon X, n and θ) is how the values $t_{(1)}$ through $t_{(M)}$ are distributed about θ. If t is to be a good estimator of θ the values of $t_{(i)}$ should be close to θ. Our first index of this closeness, bias, compares the average of the values of $t_{(i)}$, $i = 1, 2, \ldots, M$, with θ. The bias of the estimator t is given by

$$\text{Bias}(t|X, n, \theta) = (1/M)\sum_{i=1}^{M} t_{(i)} - \theta$$

Our estimator is unbiased if this bias measurement is zero; on average, the value of the estimator is equal to the value of the parameter being estimated. If the bias measurement is negative, our estimator on average is too small, underestimating θ. If the bias is positive, our t overestimates θ on average. Unbiasedness is a good quality for an estimator to have.

Standard Error of a Statistic

Bias, whether zero, positive, or negative, tells us only about the average value of t. What about their variability, sample to sample? Are the values of $t_{(i)}$ all closely clustered about their average? Or, are they rather widely dispersed? Our second index of the goodness of an estimator is the standard deviation (SD) of the sampling distribution. This SD is important enough to have a special name. We refer to it as the standard error (SE). The SE has the customary SD definition:

$$\text{SE}(t|X, n) = \sqrt{(1/M)\sum_{i=1}^{M}\left[t_{(i)} - (1/M)\sum_{i=1}^{M}t_{(i)}\right]^2}$$

the square root of the average squared distance of the $t_{(i)}$s from their average. Note that the SE of an estimator, t, does not depend on the value of the parameter being estimated, θ. In fact, the idea of a SE extends to inferential statistics other than estimators, as long as they are computed from random sample score distributions.

A small SE is desirable. This means that there is not much sample to sample variability in our estimate. We would expect t to take nearly the same value in a subsequent sample from the same population. The SE is a key concept in inferential statistics.

RMS Error of an Estimator

We've seen, then, that the t we compute from our single random sample score distribution might fail to be close to θ for either of two reasons:

1. Because t is biased, and on average the value of t is some distance from θ

2. Because t has a relatively large SE, the t computed for our particular sample might differ considerably from an average value

It can be convenient to put these two potential sources of error together in a single index. The root mean square (or, RMS) error combines the two into an easily interpreted index. The RMS error is the square root of the average squared distance of the $t_{(i)}$s from their target, the value of θ:

$$\text{RMS Error}(t|X, n, \theta) = \sqrt{(1/M)\sum_{i=1}^{M}\left[t_{(i)} - \theta\right]^2}$$

A little algebra will show just how the RMS error reflects the sizes of both the bias and the SE of an estimator:

$$\text{RMS Error}(t|X, n, \theta) = \sqrt{\text{Bias}(t|X, n, \theta)^2 + \text{SE}(t|X, n)^2}$$

Confidence Interval

Our estimator, t, provides a point estimate of the parameter θ. That is, t is a specific numeric value. This specificity overestimates the accuracy with which we can estimate θ, particularly when the bias or SE or both are substantial. A better idea is an interval estimate of θ. By widening the point into an interval we can express our uncertainty about the accuracy of t. While t is unlikely to be exactly equal to θ, t is likely to be no further away from θ than is indicated by our interval. We let the width of the interval indicate the expected accuracy of t, using a wider interval—professing less certainty about the value of θ—when the SE or bias of t is large.

The interval estimate we use is termed a confidence interval, and its definition is linked with the finer detail of the sampling distribution. A confidence interval (CI) is one in which we have a specifiable degree of confidence that the interval is one that includes θ. This degree of confidence is linked to the sampling distribution of the estimator in the following way. A 90% confidence interval is one constructed about a specific estimate, t. It is constructed by a rule that ensures that, if an interval were constructed by that same rule about each of the estimates in the sampling distribution, about each $t_{(i)}$, for $i = 1, \ldots, M$, then exactly 90% of those intervals would include the value of θ. Our 90% confidence in the particular interval we construct, then, is our confidence that ours is one of the 90% of the intervals that include the true value of the parameter.

Which 90% of the $t_{(i)}$s should we choose to have θ-including CIs? There are a variety of ways to make the choice. We will be concerned here with just two of these.

Equal Tails Exclusion $(1 - 2\alpha)100\%$ CIs

The CI of greatest interest to us is the Equal Tails Exclusion $(1 - 2\alpha)100\%$ CI. We take α to be some small proportion, usually 0.125 or smaller. Choices of α in this region translate into degrees of confidence, $(1 - 2\alpha)100\%$, of 75% or greater. The use of 2α in the definition is tied to the notion of an equal tails exclusion CI. Equal proportions of the $t_{(i)}$s, composing the two tails of the sampling distribution, are to be excluded from θ-coverage, that is, those estimates are to make up the $(2\alpha)100\%$ of the $t_{(i)}$s whose CIs do not include θ.

Figure 3.2 illustrates how we could construct an equal tails exclusion $(1 - 2\alpha)100\%$ CI if we had the sampling distribution available to us. We use the following information:

1. $F^{-1}(\alpha)$ is the α quantile of the sampling distribution of t. That is, $100\alpha\%$ of the sampling distribution is no larger than $F^{-1}(\alpha)$. Five percent of the sampling distribution is no larger than $F^{-1}(0.05)$.

2. $F^{-1}(1 - \alpha)$ is the $(1 - \alpha)$ quantile of the sampling distribution of t. Five percent of the sampling distribution is no smaller than $F^{-1}(1 - 0.05)$.

3. Between $F^{-1}(\alpha)$ and $F^{-1}(1 - \alpha)$ is the central $(1 - 2\alpha)100\%$ of the sampling distribution of t.

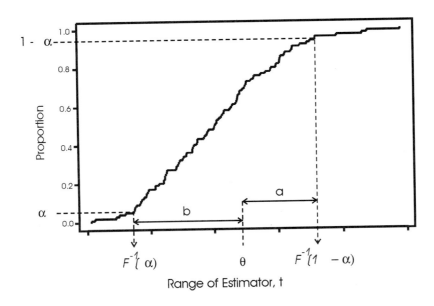

Figure 3.2:   *Constructing the Equal Tails Exclusion CI*

Our equal tails exclusion $(1 - 2\alpha)100\%$ CI should have these properties:

1. When constructed about any $t_{(i)}$ between $F^{-1}(\alpha)$ and $F^{-1}(1 - \alpha)$, the CI will include θ.

2. When constructed about any $t_{(i)}$ smaller than $F^{-1}(\alpha)$, the CI will not include θ.

3. When constructed about any $t_{(i)}$ greater than $F^{-1}(1 - \alpha)$, the CI will not include θ.

In this way our method of constructing CIs provides θ-coverage for only the central $(1 - 2\alpha)100\%$ of the possible values of t. Or, equivalently, we are assured that only the CIs for the $100\alpha\%$ smallest (5%, if $\alpha = 0.05$) and $100\alpha\%$ largest possible values of t will fail to include θ. It makes sense that the $t_{(i)}$s most distant from θ should be those excluded from θ-coverage. And often it also makes sense to split the exclusion equally between the very large and very small values of t.

Figure 3.2 illustrates how we can achieve our goal. The line segment marked a has length

$$a = F^{-1}(1 - \alpha) - \theta$$

If we subtract a from $t_{(i)}$ to form the lower bound to an interval,

$$\text{LB} = t_{(i)} - a = t_{(i)} - (F^{-1}(1 - \alpha) - \theta)$$

we are assured the interval will be wide enough to extend down to include θ for any $t_{(i)}$ between θ and $F^{-1}(1 - \alpha)$, but not far enough to reach θ from any $t_{(i)}$ greater than $F^{-1}(1 - \alpha)$.

The line segment marked b in Figure 3.2 has length

$$b = \theta - F^{-1}(\alpha)$$

If we add b to $t_{(i)}$ to form the upper bound to an interval,

$$\text{UB} = t_{(i)} + b = t_{(i)} + (\theta - F^{-1}(\alpha))$$

we are assured the interval will be wide enough to extend up to include θ for any $t_{(i)}$ between $F^{-1}(\alpha)$ and θ, but not far enough to reach θ from any $t_{(i)}$ smaller than $F^{-1}(\alpha)$.

Thus the required θ-coverage is obtained if we construct our $(1 - 2\alpha)100\%$ CI about our estimator t as

$$[t - a, t + b] = [t + \theta - F^{-1}(1 - \alpha), \, t + \theta - F^{-1}(\alpha)]$$

Unidirectional Confidence Bounds

On occasion we might want to estimate only a lower or upper bound for θ, rather than to bracket it between two values. What this means is that we want to put all of the $(2\alpha)100\%$ exclusion into one tail of the sampling distribution, rather than splitting it. If we want to estimate a lower bound for θ, very large values of $t_{(i)}$ will be uninformative and should form the excluded set. This gives a $(1 - 2\alpha)100\%$ lower bound CI of the form

$$[(t - a'), \infty] = [t + \theta - F^{-1}(1 - 2\alpha), \infty]$$

That is, we can be $(1 - 2\alpha)100\%$ confident that θ is greater than

$$[t + \theta - F^{-1}(1 - 2\alpha)]$$

Similarly we can estimate an upper bound for θ by excluding the uninformative smaller values of $t_{(i)}$. The $(1 - 2\alpha)100\%$ upper bound CI will be of the form

$$[-\infty, (t + b')] = [-\infty, t + \theta - F^{-1}(2\alpha)]$$

We can be $(1 - 2\alpha)100\%$ confident that θ is smaller than

$$[t + \theta - F^{-1}(2\alpha)]$$

Caveat

Our rules for constructing either the equal-tails or unidirectional confidence intervals are idealistic. They depend on detailed knowledge of the sampling distribution of an estimator, knowledge we never have. Instead, we must approximate the sampling distribution, the topic of Concepts 5 through 8. Confidence intervals derived from approximated sampling distributions are, in turn, only approximate confidence intervals. These approximated intervals, treated in Concepts 9 through 11, may not have exactly the correct θ-coverage.

Applications 3

Sampling Distribution Computations

Concepts 3 describes how the sampling distribution of an estimator, t, of a population characteristic, θ, is used to assess the accuracy of that estimator. First, it provides evidence of bias. An estimator is biased if the mean of its sampling distribution is appreciably smaller or larger than θ. Second, the standard deviation of the sampling distribution describes the sample to sample variability in the estimator and is known as the standard error (SE) of the estimator. These two assessments can be combined into a single measure of accuracy, the root mean square (RMS) error. Though these three accuracy indices are useful, a more important aspect to the sampling distribution is that it allows us to construct a confidence interval (CI) about the value the estimator t takes for a particular sample. The CI expresses in a standard way the uncertainty we have about our sample based estimate. We review and practice these sampling distribution computations here.

Exercises

The sampling distribution of an estimator t, though easily defined, is almost never available to us. So, to gain experience with the properties of sampling distributions, the exercises here must assume that we have a certain amount of secret knowledge about those distributions.

1. The mean, μ, of a population distribution, X, of N scores takes the value 125. We are told that the sampling distribution of the sample median, computed on the score distributions for all $M = N!/[n!(N-n)!]$ samples of size n, has a mean of 140 and a standard deviation of 15. How large is the bias of the sample median as an estimator of the population mean? How do you interpret its sign? What is the magnitude of the SE of the sample median? Compute and interpret the RMS error of the estimator.

2. We also know for the sampling distribution of the sample median that it has the following quantiles:

$$F^{-1}(0.025) = 100, \qquad F^{-1}(0.05) = 110, \qquad F^{-1}(0.095) = 200, \qquad \text{and}$$
$$F^{-1}(0.975) = 240.$$

    a. How do you interpret each of these values?

Describe the solutions for the following:

  b. a 95% equal tails exclusion CI for μ

  c. a 95% lower bound CI for μ

  d. a 95% upper bound CI for μ

  e. Show that the appropriate coverage of sample median values is ensured for each solution.

3. What key property does the 90% equal tails exclusion confidence interval have? Put another way, what does it mean to say that we have 90% confidence in a particular interval?

4. Describe how you would find a 90% equal tails exclusion CI from the sampling distribution for an estimator, t. Supply what you believe to be likely values for all the quantities involved.

Concepts 4

Testing Population Hypotheses

In Concepts 3, we developed the notion of a sampling distribution for the sample statistic, t, an estimator of a population distribution parameter θ: The sampling distribution is the collection of the values of that estimator computed from the score distributions for all possible random samples of size n drawn from that same population. We use this sampling distribution to assess the accuracy of our estimate and to communicate this accuracy in the form of a confidence interval. Now we introduce a second kind of sampling distribution, for the estimator when we assume that the population distribution parameter θ takes a specific value, θ_0. We obtain this value of θ from a statistical hypothesis known as a null hypothesis and this second sampling distribution is called the null sampling distribution of the estimator. The null sampling distribution is what the sampling distribution for the estimator t would be if the null hypothesis were correct.

Population Statistical Hypotheses

The notion of a statistical hypothesis is that it constrains, if not completely specifies, the population distribution of scores for one or more attributes. For example, we might hypothesize that the mean IQ score among students enrolling in the first grade in Seattle Public Schools is 100. Or that, among full-time university undergraduates in California, the correlation between parental income and student grade averages is zero.

Null and Alternative Hypotheses

We advance statistical hypotheses in pairs. One is termed the null hypothesis and describes one or more population distributions in which, as examples, our treatments have had no differential effects, or certain attributes of interest to us are unrelated to one another. The second hypothesis is an alternative hypothesis. This hypothesis describes one or more population distributions that provide evidence of an experimental effect on an attribute or of an observational association between attributes and is, thus, an alternative to the null, nothing is going on, hypothesis. Thus, our hypothesis of a zero correlation between university student grades and parental incomes is a null hypothesis. It hypothesizes an absence of any linear relation between these two attributes of cases. By contrast, a hypothesis that the correlation between the grades and parental income is positive would be an alternative hypothesis: It hypothesizes a linear relationship between the two.

Hypotheses and Decisions

The purpose of the paired population hypotheses is to decide between them based on our sample score distribution. More accurately, the decisions open to us are the following:

1. Reject the null hypothesis in preference to the alternative. We do this when the sample evidence is strongly inconsistent with the null hypothesis and more consistent with the alternative hypothesis.

2. Not reject the null hypothesis. We do this when the sample evidence is either inconsistent with the alternative hypothesis or not strongly inconsistent with the null hypothesis.

The sample evidence is the value of the estimator t computed from the sample distribution, x. We make our consistency judgments and, hence, our decision about the null hypothesis on statistical grounds. This statistical decision process is referred to as hypothesis testing.

Rejecting the Null Hypothesis

The alternative hypothesis is sometimes known as the scientific or experimental hypothesis. The researcher expects, or at least hopes, to find a relationship, to develop evidence that a treatment has an effect. This expectation finds expression in the alternative hypothesis. Requiring the sample evidence to be strongly inconsistent with the null hypothesis before that hypothesis can be set aside in favor of the researcher's expectation is a conservative strategy. We want to avoid claiming to have found something new when what we observe might be nothing more than an artifact of our random sampling.

Population Hypothesis Testing

The null hypothesis specifies a value for a parameter, θ, of the population distribution X. We'll refer to this null hypothesized value as θ_0. The alternative hypothesis does not fix the value of θ, but it does specify how θ differs from θ_0. The alternative hypothesis is a directional one if it specifies either that $\theta > \theta_0$ or that $\theta < \theta_0$. The alternative is a nondirectional hypothesis if it specifies only that $\theta \neq \theta_0$.

To determine if the value of our estimate, t, is strongly inconsistent with the null hypothesis, we need to know what values of the estimator we should expect to see if the null hypothesis were correct. That is, we need to know what the sampling distribution of the estimator would be if the null hypothesis were correct. We call this sampling distribution the null sampling distribution.

In the context of a particular null hypothesis, then, we can think of the estimator as having two sampling distributions. One is the true sampling distribution of Concepts 3, the one whose cdf we express as $F(t|X, n, \theta)$, where θ is the true value of the parameter in the population distribution, X, from which we actually sampled. The second, the null sampling distribution, has a potentially different cdf expressed as $F(t|X_0, n, \theta_0)$ where θ_0 is the null-hypothesis value of the population parameter under estimation. (Notice that if the null-hypothesized θ_0 is not identical to the true θ, then the null sampling distribution is made up of values of the estimator, t, computed from samples drawn, not from X, which is characterized by the parameter value θ, but from a different population distribution characterized by the parameter value θ_0. We denote this second, hypothetical population distribution X_0.)

For example, an educator might hypothesize that the mean IQ among Oregon's Head Start graduates should be above normal, that is, greater than 100. The null hypothesis, reflecting no Head Start impact, is that the mean IQ is 100. To test the expectation against this null hypothesis, the researcher randomly samples 75 graduates and collects IQ data for the sample. From the sample distribution, the researcher estimates the population distribution mean:

$$t = \text{Sample Mean IQ}$$

This value of the estimate should be close to 100 if the null hypothesis is correct, but it should be larger if the alternative, experimental hypothesis is correct.

The Null Sampling Distribution

We use the idea of a null sampling distribution to decide whether our parameter estimate is consistent with the null hypothesis or not. Values of the estimate that are in the middle of the range for the null sampling distribution are consistent with the null hypothesis. By contrast, very small or very large values of t are inconsistent with the null hypothesis.

The values of t that are more consistent with the alternative hypothesis depend on that alternative hypothesis:

1. Larger values of t are more consistent with the alternative hypothesis $\theta > \theta_0$.

2. Smaller values of t are more consistent with the alternative hypothesis $\theta < \theta_0$.

3. Both smaller and larger values of t are more consistent with the alternative hypothesis $\theta \neq \theta_0$.

To determine whether the value of t is very small or very large, we refer it to the cumulative distribution function, cdf, for the null sampling distribution, $F(t|X_0, n, \theta_0)$. Figure 4.1 illustrates a null cdf, using a more compact notation, $F_0(t)$, for the null distribution.

Let's assume that the null hypothesis behind Figure 4.1 is that of the Oregon educator; that the mean of a population IQ score distribution, X, is 100, $\theta_0 = 100$, and that our alternative hypothesis is that the mean is larger than 100, $\theta > \theta_0$. The educator computes from a (random) sample distribution, x, her population mean estimate: $t = \text{Mean}(x)$. Values of t that are close to 100 will be consistent with the null hypothesis. By contrast, larger values are more consistent with the alternative hypothesis.

The p-Value of a Test Statistic

In Figure 4.1, the educator's t is shown as taking a larger (positive) value. That result is in the direction of her alternative hypothesis, but how inconsistent is it with the null hypothesis? This is where our statistical argument begins.

We quantify the inconsistency by assessing the chances of the researcher having drawn one of the random samples that, under the null hypothesis, would give a statistic t taking a value at least this extreme, and in the direction proposed by the alternative hypothesis. This is known as the p-value of the statistic, the probability under the null hypothesis of obtaining a result at least as favorable to the alternative hypothesis as the one we observed.

The cdf for the null sampling distribution is the source of the p-value for a statistic. In our Figure 4.1 example, the null cdf evaluated for our researcher's statistic t takes the value $F_0(t) = 0.96$. Thus, there is only about a 4% chance, under the null hypothesis, of obtaining a random sample for which t would be as large or larger than the one observed. The p-value of the educator's null hypothesis test, then, is $[1 - F_0(t)] = 0.04$.

If our educator's alternative hypothesis had been, instead, nondirectional, that $\theta \neq \theta_0$, then we would have to work out not only the chances of a value of the estimate as large or larger than her greater-than-θ_0 result, but also the chances of a value as small or smaller than an equivalent smaller-than-θ_0 result as well.

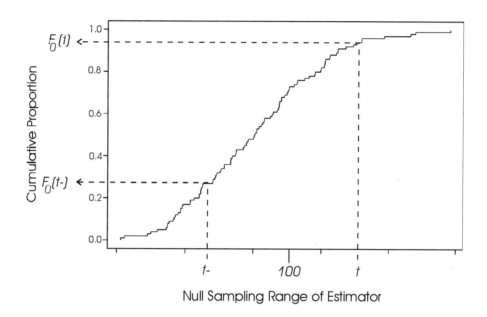

Figure 4.1:   *The Null Sampling Distribution of a Test Statistic*

The equivalent smaller-than-θ_0 result would be an estimate that takes a value as far below $\theta_0 = 100$ as the observed t was above 100. This smaller-than-θ_0 is designated, in Figure 4.1, as $t-$ and would be computed, under the null hypothesis, as

$$t- = 100 - (t - 100) = 200 - t$$

The chance, under the null hypothesis, of a value that small or smaller, $F_0(t -)$, is given in Figure 4.1 as 0.27. The p-value of the hypothesis test, for the nondirectional alternative hypothesis, would be the sum of $[1 - F_0(t)]$ and $F_0(t -)$, that is, $0.04 + 0.27 = 0.31$.

Null Hypothesis Decisions

Smaller p-values offer greater support to the alternative hypothesis and today statisticians are inclined to report the p-value for a hypothesis test as a measure of the strength of evidence against the null hypothesis. An older tradition calls for a decision to reject the null hypothesis if the p-value for the test falls below a preset level of statistical significance, say, 0.05 or 0.01. In that tradition, and at the 0.05 level, the example of Figure 4.1 would lead to a rejection of the null hypothesis if the alternative hypothesis were the directional $\theta > \theta_0$, but to a failure to reject if the alternative hypothesis were nondirectional.

Applications 4

Null Sampling Distribution p-Values

Concepts 4 builds on the notion of a sampling distribution for an estimate, t, by introducing a second, hypothetical sampling distribution. This null sampling distribution is the sampling distribution the estimator would have if the parameter under estimation were known to take the value θ_0. A null hypothesis specifies θ_0, a value that reflects, in general terms, the lack of impact of some treatment of interest. This nothing happened hypothesis is a counter to our scientific or alternative hypothesis about the value of the parameter θ. This alternative hypothesis specifies how we expect θ to differ from θ_0 and takes one of these forms: $\theta < \theta_0$, $\theta > \theta_0$, or $\theta \neq \theta_0$.

We are interested in whether t, our estimate of θ from a random sample distribution, provides evidence against the null hypothesis and in favor of the alternative. Statisticians refer to this as testing the null hypothesis against the alternative.

We test the null hypothesis about θ by determining the proportion of the null sampling distribution of our estimator that is at least as favorable to the alternative hypothesis as our sample-based estimate t. This proportion is the p-value of the hypothesis test. Small p-values offer little support to the null hypothesis.

This Applications unit illustrates hypothesis testing.

The formal definitions of true and null sampling distributions fall short of providing us sufficient information to form confidence intervals or test statistical hypotheses except for very special population distributions. This Applications unit deals with such a special distribution. Then, in Concepts 5 we consider how mathematical statisticians have made the idea of sampling distributions more useful to researchers.

In the examples and exercises here our hypotheses are about the value of a parameter θ describing the proportion of the elements of a population distribution X that are of a certain kind. The population distribution consists of N elements, N_+ of them are of the kind we are interested in (for example, Males, Democrats, Examination Passes, Successful Treatments) and $N_- = N - N_+$ of them are of a different kind or kinds.

We draw a random sample x of size n from this population distribution and observe that n_+ of them are the kind we are interested in. Our estimate of θ, the population distribution proportion, is the sample proportion, $t = n_+/n$.

We can describe the sampling distribution of t where N is large enough that the value of θ changes only infinitesimally as elements are withdrawn from X to fill out our sample, x. We then can use the cdf of a binomial random variable to develop our sampling distribution.

The p-Value of a Directional Test

The binomial random variable, $\text{Bin}(n, \theta)$ takes integer values from 0 through n. The probability that it will take a particular value, k, is given by the formula

$$P(k|n, \theta) = \frac{n!}{k!(n-k)!} \theta^k (1-\theta)^{n-k}$$

where k is the number of elements of a certain kind (generically referred to as successes) resulting when sampling, independently n times (called n independent trials) with θ the probability of obtaining a success at each trial. It provides an appropriate model for our n_+.

Most statistical packages include functions for computing such binomial probabilities. We use those provided in S-Plus.

As established in Concepts 4, a hypothesis test about θ is either directional or nondirectional, depending on the alternative hypothesis. The hypotheses $\theta < \theta_0$ and $\theta > \theta_0$ are directional alternatives to $\theta = \theta_0$. The hypothesis $\theta \neq \theta_0$ is a nondirectional alternative to the null hypothesis.

Here is an example of a directional test.

We randomly sample the names of 100 registered voters from the voter registration lists for King County (a large population) and determine that, among the sample members, 34 voted at the last primary election. Our interest in collecting the data is that we suspect voting participation was lower in that election than in the preceding one. For the earlier election, the participation rate is known to have been 41%. We shall test the null hypothesis that $\theta = 0.41$ (unchanged proportion voting) against the alternative hypothesis that $\theta < 0.41$ (our substantive hunch).

The S-Plus function `pbinom(k,n,`θ`)` provides values of the cdf for a binomial random variable, returning the probability of obtaining k or fewer successes in n trials when the success probability at each trial is θ.

What we need for our hypothesis test is the probability, under the null hypothesis (when $\theta = 0.41$), of obtaining for a random sample of 100 registered voters a proportion of primary election participants of 0.34 or smaller. Outcomes in that direction would favor the alternative hypothesis (that $\theta < 0.41$).

Equivalently, we can ask S-Plus to determine the probability of getting 34 or fewer successes in 100 trials when the success probability for each trial is 0.41:

```
> pbinom(34,100,0.41)

[1] 0.09222724
```

That is, there would be a slightly greater than 9% chance of obtaining a $t = n_+/n$ of 0.34 or smaller were θ to be equal to 0.41. The p-value of our hypothesis test is 0.09. Is that small enough to establish that our result is an unlikely one under the null hypothesis?

The p-Value of a Nondirectional Test

For the nondirectional test, we have to account for the chances of more extreme values of our estimator in both directions from the null hypothesis.

Here is such an example.

A clothespin manufacturing process is to be regulated so that it produces 2% failures. More failures than this means the company will lose customers. Fewer failures means the company is spending too much in reducing failures. The company monitors production by drawing a random sample of 300 clothespins from each day's production and recording the percentage of failures. In one such sample, there were 4% failures. Is this result different enough from 2% that we should conclude the manufacturing process is out of regulation?

We can compute, with the help of S-Plus, the probability of 12 or more failures among 300 clothespins—an observed failure percentage of 4% or greater—when the true failure percentage is 2%. To use the binomial cdf, we subtract from 1.0 the probability of observing 11 or fewer failures:

```
> 1-pbinom(11,300,0.02)
```

```
[1] 0.0189694
```

There is slightly less than 2% chance that a sample of 300 clothespins will contain 12 or more failures, if the true failure rate is just 2%.

But, that is only half the story. We also need to know, for our nondirectional test, the chances under the null hypothesis of observing a failure percentage in a sample that falls below our θ_0 of 2% by at least as much as the 4% was above that null hypothesized value. What percentage fits that description? Two possibilities suggest themselves:

1. We might think of the increase from θ_0 to 4% as the result of adding 2% to θ_0. Here, we would compute the corresponding smaller-than-θ_0 percentage by subtracting that same 2% from θ_0: $2\% - 2\% = 0\%$.

2. Or, we might think of the increase from θ_0 to 4% as the result of multiplying θ_0 by two, or doubling it. Now, we would compute the corresponding smaller-than-θ_0 percentage by dividing θ_0 by that same factor of 2, halving it: $2\%/2 = 1\%$.

To choose between these, we would need to know more about the concerns of the clothespin manufacturer. To complete the example, however, we'll choose the second definition, taking 1% to be as much smaller than 2% as 4% is larger than 2%.

A second call to the binomial cdf function in S-Plus gives us the probability of 3 or fewer failures in 300 (1% or fewer failures):

```
> pbinom(3,300,0.02)
[1] 0.1485104
```

There is almost a 15% chance of observing 1% or fewer failures in a sample of 300 when the true failure rate is 2%.

Adding the two probabilities together, we find the probability of obtaining either 4% or more failures or 1% or fewer failures in a sample of 300 clothespins when the true failure rate is 2% to be

$$0.1485104 + 0.0189694 = 0.167$$

or, just under 0.17.

Exercises

1. Repeat the primary election hypothesis test given above assuming that you took a random sample of size $n = 250$ and, again, 34% of the sample were primary election participants. Does the p-value of the hypothesis test change? In what direction? Would your substantive conclusion be different?

2. The S-Plus function $\texttt{qbinom}(\alpha, n, \theta)$ is the binomial quantile function (icdf) and returns the smallest value of k for which the binomial cdf, $F(k|n, \theta)$, exceeds α.

Use this function, or a corresponding one in your statistical package, to determine the largest number of primary election voters in a sample of 250 for which you would reject the null hypothesis ($\theta = 0.41$) in preference to the directional alternative ($\theta < 0.41$) at statistical significance level $\alpha = 0.05$. Repeat for $\alpha = 0.01$.

3. When finding the probability of 4% or more failures in the clothespin example above we subtracted from 1.0 the value of the cdf evaluated at 11. Describe why this gives the desired result.

4. We continue now with nondirectional hypothesis testing in the clothespin example. Our null hypothesis again is expressed in terms of the percentage of failures in the population, one day's production: $\theta_0 = 2\%$.

We randomly sample 500 clothespins and observe the number of failures in our sample of 25. The observed failure rate is 5%.

What is the probability, under the null hypothesis, of observing a failure rate that would diverge this much or more from 2% in either direction?

Compute this probability two ways, using the additive and multiplicative definitions of the distance from θ_0 to 5%.

Which one of the definitions do you prefer in this problem? Why? Can you think of some other situation where your preference would be different?

Concepts 5

Parametrics, Pivotals, and Asymptotics

In Concepts 3 and 4, we described both the true and null sampling distributions of an estimator, t, of a population distribution characteristic or parameter, θ. The null sampling distribution is what the sampling distribution would be if the parameter θ had a (null) hypothesized value, θ_0. Both sampling distributions are based on an idealization, that of obtaining attribute scores for all possible random samples of size n from a population consisting of N cases. Now we begin to address how statisticians cope with the impossibility of actually creating sampling distributions.

The Unrealizable Sampling Distribution

Suppose that we have carried out a study based on one random sample from an N-case population. The result, in the simplest case, is a sample distribution consisting of n scores on an attribute of interest. From this distribution, we compute a statistic, perhaps a parameter estimate.

Our computed statistic is only one of the $M = [N!/(n!(N-n)!)]$ equally likely outcomes, each tied to a different sample that might have been randomly chosen. Together, these M statistics make up the sampling distribution of our estimator. To find a second or third member of this distribution, we would have to carry out our study a second or third time on additional random samples of cases, but even one repetition could be prohibitively expensive. To repeat a study several times usually is out of the question. Generally, we can know only the one value of our estimator, the value obtained from our one random sample.

We have no prospect, then, of actually creating the sampling distribution for our statistic. Are there other ways of knowing what the sampling distribution would be like? If not, then our concepts of bias, standard error, confidence intervals, p-values, and hypothesis tests will not help us at all.

Mathematical statistics responds to the challenge to make these concepts useful by identifying those population distributions, sampling designs, statistics, and combinations of these factors for which a mathematical argument can be used to describe a sampling distribution. We now examine some of these results.

We've already seen one of these mathematical results in Applications 4: If the population distribution is large enough, the cdf for the binomial random variable describes the sampling distribution for the number of successes when we randomly sample from a mixture of successes and failures. The following is a second, widely applied mathematical result.

Sampling Distribution of a Sample Mean

We begin with the population distribution of some attribute, X, for which we know the variance, σ^2, but not the mean, μ. A random sample of n cases from the population provides a sample distribution of that attribute, x. From this sample distribution we compute the sample mean,

$$\bar{x} = (1/n)\sum_{i=1}^{n}x_i$$

to be used as an estimate of the population mean, μ.

If N and n are large enough, then the sampling distribution of the sample mean, $\bar{x}$, can be described fully:

1. The sample mean's sampling distribution has a mean of μ, the population distribution mean. Thus, the sample mean is an unbiased estimator of the population mean.

2. The standard error of the sample mean, the standard deviation of its sampling distribution, is given by

$$\text{SE}(\bar{x}) = \sqrt{\frac{\sigma^2}{n}} = \frac{\sigma}{\sqrt{n}}$$

where σ is the population distribution standard deviation.

3. The cdf of the sample mean's sampling distribution is very nearly that of a normal random variable.

The normal random variable is a mathematical construct; this variable can assume any value between $-\infty$ and ∞. By contrast, our estimator can assume, at most, $[N!/(n!(N-n)!)]$ different values. But the cdfs of the two can be close enough that the tabulated or computable values of the normal cdf provide a good basis for estimating a CI for a population distribution mean or for testing a hypothesis about that parameter.

You might recognize that these results describe the central limit theorem of statistics: Whatever the nature of the population distribution X, if the size of that population, N, is large enough, then the sampling distribution of $\bar{x}$ must converge to that of a normal random variable as the size of the sample, n, increases.

The only limitation to these results, then, is knowing when N and n are large enough for the sampling distribution of means of random samples from X to have converged closely enough to the normal distribution. In part, this depends on the population distribution X. For relatively large populations ($N > 1000$), a sample size of $n = 100$ ensures convergence in almost any instance.

The sample mean is a rare statistic, one for which the sampling distribution can be fully described, subject only to the sizes of population and sample. Even so, this description involves a second, perhaps unknown parameter, the population distribution variance, σ^2. For other statistics, we must have more information or be willing to accept additional assumptions if we are to produce sampling distributions. Figure 5.1 shows three ways in which this information can be gained through mathematical analysis.

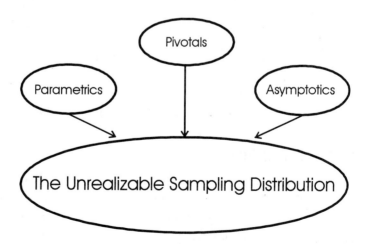

Figure 5.1:   *Mathematical Approaches to Knowledge of the Sampling Distribution*

Parametric Population Distributions

One mathematical route to specifying a sampling distribution is to require the population distribution sampled to have a particular mathematical form.

For example, consider σ^2, the variance of a population distribution. We can compute an estimate,

$$\hat{\sigma}^2 = \frac{1}{n-1} \sum_{i=1}^{n} (x_i - \bar{x})^2$$

from the n scores for a random sample of cases. If the population distribution X has the same cdf as a normal random variable, then the sampling distribution of $\hat{\sigma}^2$ can be described. In particular, the cdf for the sampling distribution of $\hat{\sigma}^2$ is proportional to that of the chi-squared random variable with $n-1$ degrees of freedom (df). The constant of proportionality is $k = [\sigma^2/(n-1)]$.

The chi-squared random variable with $n-1$ df has a mean of $n-1$. The sampling distribution of $\hat{\sigma}^2$ has a mean k-times that of the chi-squared variable,

$$k(n-1) = [\sigma^2/(n-1)](n-1) = \sigma^2$$

and our estimator is unbiased. We also can use the tabulated and computable values of the cdf of the chi-squared random variable to work out a CI for the population distribution variance.

The important point is that any statements about the bias, SE, or CI for $\hat{\sigma}^2$ based on the chi-squared random variable cdf require that the population distribution X have a normal distribution. This is never completely true, and many population distributions bear faint resemblance to the bell-shaped curve.

Pivotal Form Statistics

The sampling distribution for $\hat{\sigma}^2$ described in the previous section changes from one normal population distribution to another, as the value of the parameter being estimated, σ^2, changes. The sampling distribution of the estimator is different for $\sigma^2 = 100$ than for $\sigma^2 = 64$, and different for a null-hypothesized value, σ_0^2, than for some alternative value. To avoid the difficulties associated with a shifting sampling distribution, mathematical statisticians have developed pivotal forms for several estimators. The sampling distribution of a pivotal form does not change as we move from one population distribution to another, with a consequent change in the value of the parameter being estimated.

For the normal population distribution variance estimate, the statistic

$$v = \frac{\widehat{\sigma}^2}{k} = \frac{\widehat{\sigma}^2(n-1)}{\sigma^2}$$

is a pivotal form. The cdf for the sampling distribution of v is that of the chi-squared random variable with $n-1$ df, whatever the value of σ^2 in the normal population sampled.

Not all estimators have pivotal forms. The convenience of pivotal forms, particularly for CI estimation and hypothesis testing, has encouraged researchers to use estimators that do have pivotal forms and, thus, to focus attention on certain parameters. You might overlook parameters of potentially greater intrinsic interest if their estimators do not have pivotal forms.

Perhaps the best known pivotal form is that for the sample mean. We noted earlier that the sampling distribution for the sample mean, for n and N large enough, has a normal cdf with mean equal to μ, the mean of the population distribution sampled, and SE of $(\sigma/\sqrt{n})$, where σ^2 is the variance of the population distribution sampled. As a result, this linear transformation of the sample mean,

$$z(\overline{x}) = \frac{\overline{x} - \mu}{\text{SE}(\overline{x})} = \frac{\overline{x} - \mu}{(\sigma/\sqrt{n})}$$

will have a sampling distribution with cdf identical to the standard normal random variable, with a mean of 0 and variance of 1. That is, the sampling distribution of $z(\overline{x})$ will have this same normal cdf whatever the value of μ (and the value of σ) in the population distribution sampled.

Unfortunately, the population distribution whose unknown mean, μ, we want to estimate usually has an unknown σ^2 as well. We could estimate σ^2 from our sample distribution,

$$\widehat{\sigma}^2 = \frac{1}{n-1}\sum_{i=1}^{n}(x_i - \overline{x})^2$$

and replace the unknown variance with this estimate in $z(\overline{x})$. The result,

$$t(\overline{x}) = \frac{\overline{x} - \mu}{\widehat{SE}(\overline{x})} = \frac{\overline{x} - \mu}{(\widehat{\sigma}/\sqrt{n})}$$

is no longer a simple transformation of the sample mean; $\widehat{\sigma}$ is not constant but would change value from sample to sample. Mathematical statisticians have established that for very, very large sample sizes, as n moves toward infinity, the sampling distribution of $t(\overline{x})$ is well approximated by the standard normal cdf. For finite sample sizes, however, the approximation might not be very good.

The English statisticians Gosset (who wrote as "Student") and Fisher established that the sampling distribution of $t(\overline{x})$ would have exactly the cdf of the t-random variable with $(n-1)$ df, provided the population distribution sampled had a normal cdf. That is, $t(\overline{x})$ is a pivotal form of the sample mean when the sample is drawn from a normal population distribution; the sampling distribution of $t(\overline{x})$ has the cdf of a t-random variable, whatever the value of μ (or, of σ) in the normal population sampled.

This pivotal form is used, quite frequently, in two ways:

1. To test a null hypothesis about μ. The null hypothesized value of the population distribution mean, μ_0, is substituted for μ in the $t(\overline{x})$ statistic computation. By evaluating the t-random variable cdf at the resulting quantity, $t(\overline{x}|\mu_0)$, the researcher can determine a p-value for the hypothesis test.

2. To estimate an equal tails exclusion $(1 - 2\alpha)100\%$ CI for μ. The $t(\overline{x})$ equation is solved for the upper and lower limiting values of μ:

$$\mu_L = \overline{x} - \sqrt{\widehat{\sigma}^2/n} \times F_{t(n-1)}^{-1}(1 - \alpha)$$

and

$$\mu_U = \overline{x} + \sqrt{\widehat{\sigma}^2/n} \times F_{t(n-1)}^{-1}(\alpha)$$

where $F_{t(n-1)}^{-1}(\alpha)$ and $F_{t(n-1)}^{-1}(1 - \alpha)$ are the α and $(1 - \alpha)$ quantiles of the t-distribution with $(n-1)$ df. As these distributions are symmetric about zero,

$$F_{t(n-1)}^{-1}(1 - \alpha) = -F_{t(n-1)}^{-1}(\alpha)$$

the CI is expressed more compactly as

$$\bar{x} \pm \sqrt{\hat{\sigma}^2/n} \times F^{-1}_{t(n-1)}(\alpha)$$

We know that the cdf of the sampling distribution of $t(\bar{x})$ is that of a t-random variable only when the sample distribution, x, is drawn from a normal population distribution. Alas, all too often researchers assume that the sampling distribution of $t(\bar{x})$ has the mathematical form of a t-random variable when x is not a normal random sample.

Asymptotic Sampling Distributions

Mathematical statistics makes a third contribution to our knowledge of the unrealizable sampling distribution. This has been to work out what would be true of the sampling distributions of certain statistics asymptotically, as the size of the random sample approaches infinity. These results are then used when n is thought to be large enough for the asymptotic results to be applicable. We've seen one example of that already in the pivotal form $z(\bar{x})$, which has a standard normal cdf as the sample size approaches infinity.

The well-known Pearson chi-squared test of row-by-column independence in a two-way table of frequencies is grounded in just such an asymptotic result. In one study design, we sample cases at random from a population with a bivariate score distribution, X. That is, each case has a score on two attributes. These are categorical attributes, and the scores determine the row and column placement of the case.

Each sampled case contributes to the number in one of the cells of our two-way table. We let the number of rows and columns be I and J, respectively, and represent the number of sampled cases that are in both the i-th row category and the j-th column category as n_{ij}. The number of sample cases in the i-th row category is

$$n_{i.} = \sum_{j=1}^{J} n_{ij} \text{, for } i = 1, 2, \ldots, I$$

The number of sample cases in the j-th column category is

$$n_{.j} = \sum_{i=1}^{I} n_{ij} \text{ , for } j = 1, 2, \ldots, J$$

and the sum of all of the tabled frequencies is the sample size,

$$\sum_{i=1}^{I} \sum_{j=1}^{J} n_{ij} = n$$

Under the null hypothesis that row and column attributes operate independently of one another in classifying cases, we can model the number in the sample we would expect in the ij-th cell of the table as

$$m_{ij} = \frac{n_{i.} \times n_{.j}}{n} \text{ , for } i = 1, 2, \ldots, I \text{ and } j = 1, 2, \ldots, J$$

The Pearson chi-squared test statistic is obtained from the observed and modeled cell frequencies as

$$\chi^2 = \sum_{i=1}^{I} \sum_{j=1}^{J} \left[\frac{(n_{ij} - m_{ij})^2}{m_{ij}} \right]$$

Asymptotically, as n approaches infinity, the null sampling distribution of this statistic has the same cdf as the chi-squared random variable with $(I-1) \times (J-1)$ df. Large values of the test statistic offer evidence against the null hypothesis. Considerable work has been done by mathematical statisticians to adapt this asymptotic result to moderate sized samples, to determine how large n or the individual m_{ij}s must be for the two cdfs to be sufficiently close.

An additional assumption of the chi-squared test is that the population size, N, be large enough that drawing out n cases cannot alter the score distributions of the two attributes. If n is large relative to N, the nature of X would change when you withdraw sample scores.

Limitations of the Mathematical Approach

Unfortunately, parametric assumptions, ingenuity in devising pivotal forms of statistics, and reliance on large sample results cover only some of the data analysis questions researchers face.

We do not always have

1. Random samples of cases

2. Population distributions with known parametric forms

3. Statistics with known pivotal forms

4. Large numbers of cases, either in populations or samples

In Concepts 6 through 16, we will develop alternatives to mathematically derived sampling distributions.

Applications 5

CIs for Normal Population μ and σ^2

If the population distribution X has as its cumulative distribution function (cdf) that of a normal random variable, then we know the sampling distributions of transformations of the two estimators

$$\bar{x} = (1/n)\sum_{i=1}^{n} x_i$$

and

$$\widehat{\sigma}^2 = \left(\frac{1}{n-1}\right)\sum_{i=1}^{n}(x_i - \bar{x})^2$$

We can use this knowledge to create confidence intervals (CIs) for the population mean, μ, and variance, σ^2.

CI for a Normal Population Mean

The sampling distribution of

$$t(\bar{x}) = \frac{\bar{x} - \mu}{\sqrt{\widehat{\sigma}^2/n}}$$

has as its cdf that of the t-random variable with $(n-1)$ degrees of freedom (df), if x is a random sample of n elements from a normal population distribution with mean μ. We can invert this equation, for appropriately chosen values of t, to find a $(1 - 2\alpha)100\%$ CI for μ:

$$\bar{x} \pm \left[\sqrt{\widehat{\sigma}^2/n} \times F_{t(n-1)}^{-1}(\alpha)\right]$$

Here is an example of the use of this CI formulation.

The data in Table 5.1 are taken from Daly et al. (1995) and give the weight in grams of a random sample of 50 bags of pretzels. The bags were sampled from the output of a packaging machine thought to be set to fill the bags, on average, to a weight of 454 gm.

The population distribution of weights is known to have a normal cdf. We want a 90% confidence interval for μ, the mean of that population distribution.

Table 5.1:   *Weights (in grams) of 50 Randomly Sampled Bags of Pretzels*

464	450	450	456	452	433	446	446	450
447	442	438	452	447	460	450	453	456
446	433	448	450	439	452	459	454	456
454	452	449	463	449	447	466	446	447
450	449	457	464	468	447	433	464	469
457	454	451	453	443				

The mean weight for our sample of 50 bags is $\bar{x} = 451.22$ gm, the estimated population standard deviation is $\hat{\sigma} = 8.40$ gm.

To create our 90% equal tails exclusion CI, we need the 0.05 quantile of the t-distribution with 49 df, $F_{t(49)}^{-1}(0.05)$. The S-Plus t-random variable quantile function, qt (α, df), can be used to find this quantile:

```
> qt(0.05,49)
[1] -1.676551
```

We now have all we need to compute the limits to our CI:

$$451.22 \text{ gm} \pm \left[\left(8.40 \text{ gm}/\sqrt{50} \right) 1.677 \right] =$$

$$451.22 \text{ gm} \pm 1.99 \text{ gm} = [449.23 \text{ gm}, 453.21 \text{ gm}]$$

We are 90% confident that this interval includes μ.

CI for a Normal Population Variance

The sampling distribution of the statistic

$$v = \frac{(n-1)\hat{\sigma}^2}{\sigma^2}$$

has as its cdf that of the chi-squared random variable with $(n-1)$ df when $\hat{\sigma}^2$ is computed from a sample distribution, x, drawn from a normal population with variance σ^2. We can use this sampling distribution knowledge to form a confidence interval for σ^2. We'll show how to do this. First, though, we outline the logic of the process in some detail:

1. The α and $(1-\alpha)$ quantiles of the sampling distribution of v are the corresponding quantiles of the chi-squared random variable with degrees of freedom $(n-1)$ and we write these as:

$$F^{-1}_{\chi^2,(n-1)}(\alpha) \text{ and } F^{-1}_{\chi^2,(n-1)}(1-\alpha)$$

2. If we multiply each element of the sampling distribution of v by $[\sigma^2/(n-1)]$ the result is the sampling distribution of $\hat{\sigma}^2$. In particular, if we apply this multiplication to the α and $(1-\alpha)$ quantiles of the sampling distribution of v, we have the α and $(1-\alpha)$ quantiles of the sampling distribution of $\hat{\sigma}^2$:

$$F^{-1}_{\hat{\sigma}^2}(\alpha) = [\sigma^2/(n-1)]F^{-1}_{\chi^2,(n-1)}(\alpha)$$

and

$$F^{-1}_{\hat{\sigma}^2}(1-\alpha) = [\sigma^2/(n-1)]F^{-1}_{\chi^2,(n-1)}(1-\alpha)$$

3. If we substitute the right sides of the two equations above into the equations given in Concepts 3 for the amounts to be added to and subtracted from t to form the limits of the $(1-2\alpha)100\%$ CI,

$$a = F^{-1}_{\hat{\sigma}^2}(1-\alpha) - \sigma^2$$

and

$$b = \sigma^2 - F^{-1}_{\hat{\sigma}^2}(\alpha)$$

we see that those lengths each depend, multiplicatively, on the unknown value of σ^2:

$$a = \sigma^2 \times \left\{ \left[F^{-1}_{\chi^2,(n-1)}(1-\alpha)/(n-1) \right] - 1 \right\}$$

and

$$b = \sigma^2 \times \left\{ 1 - \left[F^{-1}_{\chi^2,(n-1)}(\alpha)/(n-1) \right] \right\}$$

4. What we have observed is that the spread of the sampling distribution of $\widehat{\sigma}^2$ depends on the value of the parameter, σ^2. The Concepts 3 solution for the CI limits, $[t - a, t + b]$, is appropriate when the location alone of the sampling distribution is sensitive to the parameter value, but this solution is not appropriate when the spread of the distribution accelerates as the parameter value changes. We need a different approach.

5. We desire an interval that if constructed around any value of $\widehat{\sigma}^2$ between

$$\widehat{\sigma}^2_L = [\sigma^2/(n-1)] \times F^{-1}_{\chi^2,(n-1)}(\alpha)$$

and

$$\widehat{\sigma}^2_U = [\sigma^2/(n-1)] \times F^{-1}_{\chi^2,(n-1)}(1-\alpha)$$

will overlap σ^2, but if constructed around a value below $\widehat{\sigma}^2_L$ or above $\widehat{\sigma}^2_U$, will fail to include σ^2.

6. The following solution depends on our knowing for the chi-squared distribution that

$$0 < F^{-1}_{\chi^2,(n-1)}(\alpha) < (n-1) < F^{-1}_{\chi^2,(n-1)}(1-\alpha)$$

In fact, the mean of the chi-squared distribution with $(n-1)$ df is $(n-1)$.

7. Let's start with $\widehat{\sigma}^2$ at the upper limit of the range we want to be covered: $\widehat{\sigma}^2_U = [\sigma^2/(n-1)] \times F^{-1}_{\chi^2,(n-1)}(1-\alpha)$. If we multiply this value of $\widehat{\sigma}^2$ by the ratio $\left[(n-1)/F^{-1}_{\chi^2,(n-1)}(1-\alpha) \right]$ the result is exactly σ^2. However, if we multiply any $\widehat{\sigma}^2$ that is larger than $\widehat{\sigma}^2_U$ by this same ratio, the product will be larger than σ^2. This gives us a formula for the lower bound of a $(1-2\alpha)100\%$ CI for σ^2,

$$\text{LB} = \widehat{\sigma}^2 \times \left[(n-1)/F^{-1}_{\chi^2,(n-1)}(1-\alpha) \right]$$

This lower bound will be larger than σ^2, excluding it from the interval, for the $100\alpha\%$ of the $\widehat{\sigma}^2$s that are at the high end, above $\widehat{\sigma}^2_U$.

8. Now, we'll take $\widehat{\sigma}^2$ to be at the lower end of the range to be covered: $\widehat{\sigma}^2_L = [\sigma^2/(n-1)] \times F^{-1}_{\chi^2,(n-1)}(\alpha)$. If we multiply this value of $\widehat{\sigma}^2$ by the ratio $\left[(n-1)/F^{-1}_{\chi^2,(n-1)}(\alpha) \right]$, the result is exactly σ^2. But if we multiply any $\widehat{\sigma}^2$

that is smaller than $\hat{\sigma}_L^2$ by this same ratio, the product will be smaller than σ^2. This gives us a formula for the upper bound of a $(1 - 2\alpha)100\%$ CI for σ^2,

$$\text{UB} = \hat{\sigma}^2 \times \left[(n-1)/F_{\chi^2,(n-1)}^{-1}(\alpha) \right]$$

This upper bound will be smaller than σ^2, excluding it from the interval, for the $100\alpha\%$ of the $\hat{\sigma}^2$s that are at the low end, below $\hat{\sigma}_L^2$.

9.   Taken together, the two results define a $(1 - 2\alpha)100\%$ equal tails exclusion CI for the variance of a normal population distribution:

$$\text{LB} = \hat{\sigma}^2 \times \left[(n-1)/F_{\chi^2,(n-1)}^{-1}(1-\alpha) \right]$$

and

$$\text{UB} = \hat{\sigma}^2 \times \left[(n-1)/F_{\chi^2,(n-1)}^{-1}(\alpha) \right]$$

Let's apply these results. The fifty bag weights of Table 5.1 are a random sample from a normal population distribution with variance σ^2. We want to construct a 90% equal tails exclusion CI for that parameter.

For this construction, we already have determined $n = 50$, $\alpha = 0.05$, and $\hat{\sigma}^2 = 70.50$ gm^2. Next, we use the S-Plus quantile function for the chi-squared random variable, qchisq(α,df), to find $F_{\chi^2,49}^{-1}(0.05) = 33.93$ and $F_{\chi^2,49}^{-1}(0.95) = 66.34$. From these results we compute lower and upper bounds to our 90% CI of

$$\text{LB} = 70.50 \text{ gm}^2 \times \left(\tfrac{49}{66.34} \right) = 52.07 \text{ gm}^2$$

and

$$\text{UB} = 70.50 \text{ gm}^2 \times \left(\tfrac{49}{33.93} \right) = 101.81 \text{ gm}^2$$

We have 90% confidence that the interval from 52.07 gm^2 to 101.81 gm^2 contains the population variance, σ^2.

Nonparametric CI Estimation

The confidence intervals for the mean and variance explored here depend on our knowing that the sample distribution was obtained from a normal population distribution. These CIs are parametric in two respects: We know the parametric form of the population distribution sampled to be normal, and that, in turn, allows us to take advantage of the mathematical derivation of the parametric forms of the sampling distributions of $t(\bar{x})$ and $v(\hat{\sigma}^2)$.

Generally we do not know the parametric form of either the population distribution we've sampled or the sampling distribution of the statistic we've computed from the sample distribution. As a result, our CIs must be nonparametric ones. Those we will discuss in Concepts 9 and 10 have the advantage of overcoming our lack of knowledge of two things important here:

1. A pivotal transformation of our statistic

2. The extent to which the spread of the sampling distribution accelerates as the parameter value changes

Exercises

The data in Table 5.2 are taken from Hand et al. (1994, Data Set 392) and give scores on the Vocabulary subtest of the WAIS-R intelligence test for 54 university students. Solely for the purposes of this exercise, we'll assume the students were randomly sampled from a large university population, and that, in that population, the vocabulary scores form a distribution with a normal cdf. In fact, there is every reason to believe that neither assumption is correct and, after learning the lessons of subsequent Concepts, we would proceed much more realistically.

Table 5.2:   *WAIS Vocabulary Scores for a University Sample*

```
14 11 13 13 13 15 11 16 10 13 14 11 13 12 10 14 10 14
16 14 14 11 11 11 13 12 13 11 11 15 14 16 12 17  9 16
11 19 14 12 12 10 11 12 13 13 14 11 11 15 12 16 15 11
```

1.   Find an equal tails exclusion 95% CI for the mean of the university population distribution of WAIS-R vocabulary scores using the sample data in Table 5.2.

Find a 95% lower bound for the mean from the same data. Interpret this bound.

2. Find a 95% equal tails exclusion CI for the variance of the population distribution of vocabulary scores sampled in Table 5.2.

3. Explain how you would find a 90% lower confidence bound for the population variance. Illustrate the process with the pretzel bag weight sample.

Concepts 6

Limitations of Parametric Inference

In Concepts 5, we outlined how mathematical analysis has helped us gain an understanding of the sampling distribution or null sampling distribution that underpins parametric population estimation and hypothesis testing. Here, we review the limitations of this approach to statistical inference.

Range and Precision of Scores

One mathematical approach to statistical inference is to conceptualize the population distribution of attribute scores as identical to that of a random variable. Random variables are characterized as discrete or continuous.

The discrete random variable can take only certain values. The binomial random variable, $\mathrm{Bin}(n, p)$, is a discrete random variable; it takes only integer values between zero and n. $\mathrm{Bin}(10, 0.5)$, for instance, models the number of Heads in 10 flips of a fair coin. We can see only values of $0, 1, 2, \ldots, 9, 10$.

In contrast, the continuous random variable can assume all possible values, at least within a certain range. The normal random variable, $\mathrm{N}(\mu, \sigma)$, is a continuous random variable; it takes all possible values between $-\infty$ and ∞, though the more frequently occurring values are those close on either side to the mean, μ. How closely the normal scores are bunched about μ is determined by the second parameter, σ, the standard deviation (SD). For example, approximately two-thirds of the scores will lie in the interval $(\mu \pm \sigma)$.

The score we obtain for a case is the result of an interaction between the strength of an attribute in that case and the characteristics of our instrument or technique for measuring that strength. The result is restricted in two ways:

1. Score values are constrained to lie within a certain range. The time taken by a college student to read four paragraphs of newspaper text, for example, certainly is bounded from below by zero and from above by 30 minutes.

2. Score values are constrained to take only certain values; that is, they are limited in their precision. Blood pressure is recorded as a positive

integer, the unit being millimeters of mercury. Values such as 120.1 or 87 plus 1/3 are not noted in any sample.

The continuous random variable cannot represent, faithfully, the population distribution of scores for any realistically measured attribute. Nonetheless, to facilitate the computation of confidence intervals (CIs) and p-values, researchers customarily assume that attributes such as cognitive task time and blood pressure have population distributions with continuous random variable cdfs, often that of the normal random variable. We'll learn to avoid such assumptions.

Size of Population

Our mathematical derivations of sampling distributions often assume that the attribute scores for our n sample cases were obtained independently of one another and from an unchanging, essentially infinitely large population distribution. Such scores are said to be independent and identically distributed or iid observations. In a case-based data analysis context, we will have iid observations if either

1. We sample the cases randomly and with replacement from our population, or

2. We sample the cases randomly and without replacement and our population is a very large one.

The first condition almost never holds. Almost no researcher would permit the same case to be included twice in a sample. For the second condition to be true, we must be satisfied that the population is so large that the population distribution of attribute scores remains constant as we withdraw cases from the population and, coincidentally, withdraw their scores from the population distribution. As a rule of thumb, we'll require here that the nonreplacement random sample contain no more than 5% of the population, that $N > 20n$.

This rule is not always met, so we shall develop appropriate ways for dealing with samples from smaller populations.

Size of Sample

We have just seen that our sample might be too large, relative to the size of the population, to justify the assumption of iid observations. That same sample also might be too small to justify the use of asymptotic results in the description of the sampling distribution of a statistic.

Typically, where we have random samples of cases, they are moderately sized samples from moderately sized populations. For realistic data analysis, we need inferential techniques that consider the actual sizes of both, in place of presumptions that either is small enough or large enough.

Roughness of Population Distributions

Any case population is defined empirically, as a circumscribed collection of, for example, patients, pupils, plots of land, or political subdivisions. These populations also are finite. We could not otherwise identify a random sample of cases.

As a result, the distribution of scores for any attribute over the population—a population distribution—also is finite and empirically defined. It is quite unlikely, therefore, that a population distribution will follow any nice mathematical form. There is little reason, for example, to assume that a histogram for a population distribution of attribute scores, were we able to create it, would show score frequencies declining smoothly and symmetrically either side of the mean or even smoothly but asymmetrically.

The normal or any other mathematically formulated smooth cumulative distribution function (cdf) might not provide a good fit to finite and empirical cdfs. Population distributions are likely to exhibit more roughness.

Parameters and Statistics of Interest

Some population parameters and their estimators are mathematically more tractable than others. This has led researchers, for example, to focus on the mean as a measure of the typical size or location of the scores in a distribution and on the variance or standard deviation as a measure of the variability of those scores.

Although the mean is so widely used—as a descriptive statistic as well as inferentially through the *t*-test, the analysis of variance, and linear regression—it is not always the best indicator of typical size.

    1. For a symmetric score distribution, the mean and median coincide. For a non-symmetric distribution, the two diverge and the median is likely a better candidate for center of the distribution than is the mean.

    2. The mean—whether of a data set or a population distribution—is notoriously nonresistant to the presence of even a single very small or very large score. Such scores can force the mean to take either an arbitrarily

small or arbitrarily large value, well away from the bulk of the data. The location of the distribution of scores would be better assessed by a more resistant measure, for example, a trimmed mean.

Similar concerns can be raised about the standard deviation as a measure of variability. The standard deviation shares with the mean a lack of resistance to the influence of an atypically large or small score.

Scarcity of Random Samples

A major limitation of statistical inference based on the sampling distribution is that it assumes our data set is the result of randomly sampling cases from a well-defined population. Random samples, however, can be difficult to obtain and are relatively uncommon in many empirical disciplines.

Approaches to inference other than the sampling distribution are needed when our data set consists of scores for available, rather than randomly sampled, cases.

Resampling Inference

In Concepts 7 through 16, we will develop three resampling approaches to statistical inference. Taken together, they overcome the limitations of classical parametric inference.

Each approach is directed at a particular class of research design. Realistic data analysis requires that the statistical inference methodology used in an analysis be appropriate to the research design from which the data were obtained. Population inference requires random sampling. Causal inference requires random treatment assignment. These principles are not mere statistical niceties. To contravene them is to risk deceiving, albeit unintentionally, those to whom you report your research findings.

As illustrated in Figure 6.1, the major considerations in the selection of a resampling approach to statistical inference are whether cases have been randomly sampled and, if not, whether the available cases were randomly assigned to different levels of treatment.

Random Samples: Bootstrap Inference

When we have a random sample of cases from a population, our goal is to draw inferences about that population. The sampling distribution approach allows us to do so. The use of bootstrap resamples extends the sampling distribution approach by overcoming the limitations of parametric inference.

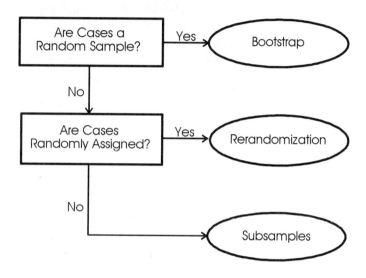

Figure 6.1:   *Resampling Strategies and Random Designs*

The bootstrap approach to statistical inference, which we introduce in Concepts 7 through 11, allows us to keep our sampling distribution development faithful to the following:

1. The size and nature of our sample

2. The size of our population

3. Our knowledge about that population

4. Our choice of estimator or test statistic

The name bootstrap alludes to the process of estimating a sampling distribution by drawing repeated resamples from the one random sample, pulling our sampling distribution up, as it were, by the bootstraps.

Random Assignment: Rerandomization

Many experiments call for dividing a set of cases among two or more treatment groups. There are advantages, from the standpoint of statistical inference, to making this division on a random basis.

1. If the cases to be allocated among treatments are a random sample from a natural case population, then their random division among K treatments yields random samples from each of K prospective populations. We can then use bootstrap methods to make inferences about the distributions of attribute scores in these K populations.

2. If the cases constitute only an available set, the random assignment of cases to treatments provides the statistical basis for a second kind of inference.

Without random sampling, we have no warrant for population inferences. But the randomization of available cases among treatments provides the statistical basis for vitally important causal inferences. If we repeatedly rerandomize treatment assignments, under an appropriate null hypothesis, we can determine how likely it is that an observed treatment response difference could be the chance result of our randomization. When a result is highly unlikely to be the result of case randomization, we can argue that the treatments themselves are responsible for or cause the treatment response difference. The basis for rerandomization inference is developed in Concepts 12 through 15.

Nonrandom Designs: Subsampling

Where the cases in a study are neither a random sample nor randomly allocated among treatments, neither population nor causal inferences are appropriate. Each depends on a random mechanism in the study design to justify inference.

In the absence of either random mechanism, scores for the available cases—which, in some instances, make up a population—can, of course, be summarized. That is, the data set can be described. This description can be simple, such as the mean score on a single attribute, or more complicated, such as a factor analysis of the intercorrelations among 30 attributes. These descriptions are, in a limited sense, inferences—inferences about the available cases and their attributes.

For such a limited inference to be useful to us scientifically, we need assurance that the description truly characterizes the available cases, that it is stable. A description is stable if it is relatively unaffected by the withdrawal of some cases. If the description changes markedly as cases are withdrawn, then we should proceed with caution in offering a description of the data.

Our approach to assessing stability is to recompute the description on a sequence of subsamples. Each subsample contains some but not all the available cases. We look at local descriptive inference based on subsamples in Concepts 16.

Applications 6

Resampling Approaches to Inference

Figure 6.1 outlines my orientation to statistical inference. We take one of three different approaches depending on how or whether a random element has been built into the case-based study:

1. If the cases are a random sample from a well-defined population of cases, then we can make estimates of or test hypotheses about properties of a population distribution of attribute scores. We call this population inference.

2. If we have only a set of available cases, not a random sample, but those cases have been randomly allocated among two or more treatment groups, then we can test hypotheses about the relative effectiveness of those treatments. We call this causal inference becasuse we are interested in whether the different treatments cause different responses.

3. If the cases have been neither randomly sampled nor randomly allocated among treatments, then neither population nor causal inference are possible. We can describe the outcomes of the study and, when that description is stable, we refer to it as a local descriptive inference.

Exercises

1. Propose a study involving random samples of cases. What population or populations would you sample? Would these be natural, prospective, or contructed populations? What attributes would you measure? What treatments, if any, would you provide? What population inferences would be of interest in your study?

2. Propose a study in which a set of available cases is randomly allocated among treatments. What is the source of your available cases? How are they randomly divided? What treatments are provided? How is the response to treatment assessed? What treatment comparisons would you make? What causal inferences are related to these treatment comparisons?

3. Propose a study in which cases are neither sampled nor allocated randomly but constitute a (small) population. What outcome descriptions would be of interest?

Concepts 7

The Real and Bootstrap Worlds

We introduce here a bootstrap world of population inference. This bootstrap world is made up of population and sample distributions and of statistics and their sampling distributions, and mirrors a real world of population inference, made up of corresponding objects. The bootstrap world is useful to us because its objects are all fully attainable whereas many of those in the real world are not. We can use these attainable bootstrap world objects, particularly the sampling distribution of a statistic, to estimate properties of the corresponding but unknowable real world objects.

The Real World of Population Inference

The real world of population inference is described by Concepts 2 through 4: It derives from a population distribution about which we would like to make one or more inferences. We begin with a review of concepts associated with this world.

At the top of Figure 7.1 is the distribution of scores on some attribute for a population of N cases. This population distribution is denoted X and has associated with it some parameter θ. We want to estimate the value of this parameter. We sample at random n cases from our population. From the sample distribution of attribute scores for these cases, x, we compute our parameter estimate, t. The sample score distribution and estimate are labeled $x(1)$ and $t(1)$ in the figure as a reminder that we have a single sample and a single sample estimate of θ.

Within the population of N cases, there are

$$M = \frac{N!}{n!(N-n)!}$$

different samples of size n that could have been drawn. At random, we obtained just one of these. The sampling distribution for our estimator is the collection of the values it takes across these M samples, $t(i)$, $i = 1, ..., M$.

The dotted circles and lines in Figure 7.1 remind us that we have only one sample and can compute only one value of t. As a result, we cannot fill out the sampling distribution at the bottom of the diagram.

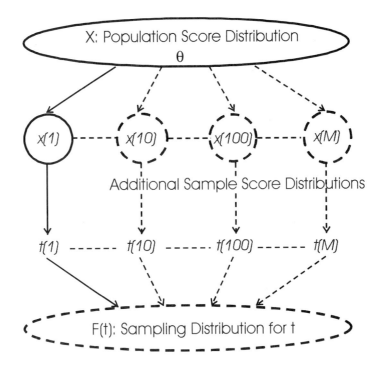

Figure 7.1:   *The Real-World Population and Sample Distributions*

We now will refer to X and x as the real-world population and sample distributions, θ as the real-world parameter, and the single sample t as the real world sample estimate of θ. The collection of values of t obtained from the M possible random samples makes up the real-world sampling distribution of t. The cumulative distribution function (cdf) of this sampling distribution is denoted as $F(t)$.

The only objects that are known to us from the real world are the population size, N; the sample size, n; the sample score distribution, x; and the value of the estimator computed from that sample distribution, t. We are limited, remember, because we cannot afford to obtain additional random sample distributions. Each such distribution would require a repetition of our experiment or observational study.

The Bootstrap World of Population Inference

The parallel bootstrap world is illustrated in Figure 7.2. At the top is a population distribution of scores on an attribute. This population distribution, which again consists of N scores, is designated $\widehat{X}$.

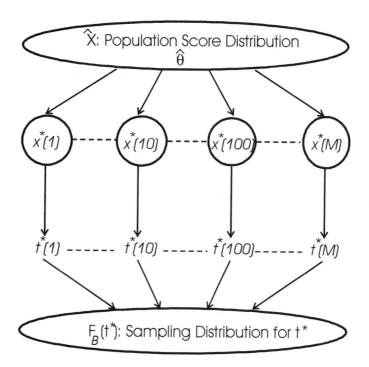

Figure 7.2:   *Elements of The Bootstrap World*

This hat ($\widehat{}$) notation identifies the bootstrap-world population distribution as an estimate of the corresponding real-world population distribution, X. We'll see shortly how we can estimate population distributions. For now, the important thing is that this estimate, $\widehat{X}$, is completely known to us; it is nothing more than a collection of N scores whose values we have specified.

Associated with this bootstrap-world population distribution is a parameter, $\widehat{\theta}$, computed from $\widehat{X}$ by the same rule we would use to compute θ from X. If θ were the mean of the population distribution X, then $\widehat{\theta}$ is the mean of the N scores making up $\widehat{X}$. Because we know all the scores in $\widehat{X}$, we can compute $\widehat{\theta}$, the bootstrap world parameter.

The key to bootstrap inference is that because the population distribution of N scores is completely known to us, we can draw from it as many random samples of size n as we want. We can, in principle, write down the sample distribution for each of the M possible random samples of size n. All that is required is a simple computing algorithm to ensure that we obtain each possible sample and the computing time to run that algorithm.

Samples from this bootstrap-world population distribution are termed bootstrap samples, and Figure 7.2 designates the score distribution for any one of these samples as $x^*(b)$, $b = 1, \ldots, M$.

Just as we calculated t, our estimate of θ, from our one real-world sample, we can calculate that same statistic for each bootstrap sample. We apply the same computational rule to the n scores in each bootstrap sample distribution that we applied to the n scores making up x. The statistic for the b-th bootstrap sample is designated $t^*(b)$.

Collecting $t^*(b)$ from each of the M samples produces the sampling distribution of t^*, a bootstrap sampling distribution. Again, because $\widehat{X}$ is simply a collection of N scores we could, given sufficient computing time, form all M sample score distributions and compute t^* from each.

In practice, M tends to be very large, and we usually approximate the bootstrap sampling distribution by computing t^* in each of a not-quite-so large number, B, of randomly chosen bootstrap sample distributions. More about this approximation appears in Concepts 8.

Real-World Population Distribution Estimates

To implement the bootstrap approach to population inference we need $\widehat{X}$, a numeric estimate of the real-world population distribution of attribute scores. Where should these N scores come from? In this section we outline four different approaches to this question. All use information from x, our single real-world sample. However, the four approaches reflect different amounts or kinds of knowledge about X. That knowledge should be considered if $\widehat{X}$ is to be our best estimate of X. In considering the four approaches it is important to distinguish knowledge from assumptions.

Nonparametric Estimate

Most likely we know only two thing about the population distribution X:

1. It consists of N scores.

2. Our random sample distribution x contains n of those N scores.

We are trying, by our experiment or observational study, to learn about X and probably the only knowledge we have about it is contained in the sample x. This would be the case, for example, if x is made up of attribute scores for the first (random) sample to be taken from our case population. Equivalently, there might have been other random samples, but ours is the first one in which this particular attribute was scored.

How should we estimate X in this situation? We know that X contains more scores than are in x, actually $(N - n)$ of them. What would they look like? In the absence of additional knowledge about the population distribution, we can only assume that the scores for the unsampled cases look exactly like those for the sampled cases. Because such an estimate does not rely on any knowledge of the form of the population distribution, it is referred to as a nonparametric estimate.

Consider this hypothetical, Number of Children, example. From a population of 1,000 female college graduates of 1992, we randomly sample ten women. We ask each the number of children she has borne. These are the results, making up our sample distribution, x:

$$0, 0, 0, 0, 1, 1, 1, 2, 2, 3$$

Our nonparametric estimate of X, the distribution of the numbers of children among the population of 1,000 women, is that it consists not only of the $n = 10$ scores in x but of $N - n = 990$ others, exactly like these ten. By that we mean that X is to be estimated as containing only the scores appearing in x $(0, 1, 2,$ and $3)$ and in the same proportions as they appear in x. Because our real-world sample distribution consists of

40% 0s, 30% 1s, 20% 2s, and 10% 3s

we create a nonparametric bootstrap-world population distribution that is also

40% 0s, 30% 1s, 20% 2s, and 10% 3s

but containing a total of $N = 1000$ scores.

More generally, the nonparametric $\widehat{X}$ consists of multiple copies of x. If $c = N/n$ is an integer, then, $\widehat{X}$ is made up of c copies of x. In our example, $c = 1000/10 = 100$ and the nonparametric $\widehat{X}$ would consist of 100 copies of each of the 10 scores in x, yielding 400 0s, 300 1s, 200 2s, and 100 3s.

If $c = N/n$ is fractional, then we'll need a slightly different approach to creating $\widehat{X}$ from x. We defer describing this, though, until later.

Parametric Estimate

In addition to the limited knowledge used for the nonparametric $\widehat{X}$, that X

1. Consists of N scores

2. Our sample distribution x contains n of those N scores

we may know as well

3. The parametric form of X

This last means that we know how to express, mathematically, the cdf of the population distribution.

To continue with our Number of Children example, we might know that the cdf of the population distribution X follows that of a Poisson random variable. The Poisson random variable takes nonnegative integer values, from zero through the positive integers to, in principle, infinity. The proportion of a Poisson population distribution taking a particular value, i, has the mathematical form

$$P(i|\lambda) = \frac{e^{-\lambda}\lambda^i}{i!} \, , \, i = 0, 1, \ldots, \infty$$

where e is the numeric constant ($e = 2.718\ldots$) and λ is the Poisson variable parameter. The mean and variance of a Poisson distribution are both equal to λ.

We might know in this example that the scores, number of children, form a Poisson distribution for the 1,000 college women making up our population, but not know the value of the parameter λ. How can we use our parametric knowledge to create a bootstrap world population distribution, $\widehat{X}$?

In general, the parametric $\widehat{X}$ is a collection of N scores following a prescribed parametric form (as closely as possible) with any unknown parameters estimated from the sample score distribution x. For our example we can estimate the unknown λ, the mean of the Poisson distribution, with the mean of our sample distribution. The mean of our ten number of children scores is

$$\text{Mean}(x) = \frac{0+0+0+0+1+1+1+2+2+3}{10} = 1$$

and when we use this as the parameter estimate, $\widehat{\lambda}$, in our mathematical formula for population proportions we have

$$P(i|\widehat{\lambda}=1) = \frac{e^{-1}1^i}{i!} \, , \, i = 0, 1, \ldots, \infty$$

which simplifies to

$$P(i|\widehat{\lambda}=1) = \frac{1}{e \times i!} \, , \, i = 0, 1, \ldots, \infty$$

as $1^i = 1$ for all possible values of i. Solving this equation for the first few, smaller values of i, we have

$$P(0|\widehat{\lambda}=1) = \frac{1}{e \times 1} = 0.367879$$

$$P(1|\widehat{\lambda}=1) = \frac{1}{e \times 1} = 0.367879$$

$$P(2|\widehat{\lambda}=1) = \frac{1}{e \times 2} = 0.183939$$

$$P(3|\widehat{\lambda}=1) = \frac{1}{e \times 6} = 0.061313$$

$$P(4|\widehat{\lambda}=1) = \frac{1}{e \times 24} = 0.015328$$

$$P(5|\widehat{\lambda}=1) = \frac{1}{e \times 120} = 0.003066$$

and

$$P(6|\widehat{\lambda} = 1) = \frac{1}{e \times 720} = 0.000511$$

Parametric forms often assume, quite unrealistically, very, very large distributions—infinite distributions. So care must be exercised in translating parametric knowledge to the estimation of a finite distribution. To match our population size, we must have 1,000 scores in $\widehat{X}$. These results suggest the following contents:

368 0s, 368 1s, 184 2s, 61 3s, 15 4s, 3 5s, and 1 6.

The important lesson here is not that we can fill in the 1,000 scores in $\widehat{X}$ when we know that X has a Poisson distribution nor that this parametric estimate differs from our earlier nonparametric one. The lesson is that to use the parametric estimate we must know the parametric form of the real-world population distribution of scores.

How would we gain parametric knowledge of a population score distribution X? Certainly not from the sample distribution x. An analysis of a sample can never establish its source. Knowledge of population distributional form must be external to the sample. This distributional-form knowledge comes only from our science, from specific, detailed knowledge of the process that generates the attribute scores making up X. Where these are scores for cases making up a finite, empirical population, such knowledge is quite unlikely. As noted earlier, such population distributions tend to be rough rather than mathematically smooth.

Much of classical parametric inference can be thought of as drawing bootstrap samples from a parametric population distribution estimate. For example, statisticians have been able to determine analytically what the bootstrap sampling distribution for the sample variance estimate would look like where $\widehat{X}$ has the same cdf as that of a normal random variable.

Smoothed Sample Estimate

The nonparametric and parametric approaches to distribution estimation occupy polar positions along a continuum. Consider this second example, based on data reported in Efron and Tibshirani (1993). Sixteen mice are randomly chosen from a laboratory population of 144 mice. The chosen mice are grouped, at random, into two groups, a control group of nine mice and a treatment group of seven mice. All 16 mice undergo a surgical procedure with the treatment group mice receiving a pre-surgical treatment intended to increase their survival times.

The survival times, in days, for the control group are given below,

$$10, 27, 31, 40, 46, 50, 52, 104, 146$$

These scores make up a sample distribution x. The nine cases contributing scores are a random sample from a prospective population, the laboratory population of $N = 144$ mice, all accorded control surgery. These 144 mice would then have a real-world population distribution of survival times, X.

A nonparametric estimate of X is easily enough defined. It would consist of c copies of x, where $c = 144/9 = 16$. The resulting nonparametric bootstrap-world population distribution, $\widehat{X}$, would contain 144 scores, but only the nine distinct values observed in x.

However, you might feel quite certain that, if you were to obtain a second sample of nine mice from this same population of mice and observe their survival times following control surgery, not all those survival times would be exactly 10, 27, 31, 40, 46, 50, 52, 104, or 146 days. Put another way, you are confident that X contains values other than the ten observed in x. But what other values?

You don't know enough about X to prescribe a parametric form. Although the nonparametric estimate incorporates less knowledge than you believe you have, the parametric estimate requires more knowledge than you have. Is there some middle ground?

There is. Instead of using the raw x values to estimate X we can use a smooth of the sample distribution. Smoothing spreads the distribution out, letting other values, those near to those appearing in x, contribute to $\widehat{X}$. For example, because 46, 50, and 52 appear in the sample distribution, you might think it quite likely that X also includes one or more values of 48, 51, and 53.

How can you smooth x? Smoothing provides a middle ground, but one that is quite wide. Many smoothing techniques appear in the statistics literature. Once you have chosen one of these, there remains a further choice of how much to smooth. The less smoothing, the closer will $\widehat{X}$ be to the nonparametric estimate. The more smoothing, the closer will $\widehat{X}$ be to some parametric form.

Figure 7.3 shows the histogram for an estimated population distribution of 144 survival times, the result of applying a particular smoothing approach known as the average shifted histogram (a.s.h.) to the control group sample distribution of survival times, repeated here:

$$10, 27, 31, 40, 46, 50, 52, 104, 146$$

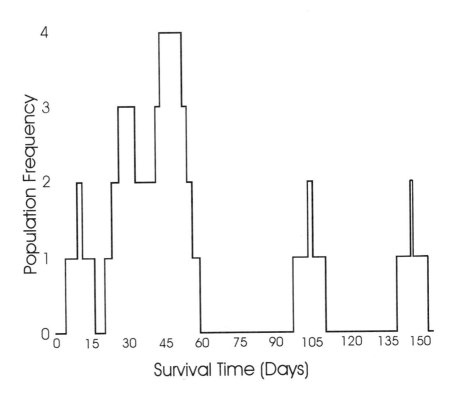

Figure 7.3:   *A Smoothed Bootstrap Population Distribution* $(N = 144)$

The a.s.h. implementation in SC used for this smooth estimate, the procedure `bincrf( )`, controls the amount of smoothing by the choice of histogram bin width. Our choice of bin width was a relatively narrow one, 10 days, and, as a result, a good deal of the roughness in the original sample distribution is preserved in $\widehat{X}$. The estimated population distribution in the figure contains, as examples, one score at each of 5 through 9 days, two scores at each of 10 and 11 days, three scores at each of 27 through 33 days, four scores at each of 44 through 52 days, zero scores at days 60 through 97, two scores at 146 days, and zero scores for any survival time exceeding 152 days. The histogram shows none of the smoothness of such parametric distributions as the normal or Poisson, but does include scores other than those in the sample distribution.

Again, we have considerable choice of type and amount of smoothing. Because, in most instances, it is not yet clear how we should be guided in our selection by any prior, external knowledge of the population distribution, we will not use smoothed bootstrap-world population distributions in subsequent Applications.

Model-Based Estimate

The fourth approach to the estimation of X requires that we know a model for generating the scores in X. The population distributions assumed in the analysis of variance provide an example. In this popular technique, the response score for the i-th member of the j-th treatment group, y_{ij}, is modeled as

$$y_{ij} = \mu_j + \epsilon_{ij}$$

where μ_j is the mean of the population distribution of response scores, all cases having been exposed to the j-th treatment level, and ϵ_{ij} is a randomly chosen error term. In the analysis of variance ϵ_{ij} is randomly chosen from the same distribution, whatever the treatment level. This error distribution is assumed to have a normal random variable cdf, to be centered at zero, and to have a variance, σ^2. We refer to this model as a treatment-effect-plus-error model; each response score is the sum of a treatment effect and a random error. The model has the errors randomly sampled rather than the scores themselves.

In the analysis of variance, the population distribution of the errors, ϵ, is estimated parametrically. The distribution is assumed to be infinitely large, normal in form, with a mean of zero and a variance that is estimated from the sum of squared deviations of the response scores about their group means,

$$\hat{\sigma}^2 = [1/(n-J)]\sum_{j=1}^{J}\sum_{i=1}^{n_j}(y_{ij}-\bar{y}_j)^2$$

where $\bar{y}_j$ is the mean response score for the n_j cases assigned to the j-th treatment level, J is the number of treatment levels and the total of all group sizes is $n = n_1 + n_2 + \ldots + n_j + \ldots + n_J$.

One or more bootstrap-world population error distributions, $\hat{\epsilon}$, can be defined nonparametrically as well. These could consist of as many copies of the n residuals:

$$e_{ij} = y_{ij} - \bar{y}_j, \; i = 1, \ldots, n_j, \; j = 1, \ldots, J$$

as are needed to fill out any one of the populations sampled.

We also could smooth the estimated error distribution, $\hat{\epsilon}$. For example, if the error distribution is known to be symmetric about zero, it could be estimated by $2n$ scores, the e_{ij}s augmented by their negatives, $-e_{ij} = -1 \times e_{ij}$.

In fact, we might create two or more estimated population error distributions if some treatment levels are known to attract errors from one distribution and other treatment levels are known to attract errors from a second. Both analysis of variance assumptions—normality of error distributions and homogeneity of error variances—can be relaxed in the bootstrap world.

In Applications 19, we illustrate the use of a model-based estimate of an error distribution. We see there how bootstrap sample observations are formed in accordance with the treatment-effect-plus-error model by adding a randomly chosen element of $\hat{\epsilon}$ to the appropriate estimate of treatment effect:

$$y_{ij}^* = \bar{y}_j + \hat{\epsilon}_{ij}$$

Nonparametric Population Estimates

Because of the paucity of our knowledge about X, parametric, sample-smoothed, and model-based estimation are more theoretical than practical alternatives to the nonparametric $\hat{X}$. For nearly all our applications of bootstrap inference, we will use nonparametric distribution estimates.

We'll follow three conventions in the formation of these nonparametric estimates, depending on sample and population sizes, n and N.

Large Population

If the population is large relative to the sample size, we can use a shortcut to formulate the bootstrap-world population distribution and to take bootstrap samples from that distribution. We know that cases are sampled without replacement from populations. However, if N is sufficiently large relative to n, then the population distribution of scores is unaffected by the sampling. As a result, we can simulate sampling without replacement by sampling with replacement. The population distribution sampled with replacement need not consist of $c = N/n$ copies of x; we can achieve the same results by sampling with replacement from a single copy of the sample distribution x. Thus, in effect, $\widehat{X}$ and x can be the same distributions. How large should N be for this shortcut? Ideally, the sample should contain no more than 1% of the population. More realistically, particularly in view of software limitations, we adopt a rule that the large population shortcut is appropriate if N is at least 20 times the size of n.

Small Population, N/n an Integer

We have described earlier how the population distribution is estimated whenever $N/n = c$ is an integer. We simply take $\widehat{X}$ to be c copies of x. The b-th bootstrap sample, x_b^*, is drawn without replacement from $\widehat{X}$.

Small Population, N/n a Fraction

If $N/n = c$ is fractional, the formation of $\widehat{X}$ is only slightly more complicated. We use an algorithm proposed by Booth, Butler, and Hall (1994). Let C be the integer part of $c = N/n$. Then, $k = N - Cn$ additional scores are needed to complete $\widehat{X}$, once C copies of x have been supplied. Initially, we complete our estimate of the population distribution by selecting k of the n sample cases at random. To balance the appearance of all sample cases in the nonparametric estimate, however, we replace these k cases with a new randomly chosen set of k after we have drawn m bootstrap samples from the current bootstrap-world population distribution. Typically, choosing $m = 100$ will ensure that the random additions are swapped frequently enough to minimize any bias in the constitution of $\widehat{X}$.

Software Considerations

In fact, many software implementations of bootstrap resampling address only the large population situation, and you may have no choice but to accept that option. The S-Plus `bootstrap` function, for example, draws bootstrap samples by sampling with replacement from a single copy of the sample distribution. The results will vary, perhaps only slightly, from what a small population analysis would yield. More often than not, however, sample sizes will not exceed 5% of the population sizes, and the large population approach will be an appropriate one.

Sample Size and Distribution Estimates

The estimation of a population distribution on the basis of a sample succeeds the closer the resemblance of the cdf for the sample distribution to that of the population distribution. This resemblance increases as the size of the sample increases. How large should n be for our nonparametric population distribution estimate to be useful? Samples of 10 or fewer cases almost always will be too small. Samples of 25 or greater will usually be large enough to use in estimating X.

As with so much in statistics, larger samples are more useful than smaller ones are. And, as in so many other texts, we use here, for convenience in exposition, examples that are counter to this principle! Remember the principle, not the textbook example.

Applications 7

Bootstrap Population Distributions

Bootstrap inference is based on building up the sampling distribution of a statistic, s^*, by computing it in each of a large number of bootstrap samples, random samples drawn from $\widehat{X}$, a fully specified distribution of N scores. This population distribution is based on a real-world sample, x, and estimates the population distribution X from which x was sampled. A critical task in bootstrap inference is forming $\widehat{X}$ from x and whatever additional knowledge we have about X.

Nonparametric Population Estimates

If all we know about X is that the scores making up x were contained in it, then our estimate, $\widehat{X}$, is said to be a nonparametric one. There are two conventions for forming a nonparametric $\widehat{X}$ and then drawing bootstrap samples from it. Which we use depends on the relative magnitudes of N, the size of the population score distribution X, and n, the size of the sample score distribution x.

1. If $(N/n) < 20$, then

    a. $\widehat{X}$ consists of as many copies of x as are necessary to provide N scores.
    b. Each bootstrap sample consists of the scores for n cases drawn randomly and without replacement from $\widehat{X}$.

2. If $(N/n) \geq 20$, then
    a. $\widehat{X}$ is represented by a single copy of x.
    b. Each bootstrap sample consists of the scores for n cases drawn randomly and with replacement from x.

Sampling with replacement from a single copy of x simulates sampling without replacement from a population distribution so large that the withdrawal of n cases has no impact on the balance of scores remaining.

Let's look now at an example.

Fifteen U.S. law schools were chosen at random from a population of 82 schools. Table 7.1 shows the average Law School Admissions Test (LSAT) scores and undergraduate grade point averages (GPA) for the classes entering those sampled schools in 1973 (Efron and Tibshirani, 1993). We are interested in forming a nonparametric estimate of the population distribution sampled.

Table 7.1: *LSAT and GPA Scores for 15 Randomly Sampled Law Schools*

	LSAT	GPA		LSAT	GPA		LSAT	GPA
1	576	3.39	6	580	3.07	11	653	3.12
2	635	3.30	7	555	3.00	12	575	2.74
3	558	2.81	8	661	3.43	13	545	2.76
4	578	3.03	9	651	3.36	14	572	2.88
5	666	3.44	10	605	3.13	15	594	2.96

The sample size here is appreciably larger than 5% of the population. Hence, we should resample without replacement from a population distribution of $N = 82$ paired scores. Because N is not an integer multiple of n, each of the 15 sampled cases cannot be represented the same number of times in any fixed estimate of the population. To have some cases represented more often than others, however, biases the population estimate. Figure 7.4 illustrates one solution (Booth, Butler, and Hall, 1994) with a Resampling Stats script.

The idea behind this approach is to randomly adjust the composition of the population estimate several times in an attempt to balance out case representation.

For purposes of illustration, the script shows the computation of the product moment correlation between LSAT and GPA scores in the real-world sample and in each bootstrap sample. Our interest, though, is in obtaining nonparametric estimates of the population originally sampled and in taking nonreplacement samples of cases (law schools) from those estimated populations. Lines 2 through 8 input the LSAT and GPA scores from the sample as the vectors L and G; and set N, NN, and BT equal to the sample size, population size, and required number of randomly chosen bootstrap samples. Line 10 computes the LSAT-GPA correlation for the real-world sample, TSTAT. Notice that the computation of this correlation and our subsequent use of the L and G vectors require the two sets of scores to be arranged in the same way relative to the sampled schools.

```
 1: ' ----- Obtain Input:
 2: copy (576 635 558 578 666 580 555 661 651 605) L
 3: concat L (653 575 545 572 594) L
 4: copy (3.39 3.30 2.81 3.03 3.44 3.07 3.00 3.43) G
 5: concat G (3.36 3.13 3.12 2.74 2.76 2.88 2.96) G
 6: copy 15 N                     'Sample Size
 7: copy 82 NN                    'Population Size
 8: copy 1000 BT                  'Resample Count (multiple of 100)
 9: ' ----- Calculate Target Statistic in Original Sample
10: corr L G TSTAT                'Target Statistic
11: ' ----- Form Estimated Population of Cases
12: copy 1,N CID                  'Case ID vector
13: divide NN N K                 'Pop'n is K multiples of sample
14: round K IK
15: if IK > K                     'Has rounded up, rather than down
16:     subtract IK 1 IK          'Want IK to be integer part of K
17: end
18: copy CID PEST                 'First copy of CID in PEST
19: subtract IK 1 IK1             'Additional copies needed
20: repeat IK1
21:     concat PEST CID PEST      'Puts IK copies of CID in PEST
22: end
23: multiply IK N IKN
24: subtract NN IKN NR            'NR cases needed to fill pop'n
25: ' ----- Begin bootstrap resample cycles:
26: divide BT 100 B100            'B100: Resample Partitions
27: repeat B100
28:     if NR > 0                 'Each 100 resamples replace NR
29:         take PEST 1,IKN PEST
30:         shuffle CID SID
31:         take SID 1,NR SID
32:         concat PEST SID PEST 'Complete PA with NR pointers
33:     end
34:     repeat 100
35:         shuffle PEST SPEST
36:         take SPEST 1,N BSAMP     'Nonreplacement case sample
37: ' ----- Calculate target statistic in resample
38:         take L BSAMP SAMPL
39:         take G BSAMP SAMPG
40:         corr SAMPL SAMPG TSTAR 'Target statistic in resample
41:         score TSTAR FBOOT        'Save the resample statistic
42:     end
43: end
44: ' ----- Resampling completed, save sampling distribution
45: write file "LAWCORR.OUT" FBOOT
```

Figure 7.4:   *Forming Nonreplacement Bootstrap Samples (Resampling Stats)*

Line 12 sets up a case identification vector, CID, containing the integers from 1 through N, the sample size. Our nonparametric estimate of the population of law schools is that it consists of equal numbers of schools the same as each of the 15 schools in the sample. That is, we want to replicate each of the cases the same number of times in the estimated population. In effect we'll do this by replicating the case identifications. Lines 13 through 17 establish the number of times, IK, that the full set of sampled cases can be replicated without exceeding the population size. For the law school example, IK takes the value 5 as the sample of 15 schools can be fully replicated five times, $5 \times 15 = 75$, but not six times; $6 \times 15 = 90$ is greater than 82. Lines 18 through 22 place IK copies of the sample case identifications, CID, into the estimated population, PEST.

Because our population size, 82, is not an integer multiple of the sample size, 15, the population estimate is not quite complete with five copies of the sample. The number of cases needed to complete the population, $82 - (5 \times 15) = 7$, is computed at lines 23 and 24 of the script as NR. The total number of bootstrap samples to be created, BT, is partitioned into B100 blocks of 100 samples at line 26. For each of these blocks, the estimated population, PEST, is completed with additional, randomly chosen NR cases by the instructions at lines 28 through 33.

In our example, each block of 100 randomly chosen bootstrap samples is drawn without replacement from a population consisting of five copies of eight of the schools and six copies of the remaining seven. The seven schools represented an extra time are randomly replaced for each block. In the Figure 7.4 computing script, the total number of bootstrap samples, BT, is set at 1,000 giving rise to ten random completions of the estimated population. These ten completions make it more likely that, over the 1,000 occasions an estimated population is sampled, each of the 15 schools will be present an equal number of times.

Although the remainder of the computing script goes beyond our present interest in forming a nonparametric population estimate, it is worth noting how the sampling from that estimate takes place. The estimated population, PEST, complete at line 33, consists of 82 case identifications, each an integer between 1 and 15 and referring back to the sample cases in Table 7.1. At lines 35 and 36, a sample is drawn from this population by shuffling or randomly reordering the population of case identifications and then taking 15 case identifications from the top of the reordering. These case identifications, in the vector BSAMP, are then used at lines 38 and 39 to obtain the LSAT and GPA scores for the sampled cases.

We see how the estimated small-population is built up and then sampled in the Resampling Stats script of Figure 7.4. The first of the 1,000 random samples from the estimated population is described by the histogram in Figure 7.5. It consists of three copies of cases 2 and 14, two copies of cases 6, 8, and 9, and one copy each of cases 3, 4, and 11.

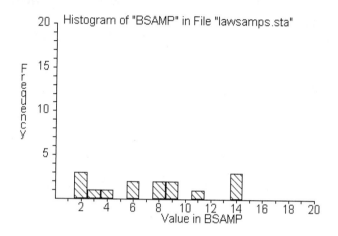

Figure 7.5:   *Histogram of Cases Selected for the First Bootstrap Sample*

Population estimation and sampling of the same form as outlined in the script of Figure 7.4 take place implicitly in the two small-population SC bootstrap procedures `fbootci2( )` and `fboot_t( )`, which you will meet in subsequent Applications.

Exercises

1. Listed below are the postsurgery survival times, in days, of seven mice given treatment before surgery designed to increase survival time (Efron and Tibshirani, 1993).

$$94, 197, 16, 38, 99, 141, 23$$

The $n = 7$ mice were randomly sampled from a laboratory population of N mice.

   a. Assume a population size of $N = 28$. Describe how you would form the nonparametric estimate of the population distribution of survival times. How would you obtain bootstrap sample distributions? Show the score distributions for four successive bootstrap samples.

b. Repeat the exercise, now assuming that the population size is $N = 144$.

2. For the postsurgery survival study of the previous exercise, describe in terms of the computational flow of Table 7.1 how estimated populations of survival times would be constructed if the laboratory population from which the seven mice were sampled consisted of 50 animals. For this particular example—$n = 7$ and $N = 50$—can you think of an alternative to randomly filling out the population from block to block of bootstrap samples? Explain.

Concepts 8

The Bootstrap Sampling Distribution

The real-world population can be sampled only once. In contrast, the bootstrap-world population distribution is nothing more than a collection of scores. In principle, we can obtain from it the score distribution for each possible sample of size n. Computing a parameter estimate for each sample distribution and collecting these gives us a sampling distribution for the estimator. This sampling distribution, of course, is based on the bootstrap-world population distribution—almost always, the nonparametric estimate $\widehat{X}$—rather than on the real-world population distribution X. What can this bootstrap sampling distribution tell us about the real one?

The Bootstrap Conjecture

To review, the real-world sampling distribution for an estimator, t, would be created in the following way. For a population of N cases, there is a population distribution of attribute scores X characterized by a parameter θ, the object of estimation. We form all M possible samples of size n from the population. Each sample will have an accompanying distribution of attribute scores x from which we compute a value of the estimator. The M estimates so computed form the sampling distribution. We name this distribution $F(t|X, n)$—the distribution of values of t when computed from all possible samples of size n taken from the real-world population distribution X. This real-world sampling distribution is unattainable; the recipe we give for its creation cannot be followed.

Now let's replace the unknown X with the nonparametric estimate described in Concepts 7. This estimate, referred to as $\widehat{X}$, is simply one or more copies of our single real-world sample distribution. $\widehat{X}$ is a bootstrap-world population distribution and has a parameter, $\widehat{\theta}$, computed from $\widehat{X}$ in the same manner that θ would be computed from X. We can now form all M possible sets of scores of size n from the N scores in $\widehat{X}$. For each of these bootstrap samples, x^*, we can compute the estimate, t^*, of $\widehat{\theta}$. The collection of the resulting M values of t^* is the bootstrap sampling distribution. We name this distribution $F_B(t^*|\widehat{X}, n)$ to indicate that it is a bootstrap distribution and specific to our choice of $\widehat{X}$ and n. Some of the ideas developed here hold as well for the parametric or smoothed $\widehat{X}$, but we shall assume throughout a nonparametric $\widehat{X}$.

The real-world sampling distribution, if we had access to it, would tell us about the accuracy of t as an estimate of θ. That is, the sampling distribution would allow us to assess the bias of our estimator,

$$\text{Bias}(t|X, n, \theta) = \text{Mean}(t|X, n) - \theta$$

its standard error (SE),

$$\text{SE}(t|X, n) = \sqrt{(1/M)\sum_{i=1}^{M}[t_i - \text{Mean}(t|, n)]^2}$$

and how close, on average, our t will be to the parameter estimated,

$$\text{RMS Err}(t|X, n, \theta) = \sqrt{(1/M)\sum_{i=1}^{M}(t_i - \theta)^2}$$

We could also use the quantiles of the sampling distribution, $F^{-1}(\alpha)$ and $F^{-1}(1 - \alpha)$, in the construction of an interval estimate of θ, one whose width would convey the extent of our confidence in the estimate.

The computations we would carry out, if we could, on $F(t|X, n)$ to find the bias, SE, RMS error, or to locate the limits to a $(1 - 2\alpha)100\%$ confidence interval (CI) for θ can be carried out on $F_B(t^*|\widehat{X}, n)$. The bootstrap conjecture, the basis for bootstrap inference, is that if these computations are carried out with sufficient attention to what is known about sampling distributions, the results can provide good estimates of the corresponding real-world sampling distribution characteristics.

Complete Bootstrap Sampling Distributions

If we were to compute t^* for every one of the

$$M = \frac{N!}{n!(N - n)!}$$

possible nonreplacement sample score distributions, x^*, of size n from among the N scores making up $\widehat{X}$, we would have the complete bootstrap sampling distribution.

The following simple algorithm ensures that we find each of these samples once and only once:

1. Assign labels of 1 through N to the cases contributing scores to $\widehat{X}$.

2. Define a sample as having positions 1 through n, to be filled with n different cases or different labels.

3. Begin by filling sample position 1 with the case labeled 1, sample position 2 with the case labeled 2, and so forth. Sample position n is filled with the case labeled n. For this first sample and for all subsequent samples, the case labels increase from sample position 1 to 2 to 3 ... to n.

4. To choose the next sample, find the first sample position—starting with position n and moving position by position towards the first position—for which the case label can be increased by one without the result exceeding $N - (n - p)$, where p is the sample position. Let P be the new label we assign to position p. Complete the sample by replacing the case labels for positions $(p+1)$ through n with labels $(P+1)$ through $[P + (n - p)]$.

5. Repeat step 4 until the case label cannot be increased at any position. The final sample will have case labels $[N - (n - 1)]$ through N in sample positions 1 through n.

A small example illustrates how the algorithm works. Let $n = 4$ and $N = 9$. The first sample consists of the cases labeled $(1, 2, 3, 4)$. The second would have cases labeled $(1, 2, 3, 5)$. The sample with cases labeled $(1, 2, 3, 9)$ would be followed by one with cases labeled $(1, 2, 4, 5)$. And the sample with cases labeled $(1, 2, 8, 9)$ would be followed by one with cases labeled $(1, 3, 4, 5)$. The last of the

$$M = \frac{9!}{4!5!} = \frac{9 \times 8 \times 7 \times 6}{4 \times 3 \times 2 \times 1} = 126$$

samples identified by our algorithm would contain the cases labeled $(6, 7, 8, 9)$.

Our algorithm is a simple one, but more often than not, M is quite a large number. In consequence, the total computing time required first to identify each possible sample, then to access its score distribution x^*, and finally to compute t^* from this score distribution can be enormous, even at currently accelerating computing speeds.

Monte Carlo Bootstrap Distributions

Because M is frequently so large, it has become customary to approximate the bootstrap sampling distribution. We form this approximation from the t^*s computed for some large number, B, of randomly chosen bootstrap samples. The size of B depends on what characteristics of the bootstrap sampling distribution we want to approximate. B can be as small as 100 if we want only a good estimate of the SE of our estimator. Values of 2,000 to 5,000, however, will be common when we want to assess bias or estimate a CI.

The larger the value of B, the closer the approximation is to the complete bootstrap sampling distribution. The approximation is referred to as a Monte Carlo approximation because of the random choice of samples to be included. Randomness enters into this process in two ways: For the Monte Carlo approximation, we obtain a random sequence of random samples of cases.

Statistical computing packages include commands that allow you to select a random sample, either sampling with or without case replacement. You will use these commands directly or indirectly when you build an approximation to a bootstrap sampling distribution. But how do these sampling commands work? How do they choose a random sample? How can they create random sequences of samples? Each of the commands calls on a more primitive feature of the statistical package, its ability to generate a sequence of random numbers. Technically these are only pseudo-random numbers as the digital computer is incapable of random behavior, but good pseudo-random number generators produce numbers that, over quite long sequences, cannot be distinguished from random numbers. Among other things, these numbers are random in their sequence; we cannot predict the next number from those that have come before. For convenience, we will refer to these pseudo-random number generators as random number generators, and refer to the numbers generated as random numbers.

Let's see how a random number generator could be used to create random sequences of random samples—first where each sample is drawn without replacement and then with case replacement.

Nonreplacement Random Samples

To create a random nonreplacement sample of n cases from a population of N cases:

 1. Direct the random number generator to assign a number to each of the N cases in the population.

 2. Sort the N cases in order, from smallest to largest random number.

 3. Select for the sample the cases assigned the n-smallest random numbers.

Steps 1 and 2 randomly shuffle the N cases. Every possible sequence of the N cases has the same chance of turning up. As a result, every possible set of n cases has the same chance of rising to the top of the shuffled order.

Steps 1 through 3 give us one random sample. How do we get a random sequence of such samples? Again, the random number generator holds the answer. Each time we carry out the first step in this algorithm, we get a new sequence of random numbers that cannot be predicted from any earlier sequences. The sequence of random samples is itself random.

Replacement Random Samples

To create a random replacement sample of n cases from a population of N cases:

 1. Direct the random number generator to assign a number to each of the N cases in the population.

 2. Sort the N cases in order, from smallest to largest random number.

 3. Include the case with the smallest random number in the sample.

 4. Repeat steps 1 through 3 an additional $n - 1$ times.

Each of the N cases has the same chance of being assigned the smallest random number in step 1 and, hence, of being chosen for the sample on Step 3. This is true each time we carry out steps 1 through 3. The algorithm ensures both that both the content of each sample is randomly determined and that the sequence of samples is a random one.

In Concepts 7, we identified the use of replacement random samples as an appropriate approximation to sampling without replacement when the population size is at least 20 times the sample size and, as a result, we have estimated X nonparametrically with a single copy of x. Our population distribution is just the n scores making up our sample distribution and, in the replacement sampling algorithm, we have $N = n$.

The Bootstrap Estimate of Standard Error

An important index of the precision of a sample-based estimate is the sample-to-sample variability of the estimator. We commonly use the standard deviation (SD) of the sampling distribution for this. And we refer to this standard deviation as the SE of the estimate.

We can compute a Monte Carlo bootstrap estimate of this SE. This estimated SE is the standard deviation estimate computed over the Monte Carlo bootstrap sampling distribution:

$$\widehat{\mathrm{SE}}_{Boot}(t|X, n) = \sqrt{[1/(B-1)]\sum_{b=1}^{B}\left[t_b^* - (1/B)\sum_{b=1}^{B}t_b^*\right]^2}$$

Because $\widehat{\mathrm{SE}}_{Boot}(t|X, n)$ is computed from the bootstrap-world sampling distribution rather than from the real-world one, it is only an estimate of the real-world SE.

The number of bootstrap samples, B, required for the SE estimate can be as small as 100. For greater stability of the estimate, however, a larger value of B should be used. Alternatively, an incremental algorithm can be used, adding bootstrap samples so long as they increase the precision of the SE estimate. Here is such an algorithm:

1. The algorithm depends on four constants. The following are reasonable choices: $B1$, the initial number of bootstrap samples, could be set equal to 100; $B2$, the incremental number of samples, would be set to some smaller value, say, 25; K, the maximum number of increments to be used, could be set to something like 10; and D, the threshold percent change in SE, might be set to 1%.

2. Collect t_b^* from $b = 1, \ldots, B1$ randomly chosen bootstrap samples. Compute and store the SD of these estimates as SE1.

3. Collect t_b^* for an additional $b = 1, \ldots, B2$ bootstrap samples. For the total number of estimates, compute the SD and store as SE2.

4a. If step 3 has been carried out K times, stop computing and report SE2 as the bootstrap SE estimate.

4b. If step 3 has been carried out fewer than K times, compare SE1 and SE2.

5a. If SE2 is between $(100 - D)\%$ and $(100 + D)\%$ of SE1, stop computing and report SE2 as the bootstrap SE estimate.

5b. If SE2 is outside these limits, replace SE1 with SE2 and return to step 3.

For the suggested values of the four constants, this algorithm would report a SE estimate based on 125, 150, 175, ..., or 350 bootstrap samples, and stop when the value of the estimate changes by less than 1% with the addition of 25 bootstrap samples.

The incremental algorithm is useful particularly where it is costly to compute t_b^* for each of $B2$ additional samples. In general, it will be easier to set B to some reasonably large value, for example, 500.

The Bootstrap Estimate of Bias

It might seem counterintuitive that it is harder to get a good estimate of bias than of the SE. After all, the average is a much simpler statistic than is the standard deviation. However, a good Monte Carlo bootstrap estimate of bias requires, when we use a nonparametric estimate of X, that each case in the sample (each score in x) be present in very nearly the same number of bootstrap samples. If some cases are picked more often than others are, we can introduce bias into the sampling distribution. The goal of equal case representation can be approached in other ways, but we'll concentrate on the effect of increasing B. As B grows larger, the chance of any appreciable case imbalance grows smaller. As a rule of thumb, we'll use values of B of the order of 2,000 or 5,000 to minimize imbalance resulting from our random choice of random samples.

The Monte Carlo bootstrap estimate of bias is given by applying the definition of bias to the bootstrap-world parameter and sampling distribution:

$$\widehat{\text{Bias}}_{Boot}(t|X, n, \theta) = (1/B)\sum_{b=1}^{B} t_b^* - \widehat{\theta}$$

In general, the bias of an estimator is a problem only if it is large relative to the SE of that estimator. Efron and Tibshirani (1993) suggest that if the absolute value of the bootstrap bias estimate is less than one-quarter the size of the bootstrap SE estimate, the bias can be ignored. That sounds like a wide margin, but recall that the more inclusive measure of estimator accuracy, the RMS error,

$$\text{RMS Err}(t|X, n, \theta) = \sqrt{(1/M)\sum_{i=1}^{M}(t_i - \theta)^2}$$

is related to the SE and bias assessments as

$$\text{RMS Err}(t|X, n, \theta) = \sqrt{\text{SE}(t|X, n)^2 + \text{Bias}(t|X, n, \theta)^2}$$

Now, if the bias is $(0.25 \times \text{SE})$ then the RMS error is

$$\text{RMS Err} = \sqrt{\text{SE}^2 + (0.25 \times \text{SE})^2} = \sqrt{1.0625 \times \text{SE}^2} = 1.0308 \text{ SE}$$

That is, if the absolute value of the bias is as large as one-quarter of the SE, the RMS error will be only about 3% larger than the SE. Biases smaller than this do not detract from the SE as an adequate measure of the accuracy of an estimator.

Simple Bootstrap CI Estimates

The most important use of the sampling distribution is the construction of a CI for the parameter we are estimating. The CI is an interval estimate of the parameter, spreading out around t to convey the inaccuracy of that point-estimate and, hence, our remaining uncertainty about the value of θ.

We complete this Concepts with a review of three of the simpler approaches to estimating an equal-tails exclusion $(1 - 2\alpha)100\%$ CI from a Monte Carlo bootstrap sampling distribution. Then in Concepts 9 and 10 we develop two more thoughtful and trustworthy bootstrap approaches to CI estimation. Except where noted, we shall assume that the $\widehat{X}$ from which the bootstrap samples are drawn is a nonparametric estimate of the population distribution X.

The Standard CI Estimate

The bootstrap standard equal-tails exclusion 90% CI ($\alpha = 0.05$) is written simply as

$$t \pm 1.645 \times \widehat{\mathrm{SE}}_{Boot}(t|X,n)$$

Where does this CI come from?

The definition has us add to and subtract from t the same distance. That is, we have $a = [F^{-1}(1-\alpha) - \theta]$ equal to $b = [\theta - F^{-1}(\alpha)]$. This implies that the sampling distribution for t is symmetric about θ. And the implicit definitions of a and b tell us that the standard CI estimate is a direct adaptation of the CI for a population mean estimated by a sample mean. Remember that the sample mean is an unbiased estimator of the population mean and has a sampling distribution with a normal cumulative distribution function (cdf). The sampling distribution for the sample mean is symmetric and centered about the population mean, μ. Furthermore, knowing that the sampling distribution follows a normal cdf tells us that the central 90% of the estimates making up that sampling distribution fall between $(\mu - 1.645 \times \mathrm{SE})$ and $(\mu + 1.645 \times \mathrm{SE})$.

Thus, the bootstrap standard CI estimate assumes that the sampling distribution for t is centered at θ and has a normal cdf. This estimate is useful because we can use the bootstrap sampling distribution to obtain good estimates of the many SEs for which we do not have handy formulae. Assuming a normal cdf for the sampling distribution, however, can be unwarranted, particularly when n is not large.

The Symmetric Percentile CI Estimate

To avoid the assumption of a normal sampling distribution cdf, Efron (1979) introduced what we'll refer to as the symmetric bootstrap percentile CI estimate.

In our quantile function notation, the $(1 - 2\alpha)100\%$ symmetric percentile CI is simply the interval

$$\left[F_B^{-1}(\alpha), \, F_B^{-1}(1 - \alpha)\right]$$

where, roughly speaking, the middle $(1 - 2\alpha)100\%$ of the B elements of the Monte Carlo bootstrap sampling distribution fall between $F_B^{-1}(\alpha)$ and $F_B^{-1}(1 - \alpha)$. These limits are the α and $(1 - \alpha)$ quantiles of the bootstrap sampling distribution.

We saw in Concepts 1 how the α and $(1 - \alpha)$ quantiles are defined for a finite distribution. Applying these ideas to the Monte Carlo bootstrap distribution, we let $t_{[1]}^*$ and $t_{[B]}^*$ designate the smallest and largest elements in the bootstrap sampling distribution. And $t_{[b]}^*$, for $b = 1, \ldots, B$, designates the b-th smallest element. Then, the α and $(1 - \alpha)$ quantiles are

$$F_B^{-1}(\alpha) = t_{[\text{int}\{\alpha \times (B+1)\}]}^*$$

and

$$F_B^{-1}(1 - \alpha) = t_{[(B+1) - \text{int}\{\alpha \times (B+1)\}]}^*$$

where $\text{int}\{\alpha \times (B + 1)\}$ is the integer part of the product of α and $(B + 1)$, the result of rounding down to the nearest integer.

For example, if $B = 2000$ and $\alpha = 0.025$, then

$$\text{int}\{\alpha \times (B + 1)\} = \text{int}\{0.025 \times 2001\} = \text{int}\{50.025\} = 50$$

and the lower and upper limits of the 95% symmetric percentile CI are given by

$$F_B^{-1}(0.025) = t_{[50]}^*$$

and

$$F_B^{-1}(0.975) = t_{[2001-50]}^* = t_{[1951]}^*$$

Our definition of the finite distribution quantiles as symmetric leads to our counting the same number of resample statistics from either end of the bootstrap sampling distribution. Here, we select the 50th smallest and 50th largest to serve as the limits of the symmetric percentile CI estimate.

What does the symmetric percentile approach to CI estimation infer about the sampling distributions of t and of t^*? In developing the CI for a location-sensitive parameter in Concepts 3, we learned that the lower limit to the CI would be given by $[t - a]$ where $a = [F^{-1}(1 - \alpha) - \theta]$, the distance separating the $(1 - \alpha)$ quantile from θ. This distance is noted in the Figure 8.1 sketch of a hypothetical bootstrap sampling distribution.

For the symmetric percentile CI estimate, however, the α quantile is used as the lower limit to the CI estimate. This is equivalent to having subtracted from t the distance from θ to the α quantile, or the distance $b = [\theta - F^{-1}(\alpha)]$ in Figure 8.1. That is, as with the standard CI estimate, a and b are assumed equal; the sampling distribution for t is assumed to be symmetric about θ. The assumption of a normal sampling distribution has been relaxed but not the assumption of symmetry.

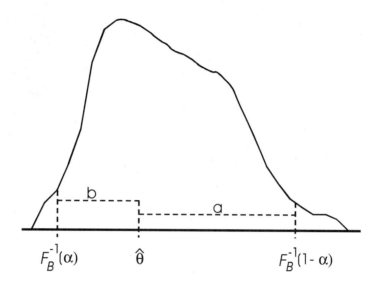

Figure 8.1:   *CI Width Determined from Bootstrap Percentiles*

The Nonsymmetric Percentile CI Estimate

In Figure 8.1, the α and $(1 - \alpha)$ bootstrap quantiles are purposely drawn not equidistant from $\widehat{\theta}$, an arrangement that we may believe to be true as well for the real-world sampling distribution. We can compute a CI estimate on that assumption. We begin by noting from the figure that for each of the central $(1 - 2\alpha)100\%$ of the B values of t^*, the interval

$$[t^* - a, \, t^* + b] = [t^* - \{F_B^{-1}(1 - \alpha) - \widehat{\theta}\}, t^* + \{\widehat{\theta} - F_B^{-1}(\alpha)\}]$$

will include $\widehat{\theta}$. This defines the CI for the bootstrap sampling distribution.

The nonsymmetric percentile CI estimate assumes that the widths a and b, correct in the bootstrap world, are good estimates of the corresponding widths for the real-world sampling distribution. That is, we can estimate our $(1 - 2\alpha)100\%$ CI for θ as

$$[t - a,\, t + b] = [t - \{F_B^{-1}(1 - \alpha) - \widehat{\theta}\},\, t + \{\widehat{\theta} - F_B^{-1}(\alpha)\}]$$

This is the nonsymmetric bootstrap percentile CI estimate.

The formulation of this CI estimate simplifies a bit when, as is often the case, t is a plug-in estimator of θ. We compute a plug-in estimate from a sample distribution x by applying the rule we would use to compute the parameter from the population distribution X. Thus, the sample mean is a plug-in estimator of the population mean. Each is computed by summing the scores in a distribution and dividing that sum by the number of scores in the distribution. The sample median also can be used as an estimate of the population mean but it is not a plug-in estimator of the population mean. The median and mean are computed by different rules.

Now if $\widehat{X}$ is a nonparametric estimate of X, as we have been assuming, and t is a plug-in estimator of θ, then

$$t = \widehat{\theta}$$

The reason for this is that t is computed over x, and $\widehat{\theta}$ is computed, by the very same rule, over an $\widehat{X}$ that is either x or multiple copies of x. Applying a particular computational rule—for example, that for the mean, median, or variance—to

$$(1, 5, 3, 8)$$

gives the same result as applying that rule to

$$(1, 5, 3, 8, 1, 5, 3, 8, 1, 5, 3, 8, 1, 5, 3, 8, 1, 5, 3, 8, 1, 5, 3, 8)$$

Where there is an equivalence of t and $\widehat{\theta}$, the nonsymmetric bootstrap percentile estimate of the $(1 - 2\alpha)100\%$ CI simplifies to

$$[t - a,\, t + b] = [2t - F_B^{-1}(1 - \alpha),\, 2t - F_B^{-1}(\alpha)]$$

With this CI estimate, we have relaxed the assumption of symmetry of our sampling distribution. However, all three of the approaches to CI estimation outlined here have assumed that a, the amount to be subtracted from the estimator to give the lower limit to the CI, would be about the same if we could compute it from $F(t)$ as when we compute it from $F_B(t^*)$. And, the same applies to b, the length of the upper arm of the CI. This is equivalent, as we'll explore in Concepts 9, to assuming that $(t - \theta)$ is a pivotal form, that is, that $(t - \theta)$ and $(t^* - \widehat{\theta})$ have the same sampling distributions.

Althought all the bootstrap sampling distributions that we work with in subsequent Concepts and Applications will be Monte Carlo approximations, based on a random sequence of B bootstrap samples, we will refer to them simply as bootstrap sampling distributions.

Applications 8

Bootstrap SE, Bias, and CI Estimates

The bootstrap sampling distribution, even a Monte Carlo approximation based on 2,000 to 5,000 randomly chosen bootstrap samples, provides a very good estimate of the standard error (SE) of a statistic, evidence about the bias of an estimator, and a starting point for estimating a confidence interval (CI) for the population distribution parameter we want to estimate, all based on our having randomly sampled cases from a well-defined, bootstrap-world population distribution.

Example

Table 5.1, reproduced below, gave the weights of 50 bags of pretzels, randomly sampled from the output of a packaging machine.

Table 5.1: *Weights (in grams) of 50 Randomly Sampled Bags of Pretzels*

464	450	450	456	452	433	446	446	450
447	442	438	452	447	460	450	453	456
446	433	448	450	439	452	459	454	456
454	452	449	463	449	447	466	446	447
450	449	457	464	468	447	433	464	469
457	454	451	453	443				

We want to estimate the median of the population distribution, using a plug-in estimator, the median of the sample. How accurately can we expect the sample median to estimate the population median?

Figure 8.2 displays an interactive SC session in which a Monte Carlo approximation to the bootstrap sampling distribution for $t_b^* = \text{Median}(x_b^*)$ is developed and aspects of that distribution are described. The first three commands retrieve the sample data, from a vector, `weights`; create a vector, `tstar`, to hold the bootstrap sampling distribution; and define a user function, `g`, to compute the median in each bootstrap sample. These three

arguments then are passed to the `boot` procedure in the fourth command:
`boot(tstar,g,weights)`.

```
> load(weights)              # Retrieve sample of pretzel bag weights
> vector(tstar,2000)           # To receive bootstrap sampling dist'n
> func g() return median($1$)   # Computes t from data

> boot(tstar,g,weights)             # Creates 2,000 bootstrap samples

> mean(tstar)
    450.65975
> sde(tstar)
    1.0996406              # The estimated SE of sample median
> median(weights)
    450                  # Our t statistic

> median(weights)-(sde(tstar)*1.645)       # Standard (Normal) CI
    448.19109
> median(weights)+(sde(tstar)*1.645)
    451.80891

> sort(tstar)
> floor(2001*0.05)
    100
> 2001-floor(2001*0.05)
    1901

> tstar[100]                 # 0.05 quantile of bootstrap dist'n
    449
> tstar[1901]                # 0.95 quantile of bootstrap dist'n
    452.5                    # Symmetric percentile CI

> 2*median(weights)-tstar[1901]      # Nonsymmetric percentile CI
    447.5
> 2*median(weights)-tstar[100]
    451
```

Figure 8.2:   *SC Computation of Bootstrap Estimates*

In this application, the `boot` procedure applied the function `g` to each of 2,000 randomly chosen bootstrap samples—each drawn randomly and with replacement from the nonparametric population distribution estimate `weight`—and stored these results in the vector `tstar`. The resampling model used is the one appropriate for large populations, where $(N/n) \geq 20$.

The mean and standard deviation (SD) of the $B = 2,000$ values of t_b^* were found to be 450.65975 gm and 1.0996406 gm, respectively. The latter is computed in estimate form—the sum of squared deviations about the mean is divided by $(B-1)$ rather than by B—and is the bootstrap estimate of the SE of the sample median:

$$\widetilde{SE}_{Boot}(t|X, n) = \sqrt{[1/(B-1)]\sum_{b=1}^{B}\left[t_b^* - (1/B)\sum_{b=1}^{B}t_b^*\right]^2}$$

Our bootstrap estimate of bias,

$$\widetilde{Bias}_{Boot}(t|X, n, \theta) = (1/B)\sum_{b=1}^{B}t_b^* - \widehat{\theta}$$

is the difference between the mean of the sampling distribution, 450.65975 gm, and the value of our parameter, the median, in the estimated or bootstrap population distribution. As our nonparametric estimate of the population distribution is just the sample distribution, $\widehat{\theta}$ is just the sample median, median(weights), of 450 gm. Thus, the bias estimate is 0.65976 gm.

How much larger will the RMS error estimate be, compared with our SE estimate?

Next, we use the estimated SE to form standard or normal theory CI estimates. The normal sampling distribution has as its 0.05 and 0.95 quantiles the values -1.645 SE and 1.645 SE. Assuming the sampling distribution is centered at the parameter value, this gives an estimated 90% equal tails exclusion CI with limits at $t \pm 1.645\,\widetilde{SE}_{Boot}(t)$. For our pretzel bag example, the normal theory or standard estimate of a 90% equal tails exclusion CI is found to have limits:

$$[448.19 \text{ gm}, 451.81 \text{ gm}]$$

To find the symmetric and nonsymmetric percentile estimates of the $(1-2\alpha)100\%$ CI, we need to find the α and $(1-\alpha)$ quantiles of the bootstrap sampling distribution. As the the distribution consisted of 2,000 elements, the two quantiles were found to be:

$$F_B^{-1}(0.05) = t^*_{[\text{int}(0.05 \times 2001)]} = t^*_{[100]} = 449 \text{ gm}$$

and

$$F_B^{-1}(0.95) = t^*_{[2001 - \text{int}((1-0.95) \times 2001)]} = t^*_{[1901]} = 452.50 \text{ gm}$$

These quantiles define the limits to the symmetric percentile estimate of a 90% CI:

$$[449 \text{ gm}, 452.5 \text{ gm}]$$

When, as in the present case, (a) our estimator is of the plug-in variety—both estimator and parameter are medians—and (b) our estimate of the population distribution is nonparametric, then the nonsymmetric percentile estimate of the 90% CI has limits

$$[t - a, \, t + b] = [2t - F_B^{-1}(1 - \alpha), 2t - F_B^{-1}(\alpha)]$$

Thus, the nonsymmetric percentile estimate of the 90%, CI for the median of the population of pretzel bag weights has lower and upper limits computed as:

$$[447.5 \text{ gm}, 451 \text{ gm}]$$

The three CI estimates are grounded in different assumptions about the form of the sampling distribution and, hence, gave somewhat different results here.

Exercises

1. Repeat the pretzel bag weight analysis using software available to you. You are certain to create a different random sequence of 2,000 random bootstrap samples. As a result, your SE, bias, and CI estimates will vary, perhaps only slightly, from those reported earlier. Comment on the magnitudes of the differences.

2. How closely Monte Carlo approximation results can be replicated depends on the number of bootstrap samples, B, we use for the approximation. This exercise shows the influence of B on the estimated SE.

Table 5.2:   *WAIS Vocabulary Scores for a University Sample*

```
14 11 13 13 13 15 11 16 10 13 14 11 13 12 10 14 10 14
16 14 14 11 11 11 13 12 13 11 11 15 14 16 12 17  9 16
11 19 14 12 12 10 11 12 13 13 14 11 11 15 12 16 15 11
```

Use the Table 5.2 data, reproduced here, giving WAIS-R Vocabulary subtest scores for 54 university students randomly chosen from a population of 7,500.

Investigate the variability in Monte Carlo bootstrap estimates of the SE of the mean. This bootstrap estimate will be given by the standard deviation of B means, each computed from a randomly chosen bootstrap sample.

a. Set B equal to 250. Estimate the SE of the mean from each of 10 successive bootstrap sampling distributions. Compute the mean and standard deviation of these 10 estimates.

b. Repeat part a. for $B = 500$.

c. Repeat part a. for $B = 1000$.

d. Repeat part a. for $B = 1500$.

e. Repeat part a. for $B = 2000$.

f. Plot the mean estimates from parts a. through e. against the values of B. Do the same for the standard deviations of the estimates from the five different values of B.

g. Interpret your results. What is happening to the mean of the 10 estimates as B increases? What is happening to the SD of the estimates? What would you expect to happen to this mean and SD as B becomes very, very large, as you move from a Monte Carlo approximation to the complete bootstrap sampling distribution?

Concepts 9

Better Bootstrap CIs: The Bootstrap-*t*

In Concepts 8, we outlined three approaches to estimating a confidence interval (CI) from a bootstrap sampling distribution. These approaches assume, correctly, that the sampling distributions for the real-world estimator t and for the bootstrap-world estimator t^* would differ in their locations. The sampling distribution for t ought to be centered about θ, the real-world population parameter estimated by t, while the sampling distribution for t^* ought to be centered about $\widehat{\theta}$, the bootstrap-world parameter estimated by the t_b^*s. The standard and symmetric percentile approaches assume that the two sampling distributions are symmetric, whereas the nonsymmetric percentile approach assumes only that the two have the same form. This leads us to estimate the lengths of the two arms of the CI about our t, $[t - a,\ t + b]$, using corresponding arm lengths obtained from the bootstrap-world sampling distribution.

The assumption of the equivalence of the forms of the two sampling distributions would be warranted if the centered estimate, $(t - \theta)$, was a pivotal form. Were that true, $(t - \theta)$ would have the same sampling distribution as $\left(t_b^* - \widehat{\theta}\right)$. In general, however, centering the estimate t is not sufficient to put it in pivotal form. In this Concepts, we review what is meant by a pivotal form and examine a bootstrap generalization of a particular pivot, the Studentized mean.

Pivotal Form Statistics

In Concepts 5, we noted the importance to mathematically based inference of having a statistic whose sampling distribution would always follow the same parametric form. Often this requires some transformation of a plug-in estimate. Thus, we noted that when we compute population mean and variance estimates, $\overline{x}$ and $\widehat{\sigma}^2$, for a sample drawn from a population distribution with a normal cumulative distribution function (cdf), then the sampling distribution for

$$t(\bar{x}) = \frac{\bar{x} - \mu}{\sqrt{\hat{\sigma}^2/n}}$$

will have the same cdf as that of the t-random variable with $n - 1$ degrees of freedom (df). This will be true, whatever the mean, μ, of the population distribution sampled. We also learned that our sample estimate, $\hat{\sigma}^2$, of the variance of a normal distribution enjoys a transformation,

$$v(\hat{\sigma}^2) = \frac{\hat{\sigma}^2(n-1)}{\sigma^2}$$

with a chi-squared sampling distribution with $n - 1$ df whatever the value of the normal population distribution variance, σ^2.

Here's a third example of a pivot. This time each case in our population has scores on two attributes, called x and y. The population distribution is bivariate; for each case the scores on the two attributes are linked together. Such a bivariate population distribution often can be characterized by a correlation parameter, ρ_{xy}. This parameter is the population analog of the sample product-moment correlation, r_{xy}. Put another way, r_{xy} is a plug-in estimator of ρ_{xy}. But what can be said about the sampling distribution of r_{xy}?

If the population distribution is that of a bivariate normal random variable (requiring, among other things, that both attributes have normal cdfs), then the Fisher transformation,

$$\hat{\phi}(r_{xy}) = (1/2)\left\{ \log\left[\frac{1 + r_{xy}}{1 - r_{xy}}\right] - \log\left[\frac{1 + \rho_{xy}}{1 - \rho_{xy}}\right] \right\}$$

has approximately a normal sampling distribution with mean of zero and variance of $[1/(n - 3)]$.

$\hat{\phi}(r_{xy})$ is almost a pivotal form statistic. This transformation of r_{xy} reduces, but does not eliminate completely the sampling distribution's dependence on the value of the parameter estimated, ρ_{xy}. The failure of this transformation to provide an exact pivot indicates how difficult it is to find pivotal forms with well-defined sampling distributions.

The pivotal forms, exact or not, that are important to mathematically based inference share three properties:

1. They have mathematically tractable sampling distributions—for example, $t(\bar{x})$, $v(\hat{\sigma}^2)$, and $\hat{\phi}(r_{xy})$ have sampling distributions with cdfs identical (or, nearly so) to those of a *t*-random variable, a chi-squared random variable, and a normal random variable, respectively.

2. These mathematical distributions are indexed by one or more parameters that control, roughly speaking, the shape of the sampling distribution—the *t*- and chi-squared cdfs vary relative to a degrees of freedom parameter, and the normal cdf depends on a variance parameter.

3. The shape of the sampling distribution of a pivotal form statistic can change with the size of the sample, n. It will not change, however, as the value of θ, the parameter under estimation, changes from one population distribution to another.

In the bootstrap world, the first two pivotal-form properties are generally unimportant. We neither require nor expect our sampling distributions to have nice mathematical forms. The third property, however, is important, particularly for estimating confidence intervals. We would like to be able to generalize properties of the sampling distribution based on repeated samples from $\hat{X}$, a population distribution with parameter $\hat{\theta}$, to those of the sampling distribution based on repeated samples from X, a population distribution with parameter θ. In general, neither t nor $(t - \theta)$ are pivotal in this sense, and the CI techniques of Concepts 8 can be inadequate.

The Bootstrap-*t* Pivotal Transformation

Mathematical statisticians have established that asymptotically—as the sample size approaches infinity—many estimators have sampling distributions that can be characterized in the following way:

$$F\left[\frac{t - \theta}{\widehat{SE}(t)}\right] \to \text{Normal}(\mu = 0, \sigma^2 = 1) \text{ as } n \to \infty$$

That is, the cdf of the standardized sampling distribution approaches that of the standard normal random variable as the sample size approaches the infinite. This result motivates the standard (or, normal) CI estimate of Concepts 8. However, asymptotic results can provide poor results for finite sample sizes.

As we have seen, Student and Fisher established that where t is the sample mean, $\bar{x}$, and the sample is drawn from a normal population distribution, this standardized sampling distribution has the cdf of a t-random variable with $(n-1)$ df:

$$F[t(\bar{x})] = F\left[\frac{\bar{x} - \mu}{\sqrt{\hat{\sigma}^2/n}}\right] \approx t\text{-random variable } (\mathrm{df} = n-1)$$

The asymptotic normal and Studentized $t(\bar{x})$ have a common form:

1. Subtract from the estimate, t, the value of the parameter, θ.

2. Divide the resulting difference by the estimated SE of the estimator.

Intuitively, the two operations should contribute to pivotality; we adjust first for the location of the sampling distribution of t, and then for the estimated variability in that distribution.

This two-step operation, $t(t) = \left[(t - \theta)/\widehat{\mathrm{SE}}(t)\right]$, has been used with estimators other than the sample mean and has come to be identified as Studentization. This results in a strongly pivotal statistic, in the sense of possessing mathematical as well as shape stability for the pair $(t = \bar{x}, \theta = \mu)$, at least when you sample from a normal population. Studentization would be very useful in bootstrap inference if it were at least mildly pivotal for a larger collection of (t, θ) pairs. By mildly pivotal we mean possessing general shape stability as the sampled population shifts from $\hat{X}$ to X and, in consequence, the value of the parameter estimated shifts from $\hat{\theta}$ to θ.

That is, we will have an improved basis for estimating CIs from a bootstrap sampling distribution whenever the bootstrap sampling distribution of

$$t(t^*) = \left[\frac{t^* - \hat{\theta}}{\widehat{\mathrm{SE}}(t^*)}\right]$$

provides a good approximation to the real-world sampling distribution of

$$t(t) = \left[\frac{t - \theta}{\widehat{\mathrm{SE}}(t)}\right]$$

CIs estimated in this way are known as bootstrap-*t* CIs. The balance of this Concepts describes how bootstrap-*t* CIs are computed and where they can be used to advantage.

The double use of a "t" in the notation $t(t)$ may appear awkward. How does it come about? I use t to denote the estimator of a parameter θ, a well-established practice in the bootstrap literature. The Studentization of a statistic is sometimes referred to as the *t*-transformation, the name stemming from the *t*-distribution that describes the sampling distribution of the Studentization of the normal sample mean. Hence, I denote the Studentization operator as $t()$.

Forming Bootstrap-*t* CIs

To form the bootstrap sampling distribution of $t(t^*)$ requires that we compute

$$t(t_b^*) = \left[\frac{t_b^* - \widehat{\theta}}{\widehat{SE}(t_b^*)} \right].$$

for each of a long sequence of randomly chosen bootstrap sample distributions, $x_1^*, \ldots, x_b^*, \ldots, x_B^*$. Note that this requires, in turn, that we compute both t_b^* and $\widehat{SE}(t_b^*)$ from each x_b^*. For now we'll assume that we know how to do that. Later in this Concepts we'll see three approaches to estimating the SE in each bootstrap sample.

The following steps describe how to compute a $(1 - 2\alpha)100\%$ equal tails exclusion bootstrap-*t* CI.

1. From the sample distribution, x, compute and save the estimate, t, and form the estimated population, $\widehat{X}$. We'll assume the latter is a nonparametric estimate.

2. Compute and save the parameter, $\widehat{\theta}$, of the estimated population. If t is a plug-in estimate of θ and $\widehat{X}$ is a nonparametric estimate, then t and $\widehat{\theta}$ will be identical.

3. For each of B bootstrap samples from $\widehat{X}$ compute t_b^*, $\widehat{SE}(t_b^*)$, and $t(t_b^*)$. Save both t_b^* and $t(t_b^*)$.

4. Compute the standard deviation (SD) of the B t_b^*s and save it as $\widehat{SE}(t)$. This result is our bootstrap estimate of the real-world SE.

5. Find the α and $(1-\alpha)$ quantiles of the bootstrap sampling distribution of $t(t^*)$. Label these $F_{t(t^*)}^{-1}(\alpha)$ and $F_{t(t^*)}^{-1}(1-\alpha)$.

6. If $t(t)$ is strongly pivotal for our t, then $F_{t(t^*)}^{-1}(\alpha)$ and $F_{t(t^*)}^{-1}(1-\alpha)$ also can be interpreted as the α and $(1-\alpha)$ quantiles of the real-world sampling distribution of $t(t)$. If $t(t)$ is only weakly pivotal, the quantiles of the sampling distribution of $t(t^*)$ still provide estimates of the corresponding quantiles of the sampling distribution of $t(t)$.

7. Multiply the $F_{t(t^*)}^{-1}(\alpha)$ and $F_{t(t^*)}^{-1}(1-\alpha)$ found in step 5 each by the $\widetilde{SE}(t)$ computed in step 4. The resulting quantities estimate the α and $(1-\alpha)$ quantiles of the sampling distribution of $[t-\theta]$. Label them $\widehat{F}_{t-\theta}^{-1}(\alpha)$ and $\widehat{F}_{t-\theta}^{-1}(1-\alpha)$.

8. The sampling distribution of $[t-\theta]$ is centered at zero rather than at θ. As a result, $\widehat{F}_{t-\theta}^{-1}(1-\alpha)$ is the estimated distance from zero to the $(1-\alpha)$ quantile of the sampling distribution of $[t-\theta]$.

$\widehat{F}_{t-\theta}^{-1}(1-\alpha)$ is also the estimated distance from θ to the $(1-\alpha)$ quantile of the sampling distribution of t. That is, $\widehat{F}_{t-\theta}^{-1}(1-\alpha)$ estimates the distance a in our formulation of the $(1-2\alpha)100\%$ equal tails exclusion CI.

By a similar argument, $-\widehat{F}_{t-\theta}^{-1}(\alpha)$ is the estimated distance between θ and the α quantile of the sampling distribution of t, and estimates the distance b in our CI formulation.

9. Thus, the $(1-2\alpha)100\%$ bootstrap-t CI for θ has a lower limit of
$$\left[t - \widehat{F}_{t-\theta}^{-1}(1-\alpha)\right]$$

and an upper limit of

$$\left[t - \widehat{F}_{t-\theta}^{-1}(\alpha)\right]$$

where t is the estimate computed in step 1 from the real-world sample distribution x and $\widehat{F}_{t-\theta}^{-1}(\alpha)$ and $\widehat{F}_{t-\theta}^{-1}(1-\alpha)$ are the quantiles computed in step 7.

The bootstrap-t algorithm is summarized in Figure 9.1. The figure links the estimation of the α and $(1-\alpha)$ quantiles of the sampling distribution of t to the computation of the corresponding percentiles of the sampling distribution of $t(t^*)$ in three stages:

1. Pivotality. The assumption that $t(t)$ is a pivotal transformation allows us to treat the α and $(1 - \alpha)$ quantiles of the sampling distribution of $t(t^*)$, which we can obtain, as if they were those of the sampling distribution of $t(t)$.

2. Scaling. Multiplying the selected quantiles of the sampling distribution of $t(t)$ by our estimate of the standard error of t, gives us estimates of the corresponding quantiles of the sampling distribution of $(t - \theta)$.

3. Shifting. If we could add the value of θ to the quantiles of $(t - \theta)$, we would have the corresponding quantiles of the sampling distribution of t. The shift is unnecessary, however, as it would preserve the distance between quantiles, and we are able to use $a = \widehat{F}^{-1}_{t-\theta}(1 - \alpha)$ and $b = -\widehat{F}^{-1}_{t-\theta}(\alpha)$ to form the $(1 - 2\alpha)100\%$ CI: $[t - a, \, t + b]$.

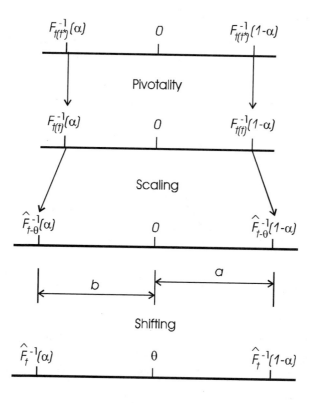

Figure 9.1:   *Stages in the Computation of the Bootstrap-t Confidence Interval*

Now that we have an algorithm for computing bootstrap-t CIs, we turn to the question of computing a SE estimate from each bootstrap sample.

Estimating the Standard Error of an Estimate

In step 3 of the bootstrap-t CI algorithm, we must compute $\widetilde{SE}(t_b^*)$ from the n elements of x_b^*. How do we do that? There are three main approaches:

1. Use a known computational formula.

2. Construct a second-level bootstrap estimate.

3. Develop and implement a smoothed relationship of $\widetilde{SE}(t_b^*)$ to t_b^*.

We'll look at each in turn.

SE Estimation by Formula

For some ts we will know a formula for $\widetilde{SE}(t)$.

Example 1. This approach is appropriate, certainly, for $t = \bar{x}$. Given a sample distribution $(x_1, \ldots, x_i, \ldots, x_n)$, we compute the sample mean as

$$\bar{x} = (1/n)\sum_{i=1}^{n} x_i$$

and estimate its SE as

$$\widetilde{SE}(\bar{x}) = \sqrt{\hat{\sigma}^2/n}$$

where

$$\hat{\sigma}^2 = [1/(n-1)]\sum_{i=1}^{n}(x_i - \bar{x})^2$$

Example 2. If we have a sample of n scores from a population distribution that is dichotomous, consisting only of a proportion, P, of successes and a proportion, $(1 - P)$, of failures, we estimate the parameter, P, as

$$\hat{P} = \frac{\#(\text{successes})}{n}$$

where $\#(x)$ is read as "The number of xs in the distribution." Our estimate of the SE of $\widehat{P}$ is obtained from the same sample as

$$\widetilde{SE}(\widehat{P}) = \sqrt{\frac{\widehat{P}(1-\widehat{P})}{n}}$$

Example 3. We have two independent samples,

$$(x_1, \ldots, x_i, \ldots, x_{n_x})$$

from a population distribution X with mean μ_X, and

$$(y_1, \ldots, y_j, \ldots, y_{n_y})$$

from a population distribution Y with mean μ_Y. We estimate the difference in population means, $(\mu_X - \mu_Y)$, with the plug-in estimator

$$\bar{x} - \bar{y} = (1/n_x)\sum_{i=1}^{n_x} x_i - (1/n_y)\sum_{j=1}^{n_y} y_j$$

whose SE is estimated as

$$\widetilde{SE}(\bar{x} - \bar{y}) = \sqrt{\frac{\sum\limits_{i=1}^{n_x}(x_i - \bar{x})^2}{n_x(n_x - 1)} + \frac{\sum\limits_{j=1}^{n_y}(y_j - \bar{y})^2}{n_y(n_y - 1)}}$$

The three examples describe computations on a real-world sample, x, or samples, (x, y). Those computational rules can be applied just as well to the b-th bootstrap sample or samples, x_b^* or (x_b^*, y_b^*).

We'll see SE estimation formulae for other estimators in the Applications. We do not know a SE estimation formula, though, for every estimator.

SE Estimation by Bootstrap Sampling

Without an explicit formula for estimating the SE, we can use the bootstrap. In Concepts 8, we learned how to estimate the SE of an estimate t using the bootstrap sampling distribution of t^*. We'll review that process.

To begin, we let x be a sample of n scores from the population distribution X with parameter θ. We compute the parameter estimate t from x. We also use x to form the nonparametric population distribution estimate, $\widehat{X}$.

We draw at random B random samples of size n from $\widehat{X}$, designating the typical bootstrap sample as x_b^*. In each of these bootstrap samples, we compute our same estimator, t_b^* from the b-th sample. Then, we have as the bootstrap estimate of the SE of t:

$$\widehat{SE}_{Boot}(t|X, n) = \sqrt{[1/(B-1)]\sum_{b=1}^{B}\left[t_b^* - (1/B)\sum_{b=1}^{B}t_b^*\right]^2}$$

We use this result at steps 4 and 7 of our bootstrap-t CI algorithm.

We can generalize the result to estimate the SE of t_b^* needed at step 3. We simply repeat the procedures described in the preceding two paragraphs, but alter the starting point.

This second time we let x_b^* be a sample of n scores from the population distribution $\widehat{X}$. As earlier, we compute from x_b^* an estimate, t_b^*, of a population distribution parameter, $\widehat{\theta}$. We also form from x_b^* the estimate of $\widehat{X}$, which we'll designate $\widehat{X}_b$. This new population estimate consists of one or more copies of x_b^*.

We draw at random $B1$ random samples of size n from $\widehat{X}_b$, designating the typical bootstrap sample as x_{ba}^*, $a = 1, \ldots, B1$. In each of these second-level bootstrap samples, we compute our same estimator, t_{ba}^* from the a-th sample. Then, we have as the bootstrap estimate of the SE of t_b^*:

$$\widehat{SE}_{Boot}(t_b^*|\widehat{X}, n) = \sqrt{[1/(B1-1)]\sum_{a=1}^{B1}\left[t_{ba}^* - (1/B1)\sum_{a=1}^{B1}t_{ba}^*\right]^2}$$

The process by which the SE of t_b^* has been estimated involves nested bootstrap sampling. Put simply, from each of the B first-level bootstrap samples, we draw a random sequence of $B1$ second-level bootstrap samples. B and $B1$ usually differ in size. B is chosen to permit good CI estimation and typically will be 2,000 to 5,000 in size. $B1$, on the other hand, need only be as large as necessary for good SE estimation, 100, perhaps.

Even with a relatively small $B1$, the nested bootstrap-*t* approach to SE estimation requires a lot of computing time. Nested bootstrap-*t* CI estimation, on the face of it, nearly multiplies by $(B1 + 1)$ the amount of computing time required for, say, the nonsymmetric bootstrap percentile CI.

SE Estimation by a Smoothed Relationship

The nested bootstrap approach to SE estimation poses a computational burden. Also, experience has shown that nested bootstrap SE estimation can behave erratically in the small-sample setting, where second-level nonparametric bootstrap samples can become quite skewed.

A remedy to both problems has been proposed by Tibshirani (1988) in the variance-stabilized bootstrap-*t*. The idea is to use a relatively small number of nested samples—B and $B1$ might be 100 and 25, respectively—as a basis for estimating a smooth relationship between t_b^* and $\widehat{SE}_{Boot}(t_b^*)$. This smoothed relation then forms the basis for a variance-stabilizing transformation of $\left(t_b^* - \widehat{\theta}\right)$, which is applied to a second sequence of $B2$ first-level bootstrap samples. The bootstrap sampling distribution for this transformation then forms the basis for CI estimation.

If $B2 = 2000$, the total number of first- and second-level bootstrap samples would be $100 + 2500 + 2000 = 4600$. This is far smaller than the number needed for the full nested solution. There, with $B = 2000$ and $B1 = 25$, we would need $2000 + 50000 = 52000$ first- and second-level bootstrap samples.

The variance-stabilized bootstrap-*t* CIs should be more accurate as well as faster to compute.

Relatively few applications of the variance-stabilized bootstrap-*t* have been reported, however, and its use at the level of this textbook would be premature.

Range of Application of the Bootstrap-*t*

The bootstrap-*t* is a reliable method for CI estimation in the nonparametric bootstrap setting whenever:

1. θ is a location parameter.

2. t is a plug-in estimator of θ.

3. We have a computational formula for $\widehat{SE}(t)$ or we are successful in estimating a variance-stabilizing transformation.

Location parameters include not only the mean and median but also the quantiles of a distribution and trimmed means (computing a mean after leaving out a proportion of the larger and smaller scores in the distribution). Location parameters also include, as we'll see in Applications 22, the weights or slope coefficients in linear regression models. Throughout the Applications, however, we shall limit our use of the bootstrap-t to estimating CIs for location parameters where we do have a computational formula for estimating the SE of the estimator. In other situations, we shall use the CI estimation approach that will be developed in Concepts 10.

Iterated Bootstrap CIs

The use of a second level of bootstrapping—drawing further bootstrap samples from an initial one—to estimate either a SE or a functional relation between estimator and SE, provides examples of iterated or double bootstrap computations. Other uses of deeper levels of bootstrapping have been thoroughly explored from a theoretical perspective. Statisticians generally agree that techniques such as the iterative bootstrap, bootstrap prepivoting, and bootstrap calibration—all involving nested bootstrapping—provide more accurate estimates than do single level bootstrap sampling techniques (for example, Shao and Tu, 1995). Their routine use must be postponed, however, until even-faster computing is widely available.

SEs and CIs for Trimmed Means

An important application of Concepts 9, one that we will use extensively, is estimating the confidence interval (CI) for a trimmed mean. As the name suggests, the trimmed mean is the mean of a distribution from which some scores have been trimmed. We are particularly interested in means computed for symmetrically trimmed distributions, those from which equal numbers of the smallest and largest scores have been trimmed.

Definition of the Trimmed Mean

A good way to define the trimmed mean is to explain how we compute it. Here is a convenient algorithm:

1. Let x be a set of n scores. This can be a sample distribution, a population distribution, or simply a data set.

2. Arrange the n scores from smallest to largest:

$$x_{[1]} \leq x_{[2]} \leq \dots \leq x_{[n-1]} \leq x_{[n]}$$

with $x_{[i]}$ being the i-th smallest score.

3. Choose a value γ from the interval $0 \leq \gamma < 0.5$.

4. Compute $g = \text{int}(\gamma \times n)$, the largest integer that does not exceed the product of γ and n.

5. Then, we compute the $(\gamma \times 100)\%$ symmetrically trimmed mean of x as

$$\overline{x}_{t,\gamma} = \left(\frac{1}{n-2g} \right) \sum_{i=(g+1)}^{(n-g)} x_{[i]}$$

The $(\gamma \times 100)\%$ trimmed mean is the mean of the middle $(1 - 2\gamma)100\%$ of the scores. We trim away both the largest $\gamma 100\%$ of the scores and the smallest $\gamma 100\%$ of the scores before averaging the remaining scores. Choosing $\gamma = 0$ gives the usual mean, no trimming. Choosing γ as close as possible to 0.50 trims away all but the middle one or middle two of the scores with the result that the trimmed mean is the median.

Importance of the Trimmed Mean

The trimmed mean provides a robust description of the location or general magnitude of a set of scores. The ordinary mean, on the other hand, lacks robustness. To appreciate what this means, consider an example of a data set consisting of ten scores. Nine of the scores are listed here:

$$100, 102, 103, 104, 104, 104, 105, 106, 108$$

The nine scores fall in an 8-point range and have a mean of $(936/9) = 104$, but the mean of the complete data set might depend very heavily on the value of the tenth score. If the tenth score is 2, the data set mean is $(938/10) = 93.8$. If the tenth score is 500, the data set mean is $(1436/10) = 143.6$. And if the tenth score is 1000, the mean becomes $(1936/10) = 193.6$. None of these values, 93.8, 143.6, or 193.6, provides a particularly good description of the location of the data set in the sense of telling us the magnitude of the bulk of the scores.

In technical terms what we have demonstrated is that the mean is not robust against the influence of a very small number (one, in our examples) of disparate scores, scores that are either very small or very large compared with the remainder of the data set. One very, very large score will make the mean very, very large as well. In contrast, our 20% trimmed mean will be uninfluenced by such outlying scores. For our examples, the data set $(2, 100, 102, 103, 104, 104, 104, 105, 106, 108)$ has a 20% trimmed mean of

$$\bar{x}_{t,0.20} = \frac{102 + 103 + 104 + 104 + 104 + 105}{6} = 103.67$$

and the data set $(100, 102, 103, 104, 104, 104, 105, 106, 108, 500)$ has a 20% trimmed mean of

$$\bar{x}_{t,0.20} = \frac{103 + 104 + 104 + 104 + 105 + 106}{6} = 104.33$$

as does the data set $(100, 102, 103, 104, 104, 104, 105, 106, 108, 1000)$. In all three instances, the trimmed mean provides a good description of the typical size of scores in the data set.

A Note on Outliers

In the random sample context, the trimmed sample mean historically has been regarded as a device for decontaminating our estimate of the population mean. It is hypothesized that we have sampled a population distribution with some simple form, for example, normal. For some reason, however, one or more of the scores appear not to have come from that population. These outliers, distinctively larger or smaller than other scores in the sample distribution, need to be discounted in our estimate of the population mean.

An Alternative View

We take a somewhat different position. First, we have no reason to believe, in general, that the population distribution of scores has any simple form. In particular, it might contain a few very small or a few very large scores, or both. Second, we have no reason to expect that some sample distribution scores will have mysterious origins and, thus, be in need of weeding-out or downplaying. Until proven to be erroneous, every sample score should be treated as a legitimate random selection from the population distribution specified in our study design.

Implications

Our alternative view of outliers has two implications:

1. Our view implies that the population mean might not be a good descriptor of the location of the population distribution. The population mean, just as the sample mean, is not robust against aberrant scores. Because we cannot inspect the population distribution, we cannot know whether or not it contains one or more small clusters of scores, well isolated from the bulk of the distribution. Our position is that we should behave as if this were true. Thus, we will tend to choose a robust location parameter, such as the 20% trimmed population mean, to be estimated from the sample.

2. Our view also implies that we should not use the sample distribution as a basis for choosing an estimation technique, for deciding which scores to attend to and which to ignore. If we throw away part of a random sample, we no longer have a random sample, and all the benefits of sampling distributions, standard errors (SEs), CIs, and p-values, are lost to us. We should decide in advance of seeing the sample distribution not only what population parameters and hypotheses interest us, but also what statistics we'll compute from the sample. In the context of random samples, our bias will be toward estimating

CIs from the nonparametric bootstrap sampling distributions of plug-in estimators.

Caveat

Our position on outliers does not mean that we should not carefully inspect our sample distributions and other data sets. Errors do occur, in the delivery of treatments, in the behavior of experimental subjects, and in the recording of data, so we should ensure that we have detected, queried, and, where possible, corrected all such errors.

Determining the Trimming Fraction

The fraction of small and large scores to be trimmed can vary from zero to almost one-half. The former gives rise to the ordinary mean, the latter to the median. Intermediate values define other location descriptors. How do we determine the trimming fraction, γ? If we choose γ too small, we let aberrant scores enter into the averaging and we lose our robustness. If we choose γ too large, the bootstrap sampling distribution of our plug-in estimator could be severely restricted. For example, if the sample size is odd, we use a nonparametric estimate of the population distribution, and choose γ to trim all but the middle score, then the trimmed mean is the median. But the median of any odd-sized bootstrap sample can take on only those distinct values appearing in the original sample. At most, there can be only n distinct values for the bootstrap sample median. Such a restricted sampling distribution can make CI estimation and hypothesis testing highly problematic. What is needed is a value of γ that is neither too small nor too large. Perhaps the best, general purpose choice for the amount of trimming is $\gamma = 0.20$ (Wilcox, 1997), and we will frequently use 20% trimmed means, each the mean of the middle 60% of a data set.

Classical statistical inference makes heavy use of the non-robust mean, both sample and population. Means are the focus of the t-test, the analysis of variance, and linear regression. The choice of the mean has been dictated by the availability of a useful mathematical structure for the sampling distribution of the mean, at least when the population distribution sampled has a normal cdf. For location measures other than the mean, no such results have been available. The use in inference of robust parameters and statistics has had to wait until statisticians and researchers could use bootstrap resampling to estimate sampling distributions and their quantiles.

Sampling Distribution of the Trimmed Mean

Our strategy, then, in estimating or testing hypotheses about the location of a population distribution is to use the 20% trimmed sample mean as a plug-in estimate of the 20% trimmed population mean. To provide a basis for these statistical inferences, we'll use nonparametric bootstrap resampling to approximate the sampling distributions of our trimmed means. As the trimmed mean is a location parameter, we can use the bootstrap-t approach to CI estimation.

Standard Error of the Trimmed Mean

The bootstrap-t approach is most reliably implemented when we can use an explicit formula to estimate the SE of a location estimate computed in the b-th bootstrap sample. We use this estimated SE in the Studentization of the estimate:

$$t(t_b^*) = \left[\frac{t_b^* - \widehat{\theta}}{\widehat{SE}(t_b^*)} \right]$$

We have such a SE estimation formula for the symmetrically trimmed mean (Wilcox, 1997). The SE of a $(\gamma \times 100)\%$ trimmed mean, computed in the b-th bootstrap sample, is estimated by

$$\widehat{SE}(\bar{x}_{t,\gamma,b}^*) = \sqrt{\frac{S_{w,\gamma}^2(x_b^*)}{(1 - 2\gamma)^2 n}}$$

where $S_{w,\gamma}^2(x_b^*)$ is the usual variance estimate, but computed from the γ-Winsorization of the bootstrap sample distribution x_b^*.

Winsorizing a data set—Charles Winsor is the statistician credited with proposing the technique—is closely related to trimming it. Trimming a proportion γ of the smallest and largest scores from the data set removes scores $x_{[1]}$ through $x_{[g]}$ and $x_{[n-g+1]}$ through $x_{[n]}$, where $g = \text{int}(\gamma \times n)$ and $x_{[i]}$ designates the i-th smallest score. This trimming leaves a data set of $(n - 2g)$ scores:

$$\left(x_{[g+1]}, \ldots, x_{[n-g]} \right)$$

Consider the earlier set of 10 scores:

$$(100, 102, 103, 104, 104, 104, 105, 106, 108, 1000)$$

ordered from smallest to largest. For $\gamma = 0.20$ we would trim away the two smallest and two largest scores leaving:

$$x_{[3]}, \ldots, x_{[8]} = (103, 104, 104, 104, 105, 106)$$

In γ-Winsorizing, instead of dropping scores $x_{[1]}$ through $x_{[g]}$, we replace each of them with $x_{[g+1]}$. And instead of dropping scores $x_{[n-g+1]}$ through $x_{[n]}$, we replace each of them with $x_{[n-g]}$. In effect, we shrink the smallest and largest scores toward the edges of the middle group, rather than drop them entirely. Our 20% Winsorized data set would contain ten, rather than six scores:

$$(103, 103, 103, 104, 104, 104, 105, 106, 106, 106)$$

As noted earlier, the $S_{w,\gamma}^2(x_b^*)$ in our equation for the estimated SE of the γ-trimmed mean of the b-th bootstrap sample, $\bar{x}_{b,\gamma}^*$, is the variance estimate obtained from the γ-Winsorization of the b-th bootstrap sample distribution. A common algebraic form for the usual variance estimate,

$$S(x_b^*)^2 = \left(\frac{1}{n-1}\right)\left[\sum_{i=1}^{n}x_{b,i}^{*\,2} - \frac{\left(\sum_{i=1}^{n}x_{b,i}^*\right)^2}{n}\right]$$

provides the framework for describing how to compute the γ-Winsorized variance:

$$S_{w,\gamma}^2(x_b^*) = \left(\frac{1}{n-1}\right)\left[c - \frac{d^2}{n}\right]$$

where

$$c = \left(g\left[x_{b,[g+1]}^{*\,2} + x_{b,[n-g]}^{*\,2}\right] + \sum_{i=g+1}^{n-g}x_{b,[i]}^{*\,2}\right)$$

and

$$d = \left(g \left[x^*_{b,[g+1]} + x^*_{b,[n-g]} \right] + \sum_{i=g+1}^{n-g} x^*_{b,[i]} \right)$$

It should be noted that although we are estimating the standard error of a symmetrically trimmed mean, we are not required to sample from symmetric population distributions (Wilcox, 1997).

Applications

Let us put these ideas into practice.

SC Example

SC has a pair of procedures, `iboot_t( )` and `fboot_t( )`, you can use to estimate bootstrap-t CIs. Both require our location estimator to be a plug-in and our estimated population distribution to be nonparametric. The first procedure assumes a large population, that is, $N/n \geq 20$, estimates the population distribution with a single copy of the sample distribution, and samples with replacement. The second procedure for small populations estimates the population distribution with an appropriate number of copies of the sample distribution and samples without replacement.

To use either of the bootstrap procedures, you must supply two additional procedures, one to compute the estimator, t^*, and the other to compute the estimated SE for that estimator. These two procedures should be written to receive three arguments:

1. A matrix with cases as rows containing the sample distribution(s)

2. A vector containing the sample size(s) and dividing the matrix by rows

3. A vector to receive m t^*_bs or m $\widetilde{SE}(t^*_b)$s, computed from x^*_b

Figure 9.2 gives the text of two procedures for computing the 20% trimmed mean and its estimated SE. The first procedure, `tmean_mn`, uses a built-in SC function, `tmean(v,gama)`, to compute the trimmed mean for each column of the input matrix (1). Each of the m trimmed means is stored as an element of the output vector, 3.

The standard error procedure, `tmean_se`, uses another built-in SC procedure, `winsor(v,pcnt)`, to replace a vector of scores with its

Winsorization. The second argument for `winsor` is given in percentage rather than proportion form.

The trimming proportion, `gama`, is set at 0.20 within each of the two procedures, but this value could be changed if a different amount of trimming were to be used.

```
proc tmean_mn(){
   local(gama,n,m,vv,g,gam2,i)
   gama= 0.20
   n= rows($1$)
   m= cols($1$)
   vector(vv,n)
   g= floor(gama*n)
   gam2= g/n
   i= 0
   repeat(m){
     i++
     getcol($1$,vv,i)
     $3$[i]=tmean(vv,gam2)
   }
}

proc tmean_se(){
   local(gama,n,m,vv,g,pcnt,h,df,i)
   gama= 0.20
   n= rows($1$)
   m= cols($1$)
   vector(vv,n)
   g= floor(gama*n)
   pcnt= 100*(g/n)
   h= (1-(2*gama))^2
   df= n*h
   i= 0
   repeat(m){
     i++
     getcol($1$,vv,i)
     winsor(vv,pcnt)
     $3$[i]= sqrt(vare(vv)/df)
   }
}
```

Figure 9.2:   *SC Procedures for Trimmed Mean and SE of Trimmed Mean*

The data in Table 9.1 are from Wilcox (1997) and consist of the time in seconds that a laboratory apparatus is kept in contact with a target by each member of a random sample of 19 university students. These times are ordered from shortest to longest.

Table 9.1:   *Time on Target for a Random Sample of University Students*

77	87	88	114	151	210	219	246	253
262	296	299	306	376	428	515	666	1310
2611								

We'll use these data to estimate a confidence interval for the 20% trimmed population mean for time on target. For purposes of illustration, we'll assume the 19 study participants were randomly chosen from a population of 100 volunteers. We want to estimate the 20% trimmed mean of the distribution of time on target scores for this population of 100 students.

The SC procedure `fboot_t` produces bootstrap-*t* CIs for parameters of finite (small) populations. It takes eight or more arguments:

$$fboot\_t(p\_t,p\_se,M,nn,NN,m,B,F,CI,\ldots)$$

p_t: Procedure to compute one or more plug-in estimate
p_se: Procedure to estimate the SE for each estimate
M: Data matrix, cases as rows
nn: Vector of sample sizes, blocking the rows of M
NN: Vector of population sizes, same length as nn
m: The number of parameters estimated by p_t
B: The number of bootstrap samples to be drawn
F: Name of a B × m matrix to hold the *t*(*t**) results
CI: Optional, one or more CI level, e.g., 75 or 90, 95

Figure 9.3 reproduces an interactive SC session in which `fboot_t` is used. The initial command retrieves three objects that had previously been stored, a matrix TTARG with 19 rows and 1 column holding the sample time on target scores shown in Table 9.1, and the single-element vectors nn and NN containing the sample and population sizes. The procedures tmean_mn and tmean_se are described in Figure 9.2. They are user-written procedures, rather than procedures supplied with SC, so they must be declared before use. We use the commands `proc tmean_mn(){}` and `proc tmean_se(){}` to declare the procedures.

```
> load(wilcox31)
  [TTARG nn NN]
> nn
  19
> NN
  100
> proc tmean_mn(){}
> proc tmean_se(){}
> fboot_t(tmean_mn,tmean_se,TTARG,nn,NN,1,2500,FBOOT,90)

_____thetahat:
  282.69231

_____means of  2500  bootstrap estimates:
  295.76514

_____SDs of  2500  bootstrap estimates:
  61.361895

    Parameter: 1   90 % CI:    210.64605 to   394.48682
    Total fboot_t() running time:  17.2
```

Figure 9.3: *Estimating the 90% CI for a Trimmed Mean with fboot_t( ) in SC*

When you invoke `fboot_t` with the arguments given here, you are specifying that the user-written `tmean_mn` returns only one value, that 2,500 bootstrap samples are to be drawn (each without replacement from an estimated population distribution of 100 elements), that the 2,500 values of

$$t(\overline{x}^*_{t,0.2,b}) = \left[\frac{\overline{x}^*_{t,0.2,b} - \widehat{\mu}_{t,0.2}}{\widehat{SE}(\overline{x}^*_{t,0.2,b})} \right]$$

are to be saved as the single column of a matrix `FBOOT`, and that a 90% equal tails exclusion CI for $\mu_{t,0.2}$ is to be estimated from the quantiles of `FBOOT` and reported.

The 20% trimmed sample mean, given our nonparametric estimate of the population distribution, is also the 20% trimmed mean of the estimated distribution: $\widehat{\mu}_{t,0.2} = 282.69$ sec (`thetahat`). The mean and SD of the bootstrap sample 20% trimmed means, for this random sequence of 2,500 random samples, were 295.77 sec and 61.36 sec, respectively. Finally, the estimated 90% CI for the population 20% trimmed mean extends from 210.65 sec to 394.49 sec.

The SC procedure `iboot_t`, which we will use in subsequent applications, is the large population version of `fboot_t`. It requires the same arguments; except the vector of population sizes, NN, is omitted because the populations are assumed to be so large that the bootstrap samples can be drawn with replacement from the sample distribution.

Resampling Stats Example

Next, we examine a Resampling Stats script for this same problem. Because it must accomplish the complete process without using higher level building blocks, such as those used in the SC solution, the script is quite long. For convenience, it has been broken up here and reproduced as Figures 9.4 through 9.7.

```
maxsize default 10
maxsize sampdist 100 trimdist 100 windist 100 addon 100
maxsize bootsamp 100
maxsize popdist1 2500 popdist2 2500
maxsize ftstar 5000 fxstar 5000
copy 0.20 gama              'trimming proportion
copy 0.05 alfa             'CI alpha level
copy 25   B1               'number of population rebalances
copy 100 B2                'resamples from each rebalance
'
copy (77 87 88 114 151 210 219 246 253 262 296) sampdist
concat sampdist (299 306 376 428 515 666 1310 2611) sampdist
copy 19 sampsiz             'sample size
copy 100 popsiz             'population size
'
```

Figure 9.4:   *Part 1, Resampling Stats Script for Trimmed Mean CI Estimation*

Part 1 of the script (Figure 9.4) handles input for the problem. The `maxsize` commands, when used, occur at the beginning of a script and control memory allocation for variables. Next the trimming proportion, γ, and the level of the $(1 - 2\alpha)100\%$ CI are declared, as are the number of bootstrap samples to be drawn. B1 specifies the number of times the finite population distribution estimate is to be adjusted for any imbalance in sample score representation, and `B2` specifies the number of resamples to be drawn from each adjusted population distribution; overall, `B1` × `B2` bootstrap samples will be drawn.

```
multiply gama sampsiz rawcut
round rawcut trim
if trim > rawcut
  subtract trim 1 trim        'number to trim from either end
end
add trim 1 lo
subtract sampsiz trim hi      'range of ordered scores for mean
'
multiply gama 2 x
subtract 1 x x
multiply x x x
multiply x sampsiz varden     'denominator of variance estimate
'
sort sampdist sampdist
take sampdist lo,hi trimdist
mean trimdist trimmean        'the gamma trimmed mean
'
divide popsiz sampsiz x
round x reps
if reps > x
  subtract reps 1 reps        'number of complete replicates
end                           'of sample in population
multiply reps sampsiz whol
subtract popsiz whol rem      'remainder needed to complete
subtract reps 1 rm1
copy sampdist popdist1
repeat rm1
  concat popdist1 sampdist popdist1
end
'
```

Figure 9.5: *Part 2, Resampling Stats Script for Trimmed Mean CI Estimation*

The final four lines of Figure 9.4 are specific to the time on target data. The sample distribution of scores is copied to the vector `sampdist` and the sample and population sizes declared. These four lines are the only ones, with the possible exception of one or more of the `maxsize` declarations, that you would have to modify to use this computing script with another one-sample problem.

Part 2 of the script, Figure 9.5, achieves the following:

> 1. Determines from γ and the sample size the number of scores to be trimmed from each end of the sample (and each bootstrap resample) score distribution before a mean is computed, `trim`

2. Computes the denominator term of the estimate of the sampling variance of the trimmed mean, $(1 - 2\gamma)^2 n$ or `varden`

3. Finds the γ-trimmed mean of the sample distribution, `trimmean`. This quantity is both our estimate of the γ-trimmed mean of X and the γ-trimmed mean of our $\widehat{X}$

4. Computes the number of complete copies of the sample distribution that can be included in the estimated population distribution, `reps`

5. Determines the number of additional cases needed to complete the estimated population, `rem`

6. Forms an incomplete estimate of the population distribution by transferring `reps` copies of `sampdist` to the vector `popdist1`

Figure 9.6 presents Part 3 of the Resampling Stats script, the bootstrap portion of the CI estimation task. In each of the `B1` outer loops through these instructions, the incomplete population distribution estimate, `popdist1`, is completed by concatenating to it `rem` additional scores, randomly chosen from `sampdist`.

For each of these completed population distribution estimates, an inner loop of instructions is executed `B2` times. On each pass through this inner loop:

1. A nonreplacement random sample, a bootstrap sample, is drawn from the estimated population distribution, `bootsamp`.

2. The γ-trimmed mean of the bootstrap sample, $\overline{x}^*_{t,0.2,b}$, is computed as `xstar` and stored as an element of the bootstrap sampling distribution `fxstar`.

3. The bootstrap sample is γ-Winsorized, and the result is stored as the vector `windist`.

4. The γ-Winsorized variance of the bootstrap sample is found, `winvar`, and used together with the earlier computed $(1 - 2\gamma)^2 n$ or `varden` to calculate the estimated SE for the bootstrap trimmed mean, `tden`.

5. The Studentization of the bootstrap trimmed mean,

$$t(\overline{x}^*_{t,0.2,b}) = \left[\frac{\overline{x}^*_{t,0.2,b} - \widehat{\mu}_{t,0.2}}{\widehat{SE}(\overline{x}^*_{t,0.2,b})} \right]$$

is computed as `tstar` and is stored as an element of the bootstrap sampling distribution `ftstar`.

```
repeat B1
  copy popdist1 popdist2
  if rem > 0
    shuffle sampdist sampdist
    take sampdist 1,rem addon
    concat popdist1 addon popdist2
  end
  repeat B2
    shuffle popdist2 popdist2
    take popdist2 1,sampsiz bootsamp    'sample non-replacement
    sort bootsamp bootsamp
    take bootsamp lo,hi trimdist
    mean trimdist xstar                 'bootstrap trimmed mean
    score xstar fxstar                  'saved
    take bootsamp lo bot                      'forming the
    take bootsamp hi top                      'winsorized
    concat trimdist trim#bot trim#top windist  'bootstrap samp
    variance windist winvar
    divide winvar varden x
    sqrt x tden                         'the SE estimate
    subtract xstar trimmean tnum
    divide tnum tden tstar              'the t-transformation
    score tstar ftstar                  'saved
  end
end
```

Figure 9.6:   *Part 3, Resampling Stats Script for Trimmed Mean CI Estimation*

At the end of Part 3 of the script, the bootstrap sampling distributions `fxstar` and `ftstar` consist of $B1 \times B2 = 2500$ values of $\overline{x}^*_{t,0.2,b}$ and $t(\overline{x}^*_{t,0.2,b})$, respectively.

In the final part of the script, Figure 9.7, a $(1 - 2\alpha)100\%$ CI for the population γ-trimmed mean is estimated from these two bootstrap sampling distributions. Using the notation of Concepts 9, the lower and upper limits for this CI can be expressed as

$$t - F^{-1}_{t(t^*)}(1 - \alpha) \times \widehat{\text{SE}}(t)$$

and

$$t - F_{t(t^*)}^{-1}(\alpha) \times \widetilde{SE}(t)$$

respectively. In our Resampling Stats script, (a) the plug-in estimator t is called `trimmean`; (b) the bootstrap SE estimate, $\widetilde{SE}(t)$, is the standard deviation of `fxstar`, called `seout`; and (c) $F_{t(t^*)}^{-1}(\alpha)$ and $F_{t(t^*)}^{-1}(1 - \alpha)$, the α and $(1 - \alpha)$ quantiles of `ftstar`, are referred to as `lim1` and `lim2`, respectively.

```
multiply B1 B2 x
add x 1 BP1
multiply alfa BP1 x
round x qlo
if qlo > x
   subtract qlo 1 qlo
end
subtract BP1 qlo qhi            'the alpha quantile
variance fxstar x               'the (1 - alpha) quantile
sqrt x seout
sort ftstar ftstar              'bootstrap SE estimate
take ftstar qlo lim1
take ftstar qhi lim2            'computing the
concat lim2 lim1 lims           'bootstrap-t
multiply lims seout lims        'confidence interval
subtract trimmean lims ci       'limits
print trimmean
print seout
print ci
```

Figure 9.7:   *Part 4, Resampling Stats Script for Trimmed Mean CI Estimation*

One run of the script of Figures 9.4 through 9.7 produced the results shown in Figure 9.8. The CI estimate, which based on a second series of 2,500 randomly chosen bootstrap samples, is close to that obtained with the SC procedure `fboot_t`. Based on the performances of 19 randomly chosen volunteer subjects, we are 90% confident that the interval from 210 sec to 394 sec contains the mean of the middle 60% of the time on target scores (the 20% trimmed mean) for the entire population of 100 volunteers.

```
TRIMMEAN  =       282.69

SEOUT    =        63.007

CI       =        210.38       393.07

Successful execution. (14.2 seconds)
```

Figure 9.8: *Resampling Stats Estimate of Time on Target Trimmed Mean CI*

S-Plus Example

To estimate bootstrap-t CI limits in S-Plus, we must arrange for the `bootstrap` function to create bootstrap sampling distributions both for t^* and for $t(t^*) = (t^* - t)/\widehat{SE}(t^*)$. Figure 9.9 shows how we can apply this approach to the time-on-target problem.

The user-supplied function `ttmean2` displayed in Figure 9.9 uses a vector X, containing a bootstrap sample x^*, and two constants—`gama`, the trimming proportion γ, and `tmuhat`, the γ-trimmed mean of our real-world sample—to compute and return two quantities:

$t^* = \bar{x}_{t,\gamma}^*$, the γ-trimmed mean of the vector X

and

$t(t^*) = (\bar{x}_{t,\gamma}^* - \hat{\mu}_{t,\gamma})/\widehat{SE}(\bar{x}_{t,\gamma}^*)$, the Studentized $\bar{x}_{t,\gamma}^*$

The `bootstrap` function requires two arguments:

1. The name of a data structure containing the real-world sample x. Here, that is the vector `ttarg`.

2. The name of a user-provided function that will return t^* from bootstrap sample. Here, our `ttmean2` returns both t^* and $t(t^*)$.

The `bootstrap` function samples with replacement from the random sample distribution. That is, the `bootstrap` function assumes that our nonreplacement random sample was drawn from a very large population and estimates the population distribution nonparametrically with a single copy of the sample distribution. The Monte Carlo approximation to the bootstrap sampling distribution will be based on 1,000 random samples unless another value of B is specified. Here, for consistency with the SC and Resampling Stats solutions, we set B at 2,500.

```
> ttarg
[1]    77    87    88  114  151  210  219  246  253  262  296
[12]  299  306  376  428  515  666 1310 2611
> gama<-0.20
> tmuhat<-mean(ttarg,trim=gama)
> tmuhat
[1] 282.6923
> ttmean2<-
function(X)
{
    x <- sort(X)
    n <- length(x)
    a <- mean(x, trim=gama)
    den <- n*(1 - 2*gama)^2
    t <- floor(n*gama)
    l <- x[t-1]
    u <- x[n-t]
    wv <- c(rep(l,times=t),x[(t+1):(n-t)],rep(u,times=t))
    b <- (a-tmuhat)/sqrt(var(wv)/den)
    c(a,b)
}
> boot_t<-bootstrap(ttarg,ttmean2,B=2500)
> print(boot_t)
Summary Statistics:
          Observed      Bias      Mean        SE
ttmean21     282.7   17.9690  300.6613  77.6219
ttmean22       0.0   -0.1107   -0.1107   0.9619
> setmn<-77.6219
> limits.emp(boot_t)
                 2.5%          5%         95%       97.5%
ttmean21   201.453846  212.457692  443.269231  503.736538
ttmean22    -2.365723   -1.914523    1.227503    1.396336
> finv05<- -1.914523
> finv95<- 1.227503
> tmuhat-(setmn*finv95)
[1] 187.4112
> tmuhat-(setmn*finv05)
[1] 431.3012
```

Figure 9.9:   *S-Plus Estimation of Bootstrap-t CI Limits*

In the lower portion of Figure 9.9, we obtain what we need for the bootstrap-*t* CI estimation from the two bootstrap sampling distributions:

1. The `print(boot_t)` command gives summary statistics for the two bootstrap sampling distributions. From this, we extract the standard deviation of the $t^* = \bar{x}_{t,\gamma}^*$ distribution, $\widehat{SE}(\bar{x}_{t,\gamma})$: `setmn`.

2. The `limits.emp(boot_t)` command gives selected quantiles of the two bootstrap sampling distributions. For our 90% CI estimation, we need the 0.05 and 0.95 quantiles of the sampling distribution of $t(t^*) = (\bar{x}_{t,\gamma}^* - \hat{\mu}_{t,\gamma})/\widehat{SE}(\bar{x}_{t,\gamma}^*)$. These are saved as `finv05` and `finv95`.

Finally, we compute the limits for our 90% bootstrap-t CI estimate :

1. The lower limit: $\hat{\mu}_{t,\gamma} - F^{-1}_{t(x_{t,\gamma}^*)}(0.95) \times \widehat{SE}(\bar{x}_{t,\gamma})$

2. The upper limit: $\hat{\mu}_{t,\gamma} - F^{-1}_{t(x_{t,\gamma}^*)}(0.05) \times \widehat{SE}(\bar{x}_{t,\gamma})$

The estimated CI reported here, [187, 431], is considerably wider than the intervals obtained by SC, [211, 394], and Resampling Stats, [210, 393]. This is because the earlier two estimates account for the small size of the population sampled, $N = 100$. The S-Plus solution, in contrast, assumes the population was considerably larger. Knowing the correct population size is clearly important to CI estimation.

Exercise

Kleinbaum, Kupper, and Muller (1988) report the Cornell Medical Index (CMI) scores for two random samples of 25 women from two different neighborhoods as reproduced in Table 9.2. The Cherryview women were sampled from 98 households, and the Morningside women were sampled from 211 households. For this exercise we have random samples drawn from quite small populations.

Table 9.2:   *CMI Scores for Two Samples of Women*

Cherryview:	49 12 28 24 16 28 21 48 30 18 10 10
	15 7 6 11 13 17 43 18 7 10 9 12 12
Morningside:	5 1 44 11 4 3 14 2 13 68 34 40
	36 40 22 25 14 23 26 11 20 4 16 25 17

Estimate 95% CIs for the 20% trimmed mean in each of the two populations, ideally taking the population sizes into account. How great is the overlap between the two CIs?

Concepts 10

Better Bootstrap CIs: BCA Intervals

The bootstrap-t confidence interval (CI) developed in Concepts 9 uses an explicit pivoting, Studentization, as a way of closing the gap between the sampling distributions of $[t^* - \widehat{\theta}]$ and $[t - \theta]$. That approach works well where t is a plug-in estimator of a location parameter. In that situation, we often have a closed-form estimate of the standard error (SE), that is, we can estimate the SE of t and, more important, of t_b^* by formula.

For other estimator-parameter-SE triads, $\{t, \theta, \widehat{SE}(t)\}$, we might neither know an explicit pivotal transformation nor be able to depend on Studentization. However, not knowing a pivotal transformation does not mean that there is not one. And our second improved bootstrap CI approach, the bias-corrected and accelerated (BCA) CI estimate (Efron and Tibshirani, 1993) assumes only that there exists some transformation for t and for t^* that normalizes their sampling distributions. We need not know how to carry out this transformation to a distribution with a normal cdf. The BCA approach uses an implicit pivotal form. We can take advantage of the normalization without ever learning how to carry it out.

Bias-Corrected and Accelerated CI Estimates

We learned in Concepts 9 that estimating a CI based on an explicit transformation to pivotal form is a two-step process. We first find the upper and lower CI limits for the transformed estimator, now in pivotal form. Then, we back-transform these results so they are phrased in terms of the original estimator. A similar two-step process lies behind BCA CI estimates as well.

The Normalized Sampling Distribution

The BCA approach hypothesizes a transformation of the estimator t, expressed as $\phi(t)$, such that the sampling distribution of $\phi(t)$ has these properties:

1. A normal cdf

2. A mean, $\mu_{\phi(t)} = \phi(\theta) - b[1 + a\phi(\theta)]\tau$

3. A standard deviation (SD), $\sigma_{\phi(t)} = [1 + a\phi(\theta)]\tau$

In the definitions of the mean and SD, $\phi(\theta)$ is the transformed value of the parameter θ, and b, a, and τ are constants, whose roles need explaining.

Note that the first property establishes that the $\phi(t)$ transformation is a normalizing one. However, the next two properties tell us that the transformation might not be pivotal. The mean and SD of the transformed, normal $\phi(t)$ might not be constant; either can depend on the value of θ.

The first of the constants, b, describes the bias of the transformed sampling distribution. If, in the second property, we set $b = 0$, then the mean of the sampling distribution for $\phi(t)$ is $\phi(\theta)$. This tells us that the sampling distribution of the transformed estimator will be unbiased whenever $b = 0$.

The second of the constants, a, is the acceleration constant. It describes how the SD of the sampling distribution changes with the value of θ—more precisely, a reflects the strength of the linear dependence of the sampling distribution SD on the value of $\phi(\theta)$. If, in the third property, we set $a = 0$, then the SD of the sampling distribution of $\phi(\theta)$ is τ, whatever the value of θ. The constant a is termed an acceleration constant because it measures the rate of change in the SD as $\phi(\theta)$ changes.

The third constant, τ, is a scaling constant, reflecting the units in which our attribute was measured when we calculated t, for example, feet, inches, or centimeters. The value of τ cancels out in the computational solution and τ plays no further role in our discussion of the BCA approach.

The BCA Solution

The BCA approach to CI estimation takes its name from the goal of correcting the percentile CI estimates both for any bias in the sampling distribution and for any tendency for the sampling distribution spread to accelerate as the value of θ changes. The constants b and a clearly will play roles in that correction. Bias, we already know, can be negative as well as positive. This is also true of acceleration; the spread can decrease rather than increase as θ increases.

The BCA solution depends on two other values as well:

$$p_i = F(t_i)$$

and

$$\pi = F(\theta)$$

Both are defined in terms of the cdf for the real-world sampling distribution for t. That is, p_i is the proportion of the sampling distribution that is as small or smaller than a particular t_i, and π is the proportion of the sampling distribution of t that is as small or smaller than θ. Like b and a, these two quantities are unknown to us.

Standard Normal Limits by Transformation

To review, the normalizing transformation $\phi(t)$ might not be quite pivotal; the mean and SD of the resulting sampling distribution can depend on the value of θ. However, if we were to transform t a second time by standardizing $\phi(t)$,

$$\Phi(t) = \frac{\phi(t) - \mu_{\phi(t)}}{\sigma_{\phi(t)}}$$

the resulting $\Phi(t)$ would be in pivotal form; $\Phi(t)$ has a normal cdf with a mean of zero and a SD of one—whatever the value of θ.

$\Phi(t)$ is still an implicit pivotal form as we do not know how to carry out the first, normalizing transformation, $\phi(t)$. But both the normalizing and the standardizing transformations are order-respecting. For example, if we pick the 0.05 quantile, $F^{-1}(.05)$, as an element in the sampling distribution of t, then the normalizing transformation of that t will be the 0.05 quantile of the sampling distribution of $\phi(t)$,

$$\phi\big(F^{-1}(.05)\big) = F_{\phi(t)}^{-1}(.05)$$

and the further, standardizing transformation of that t,

$$\Phi\big(F^{-1}(.05)\big) = \frac{F_{\phi(t)}^{-1}(.05) - \mu_{\phi(t)}}{\sigma_{\phi(t)}} = F_{\Phi(t)}^{-1}(.05)$$

will be the 0.05 quantile of the sampling distribution of $\Phi(t)$.

The lower and upper limits to a $(1 - 2\alpha)100\%$ CI for $\Phi(\theta)$ can be shown to be equal to

$$z_{lo} = z_{[p_i]} - \frac{\left(z_{[1-\alpha]} - z_{[\pi]}\right)}{1 + a\left(z_{[1-\alpha]} - z_{[\pi]}\right)}$$

and

$$z_{up} = z_{[p_i]} + \frac{\left(z_{[\pi]} - z_{[\alpha]}\right)}{1 - a\left(z_{[\pi]} - z_{[\alpha]}\right)}$$

In these equations, $z_{[q]}$ indicates the q quantile of the standard normal distribution:

1. $z_{[\alpha]}$ is the standard normal z-score that separates the lowest $\alpha 100\%$ of the distribution from the remainder. If $\alpha = 0.05$, then $z_{[\alpha]} = -1.645$.

2. $z_{[1-\alpha]}$ is the standard normal z-score that separates the highest $\alpha 100\%$ of the distribution from the remainder. If $\alpha = 0.05$, then $z_{[1-\alpha]} = 1.645$.

3. $z_{[\pi]}$ is the standard normal z-score that separates the lowest $\pi 100\%$ of the distribution from the remainder. π, in turn, is the proportion of the sampling distribution of t that is as small or smaller than θ. If $\pi = 0.50$, then $z_{[\pi]} = 0.0$.

4. $z_{[p_i]}$ is the standard normal z-score that separates the lowest $p_i 100\%$ of the distribution from the remainder.

The acceleration constant, a, plays a role in these CI limits, but the bias constant would appear not to. In fact, $b = z_{[\pi]}$ and, under this new name, the bias constant appears prominently in the limits.

You might notice that if $a = 0$—meaning that the variance of the (implicit) normal transformation of the sampling distribution of t is a constant, uninfluenced by the value of θ—then z_{lo} and z_{up} have exactly the structure laid out for the upper and lower limits of a $(1 - 2\alpha)100\%$ equal tails exclusion CI; the lower limit is the result of subtracting from t the distance separating the $(1 - \alpha)$ quantile and θ, and the upper limit is the result of adding to t the distance separating θ and the α quantile. The only wrinkle is that these quantities and distances have been replaced here by their standard normal z-score equivalents.

If both a and $z_{[\pi]}$ are equal to zero, as would be the case if the normal transformation of t has a constant variance and is always centered at θ, then our lower and upper limits have much the same form as we found for the CIs for the centered-at-zero $t(t)$ transformation in Concepts 9.

The derivation of z_{lo} and z_{up} is based on two successive transformations of the real-world sampling distribution of t. The first implicitly transforms the sampling distribution to a normal one, with the mean and variance depending on the value of θ. The second transformation standardizes this normal distribution, giving it a mean of zero and variance of one.

Bootstrap-World Standard Normal Limits

If $\pi = F(\theta)$, $p = F(t)$, and the acceleration constant a were known to us, we could solve the two prevoious equations above for z_{lo} and z_{up}. They aren't. However, if we assume that the bootstrap-world sampling distribution of t^* can be transformed in the same way—first to a normal distribution with mean and variance dependent on $\widehat{\theta}$, and then to a standard normal distribution—then the z_{lo} and z_{up} equations given earlier are pivotal. That is, the right sides of the two equations can be defined, just as well, relative to the bootstrap-world sampling distribution:

$$\widehat{z}_{\text{lo}} = z_{[p]} - \frac{\left(z_{[1-\alpha]} - z_{[\widehat{\pi}]}\right)}{1 + \widehat{a}\left(z_{[1-\alpha]} - z_{[\widehat{\pi}]}\right)}$$

and

$$\widehat{z}_{\text{up}} = z_{[p]} + \frac{\left(z_{[\widehat{\pi}]} - z_{[\alpha]}\right)}{1 - \widehat{a}\left(z_{[\widehat{\pi}]} - z_{[\alpha]}\right)}$$

On the left sides of these equations, $\widehat{z}_{\text{lo}}$ and $\widehat{z}_{\text{up}}$ are our bootstrap-world estimates of z_{lo} and z_{up}, estimates that are based on the assumption of the pivotal-equivalence of the two standard normal transformations.

On the right sides of these equations, $z_{[\alpha]}$ and $z_{[1-\alpha]}$ are defined as before, standard normal z-scores that cut off the lowest and highest $\alpha100\%$ of the distribution. The two other z-scores are estimates, obtained from the bootstrap sampling distribution:

1. $z_{[\widehat{\pi}]}$ is the standard normal z-score that separates the lowest $\widehat{\pi}100\%$ of the distribution from the remainder. $\widehat{\pi}$, in turn, is the proportion of the bootstrap sampling distribution of t^* that is as small or smaller than $\widehat{\theta}$. This z-score is our estimate of the bias constant, $\widehat{b} = z_{[\widehat{\pi}]}$.

2. $z_{[p]}$ is the standard normal z-score that separates the lowest $p100\%$ of the distribution from the remainder. p is the proportion of the bootstrap sampling distribution of t^* that is as small or smaller than t, our real-world sample estimate.

3. Recall that if t is a plug-in estimator of θ and $\widehat{X}$ is a nonparametric estimate, then $t = \widehat{\theta}$ and $z_{[p]} = z_{[\widehat{\pi}]}$.

The one remaining term on the right sides of the equations for $\widehat{z}_{lo}$ and $\widehat{z}_{up}$ is $\widehat{a}$, which is our estimate of the acceleration constant. To complete the development of the BCA confidence interval we'll assume that we have such an estimate. Then, we'll see how to compute the estimate.

From Standard Normal to Bootstrap Limits

The two z-scores, $\widehat{z}_{lo}$ and $\widehat{z}_{up}$, estimate the lower and upper limits to a confidence interval for $\Phi(\theta)$, an implicit double transformation of θ that first normalizes the sampling distribution of t, and then standardizes the spread and location of that normal distribution.

What we want are estimates of the CI limits for the untransformed θ. We use the order-preserving properties of the two transformations to find these. Let q_{lo} be the proportion of the standard normal distribution falling below $\widehat{z}_{lo}$ and q_{up} be the proportion of the standard normal distribution falling below $\widehat{z}_{up}$. You can obtain these proportions from a standard normal distribution table.

The limits of the $(1 - 2\alpha)100\%$ BCA confidence interval estimates are $t^*_{[lo]}$ and $t^*_{[up]}$, which are the q_{lo} and q_{up} quantiles of the bootstrap sampling distribution.

Estimation of the Acceleration Constant

The last stumbling block is estimating the acceleration constant. Efron and Tibshirani (1993) advocate an estimate of a that can be adapted to any plug-in estimator. It is computed as

$$\widehat{a} = \frac{\sum_{i=1}^{n}(U_i/n)^3}{6\left[\sum_{i=1}^{n}(U_i/n)^2\right]^{3/2}}$$

where

$$U_i = (n-1)\left[t_{(.)} - t_{(-i)}\right]$$

$$t_{(-i)} = t[x, \text{ omitting the } i\text{-th case}]$$

and

$$t_{(.)} = (1/n)\sum_{i=1}^{n} t_{(-i)}$$

$t_{(-i)}$ is an omitted-case parameter estimate, the estimate computed without including the score or scores for the i-th case. $t_{(.)}$ is the mean of these omitted-case estimates. The difference between the omitted-case estimate for the i-th case and the average of the omitted-case estimates,

$$(U_i/n) = [(n-1/n)]\left[t_{(.)} - t_{(-i)}\right]$$

assesses the influence of the i-th case on the estimate. If (U_i/n) is large, either positive or negative, then the i-th case has a large influence, relative to that of the other cases, on the size of the estimate t.

Our estimated acceleration constant, $\hat{a}$, measures the skewness of the distribution of the n influence measures. The acceleration will be zero if the distribution of influence statistics is symmetric about zero. The acceleration will be positive if the distribution of influence statistics is skewed positively, by the presence in the sample of cases whose scores drastically increase the size of the parameter estimate. Or the acceleration will be negative, if our sample includes cases whose scores dramatically decrease the size of t.

The relation of this estimate to the concept of acceleration is not obvious. That linkage can be shown, although at more length than is appropriate here. In particular, be assured that $\hat{a}$ is an acceleration measure in the bootstrap world, at least, when our estimator is a plug-in one and our population distribution is the infinite nonparametric bootstrap-world one. Because this is so, the skewness-of-influence $\hat{a}$ provides an appropriate estimate of the real-world acceleration constant, a.

We have seen in earlier Concepts that the nonparametric population distribution estimate is the most realistic one given the limited knowledge we have about population distributions. We also have a preference for plug-in estimators. Their sampling distributions characteristically display little bias and relatively small SEs. The combination of a nonparametric distribution estimate and a plug-in estimator is common in bootstrap inference.

Computation of BCA Confidence Intervals

We now can describe an algorithm for finding the limits to a $(1 - 2\alpha)100\%$ BCA CI. The key stages are summarized in Figure 10.1.

1. From the sample distribution x and from the bootstrap sampling distribution of t^* estimate the bias and acceleration constants, b and a.

2. Use these estimated constants and the predetermined confidence coefficient α to compute the lower and upper limits to the CI for the transformed and standardized parameter.

3. Use the bootstrap sampling distribution of t^* to back-transform these limits to those appropriate to the untransformed parameter.

Figure 10.1: *Key Stages in Computing a BCA Confidence Interval*

We describe here the BCA CI algorithm for the plug-in estimator and nonparametric bootstrap population distribution.

1. Compute the plug-in estimate, t, from the sample distribution, x.

2. Compute the n omitted-case estimates, $t_{(-i)}$.

3. Compute the average of the omitted-case estimates, $t_{(.)}$.

4. Compute n influence statistics, $(U_i/n) = [(n - 1/n)][t_{(.)} - t_{(-i)}]$.

5. Compute the acceleration estimate, $\hat{a} = \dfrac{\sum\limits_{i=1}^{n}(U_i/n)^3}{6\left[\sum\limits_{i=1}^{n}(U_i/n)^2\right]^{3/2}}$.

6. Form $\hat{X}$ from one or more copies of x.

7. Set $\hat{\theta}$ equal to the t of step 1.

8. Draw a bootstrap sample, x_b^*, from $\hat{X}$.

9. Compute the estimate, t_b^*, from the bootstrap sample.

10. Repeat steps 8 and 9 a total of B times, forming the bootstrap sampling distribution of t_b^*.

11. Compute $\hat{\pi}$, the proportion of the bootstrap sampling distribution smaller than $\hat{\theta}$.

12. Select α, the confidence level for the $(1 - 2\alpha)100\%$ CI.

13. Use the tabled distribution of the standard normal distribution to determine $z_{[\hat{\pi}]}$, $z_{[\alpha]}$ and $z_{[1-\alpha]}$. These are the z-scores that cut off the lower $\hat{\pi}$, α, and $(1 - \alpha)$ proportions of the standard normal distribution.

14. Use the acceleration estimate from step 5 and the z-scores from step 13 to compute

$$\hat{z}_{lo} = z_{[\hat{\pi}]} - \frac{\left(z_{[1-\alpha]} - z_{[\hat{\pi}]}\right)}{1 + \hat{a}\left(z_{[1-\alpha]} - z_{[\hat{\pi}]}\right)}$$

and

$$\hat{z}_{up} = z_{[\hat{\pi}]} + \frac{\left(z_{[\hat{\pi}]} - z_{[\alpha]}\right)}{1 - \hat{a}\left(z_{[\hat{\pi}]} - z_{[\alpha]}\right)}$$

15. Use the tabled distribution of the standard normal distribution to find q_{lo} and q_{up}, the proportions of the distribution falling below the two z-scores computed at step 14, $\hat{z}_{lo}$ and $\hat{z}_{up}$.

16. Use the two step 15 proportions and the B of step 10 to compute $lo = int[q_{lo} \times (B + 1)]$ and $up = (B + 1) - int[(1 - q_{up}) \times (B + 1)]$.

17. Sort the bootstrap sampling distribution of step 10 in order from the smallest element, $t_{[1]}^*$, to the largest element, $t_{[B]}^*$.

18. Use the integers computed in step 16 to select $t_{[lo]}^*$ and $t_{[up]}^*$ from the sorted bootstrap sampling distribution. The selected $t_{[lo]}^*$ and $t_{[up]}^*$ are the q_{lo} and q_{up} quantiles of the bootstrap sampling distribution, just as $\hat{z}_{lo}$ and $\hat{z}_{up}$ were the q_{lo} and q_{up} quantiles of the standard normal distribution.

19. Steps 15 through 18 describe the translation of the CI limits from the standard normal to the bootstrap sampling distribution. The lower and

upper limits of our $(1 - 2\alpha)100\%$ BCA confidence interval are given by $t^*_{[\text{lo}]}$ and $t^*_{[\text{up}]}$.

The solution for BCA estimates appears as complicated as its rationale is technical. Fortunately, two of the three statistical packages we use to demonstrate resampling, SC and S-Plus, have built-in functions for computing BCA estimates. Thus, estimates can be produced automatically. And in Applications 10 we show how the third package, Resampling Stats, can be programmed to do all the work except steps 13 and 15, which require knowledge of values of the standard normal cdf and quantile function. For those steps we simply look up the required tabled values for the standard normal distribution.

Steps 13 and 15, of course, can be built into a BCA computing routine if your statistical computing package has commands to compute z-scores and tail proportions for the standard normal distribution.

Applications of the BCA CI

The BCA approach to CI estimation has proven to have wide applicability, for example, Efron and Tibshirani, 1993. We'll use this implicit pivotal approach in our random sample applications whenever θ is not a location parameter for which we can use the explicit pivot of Studentization to develop bootstrap-t confidence intervals.

Better Confidence Interval Estimates

We have noted that the bootstrap-t and BCA approaches provide better CI estimates than do the simpler standard and percentile solutions of Concepts 8. How are they better? Recall that a 90% equal-tails exclusion CI estimation procedure should fail to cover the parameter, θ, for the 5% largest and 5% smallest estimates provided by random samples drawn from a population. How well an estimation procedure attains this goal is its coverage accuracy. The bootstrap-t and BCA approaches have been shown both theoretically and empirically to have coverage accuracies superior to those of the standard and uncorrected percentile methods (Davison and Hinkley, 1997; Efron and Tibshirani, 1993; Shao and Tu, 1995).

Applications 10

Using CI Correction Factors

Concepts 10 describes the estimation of bias-corrected and accelerated (BCA) confidence intervals (CI). The computations, beyond the Monte Carlo approximation of a bootstrap sampling distribution, can appear daunting, and in this Application we look at how the BCA approach has been implemented in three statistics packages.

Requirements for a BCA CI

We'll restrict our interest to those settings in which BCA estimates have been most fully developed:

1. Our estimator, t, is a plug-in estimator, computed over a sample distribution by the same rule we would use to extract the parameter, θ, from the population distribution, and

2. Our population distribution estimate, $\widehat{X}$, is fully nonparametric, consisting of one or more copies of a sample distribution, x.

To obtain BCA CIs we need the following:

1. A bootstrap sampling distribution of t^*

2. An estimate of the median bias of t, the extent to which the median of the sampling distribution of t diverges from θ

3. An estimate of the variance acceleration, the extent to which the sampling variance increases or decreases as the value of θ increases

4. A choice of α, determining the confidence level of the $(1 - 2\alpha)100\%$ CI

5. Access to the cumulative distribution function (cdf) and its inverse, the quantile function, of the standard normal distribution

We'll follow the Concepts 9 algorithm in estimating acceleration by the skewness of the influences of the n cases on the statistic t, and in estimating median bias by determining the proportion of the bootstrap sampling distribution taking values smaller than our estimator, t.

Implementations of the BCA Algorithm

Both SC and S-Plus provide commands that isolate the user from the details of estimating BCA confidence limits. Our Resampling Stats script, however, spells out the steps involved and, indeed, requires the user to make a side computation.

Estimating a CI for the Population Variance

As a working example, we'll use the sample data of Table 10.1, taken from Efron and Tibshirani (1993). These are scores on a test of spatial ability for a random sample of 26 children.

Table 10.1:   *Spatial Ability Scores for a Random Sample of 26 Children*

48	36	20	29	42	42	20	42	22
41	45	14	6	0	33	28	34	4
32	24	47	41	24	26	30	41	

We want to estimate a 90% CI for the population variance of these scores. There would be a parametric solution if we were to know that the population distribution enjoyed a normal cdf, but we have no such knowledge.

SC Procedures for BCA Intervals

We will use two SC procedures, `ibootci2` and `fbootci2` in this and subsequent Applications for estimating BCA CIs. The first presumes a large, relative to sample size, population and forms bootstrap samples by sampling with replacement from the sample distribution. The second procedure is a small population version, sampling without replacement from a population distribution estimate formed from an appropriate number of copies of the sample distribution.

The arguments required by these procedures are quite similar to those required by the bootstrap-*t* procedures introduced in Applications 9. For example, `fbootci2` is called with a sequence of eight or more arguments:
`fbootci2(p1,p2,M,nn,NN,m,B,FSTAR,CI,...)`

p1   A user-written procedure to compute *m* estimates *t*

p2	User-written procedure, computing m bootstrap $\widehat{\theta}$s
M	Data matrix, cases as rows, grouped into samples
nn	Vector of sample sizes, blocking the data matrix
NN	Vector of population sizes, length as for nn
m	Number of estimates computed by p1
B	Number of bootstrap samples to be drawn
FSTAR	Name of a B $\times$ m matrix to receive t^*s
CI	Optional, one or more CI levels, for example, 90.

The calling sequence is the same for ibootci2 except that the population size argument is omitted.

The two user-written procedures should be written to receive three arguments: p1(DM,gs,out)

DM	Data matrix, cases as rows, grouped into samples
gs	Vector of sample sizes, blocking the data matrix by rows
out	Vector to receive estimates or parameters

If, as in our work, the estimator is a plug-in type, then the two procedures p1 and p2 will be identical, and p1 should be substituted for p2 when you invoke either ibootci2 or fbootci2.

Figure 10.2 shows an interactive SC session in which a BCA CI estimate is obtained for the variance of the population distribution of spatial ability scores. Here, we assume that that population is large relative to the sample size, that is, larger than $20 \times n = 520$, and we use ibootci2.

First, the data are read into a vector, TESTA, and then set into the first and only column of a matrix MT. We do this because ibootci2 wants data in the form of a matrix. A procedure usrvar is created that will compute the variance of the scores in each column of a matrix (1, its first argument) and store those as the elements of a vector (3, its third argument). The sample size vector, the second argument passed to usrvar, is irrelevant to that procedure in this one-sample application. Such a vector is needed, however, by ibootci2, so we create the one-element vector g before we call the BCA procedure.

In the call to ibootci2, we specify that our procedure usrvar is to be used to compute both t and $\widehat{\theta}$, that the procedure returns a single value, that we want our Monte Carlo approximation to the sampling distribution of t^* to be based on 1,000 bootstrap samples, and that this sampling distribution is to be stored in the matrix FBOOT. Finally, we request a 90% CI estimate.

```
> vector(TESTA,26)
> read(TESTA)
48 36 20 29 42 42 20 42 22 41 45 14 6
0 33 28 34 4 32 24 47 41 24 26 30 41
     26
> matrix(MT,26,1)
> setcol(MT,TESTA,1)
> proc usrvar(){
   $3$= var($1$')
   }
> vector(g,1)
> read(g)
26
    1
> ibootci2(usrvar,usrvar,MT,g,1,1000,FBOOT,90)
_____tstat:
  171.53402
_____thetahat:
  171.53402
_____means of  1000  bootstrap estimates:
  162.43828
_____SDs of  1000  bootstrap estimates:
  42.213234
_____z0 (bias corrections):
  0.20957422
_____a (acceleration constants):
  0.06124012
  Theta 1  90 % Standard (Norm) CI:   102.09943   to 240.96861
  Theta 1  90 % Percentile (Sym) CI:   94.08284   to 232.98225
  Theta 1  90 % Percentile (Inv) CI:  110.0858    to 248.98521
  Theta 1  90 % Bias Adjusted CI:     103.17858   to 242.07799
  Theta 1  90 % Bias Corrected CI:    108.70562   to 251.39053
  Theta 1  90 % BC & Accelerated CI:  115.7929    to 260.78698
      Total ibootci2() running time:  2.31
```

Figure 10.2: *SC Estimation of a BCA CI for Variance of Spatial Ability Scores*

The output from `ibootci2` consists of

`tstat`	The estimate, t, computed from the sample
`thetahat`	The parameter in the bootstrap population, $\widehat{\theta}$
	The mean of the t_b^*, $b = 1, \ldots, B$
	The standard deviation of the t_b^*, $b = 1, \ldots, B$
`z0`	The bias correction, a standard normal score
`a`	The acceleration estimate

as well as several estimated CIs. We are interested in the final one, the BCA CI. We are 90% confident that computed interval, from 115 to 261, contains the variance of the population distribution of spatial ability scores. Notice that the interval is not symmetric about our point estimate of 171.5. This is characteristic of variance CIs.

Finding a Resampling Stats BCA Interval

The SC procedures `ibootci2` and `fbootci2` hide the details of first estimating the bias and acceleration constants and then using these to define a BCA interval. In using Resampling Stats for this same purpose, we see these spelled out. Indeed, because Resampling Stats does not have access to standard normal cdf and icdf tables, we divide the solution into three phases.

Figures 10.3 through 10.7 show a script that carries out the first phase of the BCA computations. The text in Figure 10.3 describes the goal of this phase.

```
'  Bias Corrected and Accelerated (BCA) bootstrap confidence
'  interval determination, Finite Population Sampling, Part I.
'
'  Computes the acceleration constant (AA) and (BB) proportion of
'  THETASTARs less than THETAHAT (needed for bias correction)
'  as well as saving the bootstrap distribution of THETASTARs
'  in BCA_PT1.OUT. A side computation is needed to convert
'  AA and BB into the ALF1 and ALF2 needed for CI determination
'  in BCA_PT2.STA.
'
'  Variance is the univariate statistic being bootstrapped.
'
```

Figure 10.3: *Part 1 Phase 1, BCA Limits via Resampling Stats*

The commands constituting the second part of the script, shown in Figure 10.4, set the sizes of vectors, input the data vector, provide sample and population sizes, establish the number of bootstrap samples to be drawn, calculate t (and $\widehat{\theta}$) from the sample, and then estimate the population distribution as closely as possible with copies of the sample distribution.

In Figure 10.5, the third part of first phase computations, we see the bootstrap sampling.

Notice that in addition to accumulating the Monte Carlo bootstrap sampling distribution (in the vector BOOT), a count is kept in BB of the number of t_b^*s that are smaller than $\widehat{\theta}$.

We will need this quantity for the bias correction calculation.

In the fourth part of first phase computations, displayed in Figure 10.6, we see the bias correction and acceleration computations. After converting BB to a proportion, a loop of instructions computes the n omitted-case estimates, $t_{(-i)}$, and stores them in the vector SA.

```
maxsize default 1
'
' ----- Obtain Input:
'
maxsize A 26 SA 26 A2 26 A3 26 AA 26    'Set to Sample Size
maxsize SPA 126 PA 126                  'Set to Pop Size +1
maxsize BOOT 1000                       'Set to Resample Count
copy (48 36 20 29 42 42 20 42 22 41 45 14 6) A
concat A (0 33 28 34 4 32 24 47 41 24 26 30 41) A
copy 26 N                               'Sample Size
copy 125 NN                             'Population Size
copy 1000 B                             'Resample Count (mult of 100)
'
' ----- Calculate Target Statistic in Original Sample
'
variance divn A THAT                    'Target Statistic
'
' ----- Form Estimated Population from Sample
'
divide NN N K                           'Pop'n is K mults of sample
round K IK
if IK > K
   subtract IK 1 IK                     'IK to be integer part of K
end
copy A PA                               'Initialize Population vector
subtract IK 1 IK1
repeat IK1
   concat PA A PA                       'Puts IK copies of A in PA
end
multiply IK N IKN
subtract NN IKN NR                      'NR obs'ns needed to fill pop'n
'
' ----- Estimated Population, except for sample fraction (NR)
```

Figure 10.4:   *Part 2 Phase 1, BCA Limits via Resampling Stats*

From the omitted-case estimates, the following quantities are computed in turn

A2   the n quantities: $\left(t_{(.)} - t_{(-i)}\right)^2$

A3   the n quantities: $\left(t_{(.)} - t_{(-i)}\right)^3$

ATOP $\displaystyle\sum_{i=1}^{n}\left(t_{(.)} - t_{(-i)}\right)^3$

B1 $\displaystyle\sum_{i=1}^{n}\left(t_{(.)} - t_{(-i)}\right)^2$ and

ABOT $6\left[\displaystyle\sum_{i=1}^{n}\left(t_{(.)} - t_{(-i)}\right)^2\right]^{3/2}$

```
' ----- Begin bootstrap resample cycles:
divide B 100 B100              'B100: Resample Partitions
copy 0 BB                      'Initialize bias count
repeat B100
   if NR > 0                   'Each 100 resamples replace NR
      take PA 1,IKN PA
      shuffle A SA
      take SA 1,NR SA
      concat PA SA PA          'Complete PA with NR obs'ns from A
   end
   repeat 100
      shuffle PA SPA
      take SPA 1,N SA          'Nonreplacement sample from pop'n
' ----- Calculate target statistic in resample
      variance divn SA TSTAR        'Target statistic in resample
' ----- Continue resample processing
      if TSTAR < THAT
         add BB 1 BB                 'Count for bias adjustment
      end
      score TSTAR BOOT          'Bootstrap t statistics saved
   end
end
```

Figure 10.5: *Part 3 Phase 1, BCA Limits via Resampling Stats*

The ratio of ATOP to ABOT is the estimated acceleration constant, AA. Although the computational flow differs from that outlined in Concepts 10, the AA computed in Figure 10.6 is the acceleration estimate described earlier:

$$\widehat{a} = \frac{\sum\limits_{i=1}^{n}(U_i/n)^3}{6\left[\sum\limits_{i=1}^{n}(U_i/n)^2\right]^{3/2}}$$

```
' ----- Resampling completed, Compute bias adjustment
divide BB B BB                     'Bias adjustment factor
' ----- Begin acceleration factor computation
clear SA
subtract N 1 NM1
repeat N                            'Drop one case at a time
    take A 1,NM1 A2
    take A N A1                     'last case held out
' ----- Calculate target statistic in jackknife sample
    variance A2 VI                  'Jackknifed target statistic
' ----- Finish loop
    score VI SA
    concat A1 A2 A                  'Rotate held out case to front
end
' ----- Summarize for acceleration factor
mean SA MDOT                        'Mean of jackknifed stats
subtract MDOT SA SA                 'Deviations from mean
multiply SA SA A2                   'Squared deviations
multiply A2 SA A3                   'Cubed deviations
sum A3 ATOP                         'Sum of cubed deviations
sum A2 B1                           'Sum of squared deviations
multiply B1 B1 B2
multiply B2 B1 B2                   'Cube of sum of squared dev'ns
sqrt B2 B1
multiply B1 6 ABOT                  '(6)sqrt(cube sum of squares)
divide ATOP ABOT AA                 'Acceleration factor
```

Figure 10.6:   *Part 4 Phase 1, BCA Limits via Resampling Stats*

We have simply omitted two scalings of the $\left(t_{(.)} - t_{(-i)}\right)$s, first by $(n-1)$ to produce the U_is and then by $1/n$. These scalings affect the numerator and denominator of $\widehat{a}$ in the same way and cancel out in the formation of the ratio.

The first phase script is completed in Figure 10.7 with the reporting of $\widehat{\theta}$, the acceleration coefficient estimate, the proportion of the bootstrap sampling distribution falling below $\widehat{\theta}$, and the saving of the bootstrap sampling distribution to a disk file, BCA_PT1.OUT. We save the distribution because we will need it in the third phase of the BCA solution.

```
'  ----- Report results of run:
print THAT                'Target statistic
print AA                  'The acceleration factor.
print BB                  'Bias factor: Needed for lookup of Z0
write file "BCA_PT1.OUT" BOOT  'Resample statistics to disk file
```

Figure 10.7:   *Part 5 Phase 1, BCA Limits via Resampling Stats*

Running the computational script making up Figures 10.3 through 10.7 produced the results displayed in Figure 10.8.

The variance in our estimated population distribution, $\hat{\theta}$ (or, THAT), is 171.53; the estimated acceleration coefficient, $\hat{a}$ (or, AA), is 0.06124; and the proportion of the bootstrap sampling distribution smaller than $\hat{\theta}$ was 0.562, $\hat{\pi}$ (or, BB). As they should be, the first two results are identical with what we saw in Figure 10.2 for the use of the SC procedure ibootci2. Both results depend only on the sample distribution. Had ibootci2 reported a $\hat{\pi}$ value, it might have differed slightly from 0.562 because it would have been based on a different, randomly chosen sequence of bootstrap random samples.

```
THAT      =       171.53
AA        =      0.06124
BB        =       0.562
Line 118: 1000 records (0 missing values) written to
C:\RESAMP\STA\BCA_PT1.OUT
```

Figure 10.8:   *Phase 1, Resampling Stats Script Output, Spatial Ability Variance*

Resampling Stats does not have access to the standard normal cdf and icdf. So we must carry out the second phase of the BCA solution by hand. This phase carries the solution from step 12 through step 15 of the BCA algorithm presented in Concepts 10:

12. Select α, the confidence level for the $(1 - 2\alpha)100\%$ CI.

For a 90% CI we have $\alpha = 0.05$

13. Use the tabled distribution of the standard normal distribution to determine $z_{[\hat{\pi}]}$, $z_{[\alpha]}$ and $z_{[1-\alpha]}$. These are the z-scores that cut off the lower $\hat{\pi}$, α, and $(1 - \alpha)$ proportions of the standard normal distribution.

Our choice of α means that $z_{[\alpha]} = -1.65$ and $z_{[1-\alpha]} = 1.65$. The first phase results give us $\hat{\pi} = 0.562$. This ought to give us a $z_{[\hat{\pi}]}$ that is slightly greater than zero; in fact, $z_{[\hat{\pi}]} = 0.156$.

14. Use the acceleration estimate from step 5 and the z-scores from step 13 to compute

$$\widehat{z}_{\text{lo}} = z_{[\hat{\pi}]} - \frac{\left(z_{[1-\alpha]} - z_{[\hat{\pi}]}\right)}{1 + \widehat{a}\left(z_{[1-\alpha]} - z_{[\hat{\pi}]}\right)}$$

and

$$\widehat{z}_{\text{up}} = z_{[\hat{\pi}]} + \frac{\left(z_{[\hat{\pi}]} - z_{[\alpha]}\right)}{1 - \widehat{a}\left(z_{[\hat{\pi}]} - z_{[\alpha]}\right)}$$

Substituting into these equations, we have

$$\widehat{z}_{\text{lo}} = 0.156 - \frac{1.65 - 0.156}{1 + 0.06124(1.65 - 0.156)} = -1.213$$

and

$$\widehat{z}_{\text{up}} = 0.156 + \frac{0.156 + 1.65}{1 - 0.06124(0.156 + 1.65)} = 2.187$$

15. Use the tabled distribution of the standard normal distribution to find q_{lo} and q_{up}, the proportions of the distribution falling below the two z-scores computed at step 14, $\widehat{z}_{\text{lo}}$ and $\widehat{z}_{\text{up}}$.

The z-scores $\hat{z}_{lo} = -1.213$ and $\hat{z}_{up} = 2.187$ are the 0.113 and 0.986 quantiles of the standard normal distribution: $q_{lo} = 0.113$ and $q_{up} = 0.986$.

```
'-- BCA bootstrap confidence intervals, Phase 3
maxsize default 2
maxsize BOOT 5000
copy 1000 B                       'Size of Bootstrap Distribution
copy 0.113 QLO                    'Lower Adjusted Quantile
copy 0.986 QHI                    'Upper Adjusted Quantiile
read file "BCA_PT1.OUT" BOOT
sort BOOT BOOT
'
add B 1 BP1
multiply BP1 QLO X
round X FLO
if FLO > X
   subtract FLO 1 FLO
end                               'Position of lower limit
subtract 1 QHI X
multiply BP1 X X
round X FHI
   if FHI > X
   subtract FHI 1 FHI
end
subtract BP1 FHI FHI              'Position of upper limit
concat FLO FHI LIMS
take BOOT LIMS CI
print CI
```

Figure 10.9: *Phase 3, BCA Limits via Resampling Stats*

For the third and final phase of the BCA CI solution, we need only find the corresponding quantiles of the bootstrap sampling distribution saved in the first phase.

The Resampling Stats script of Figure 10.9 does the work. It reads in the saved bootstrap distribution and sorts the elements from smallest to largest. Then the script works out the positions, FLO and FHI, of q_{lo} and q_{hi} in the sorted t^* distribution, and extracts those two values from the distribution. These are then displayed as the limits to the BCA CI.

Running the script of Figure 10.9 produced CI limits of 116.75 and 255.99 for the spatial ability population variance.

Finding BCA Confidence Limits with S-Plus

The statistical package S-Plus can compute BCA limits for a population parameter when the estimator is a plug-in type and the population distribution is large in size. The key function is `bootstrap`. This function can be modified in a variety of ways. In its most basic form, we need only specify a sample distribution and a plug-in estimator. The `bootstrap` function samples with replacement from the sample distribution to form bootstrap samples. $B = 1000$ samples are drawn unless we specify another value for B. In subsequent applications, we will explore modifications of this and other optional `bootstrap` arguments.

Figure 10.10 carries forward our example, finding BCA 90% confidence limits from a bootstrap distribution of 1,000 plug-in variance estimates. The 26 spatial ability scores have been entered previously as a vector, `spab`.

```
>fulln<-list(unbiased=F)
>temp<-bootstrap(spab,var,args.stat=fulln)
>limits.bca(temp,details=T)
```

Figure 10.10: S-Plus Determination of BCA Limits

The basic bootstrap command is on the second line of the display. The first argument is the name of the vector containing our data, the second is the name of the function used to compute t_b^* in each bootstrap sample. In this application, the function `var` will compute the variance. By default, however, `var` computes an unbiased variance estimate, dividing the sum of squared deviations about the mean by $n - 1$. We want it to compute a plug-in estimator of the variance, dividing by n instead. To do so, we must specify an additional argument to the variance function, `unbiased=F`. We provide this additional argument in our call to the `bootstrap` command by including a third argument, `args.stat`, that supplies a list of additional arguments needed by the statistics function. This additional arguments list, `fulln`, is created by the first S-Plus command displayed in Figure 10.10.

The results of the `bootstrap` are saved in an object named `temp` and the third command requests that BCA confidence limits, including those for a 90% interval, be computed from these results. As part of the output to this command, we learn that $\hat{z}_{[\hat{\pi}]}$ was estimated in this sequence of 1,000 bootstrap samples to be 0.176, and that the acceleration coefficient, as in the SC and Resampling Stats solutions, was estimated at 0.061240. The lower and upper limits of the 90% BCA interval were estimated at 114.27 and 261.54.

Exercise

1. The data of Table 10.2 are taken from Hand et al. (1994, Data Set 286). A sample of 318 12- and 13-year-old girls are classified by socioeconomic class of parents and degree of physical development.

Table 10.2:   *Physical Development and Family Socioeconomic Class of 318 Girls*

Socioeconomic	Developmental Stage				
Class	5	4	3	2	1
A	18	40	28	14	2
B	9	25	25	21	1
C	2	12	12	12	1
D	6	33	34	17	6

Both classifications are ordered. Socioeconomic class A is the highest, D is the lowest. Developmental stage 5 is the most mature, stage 1 is the least mature. Of interest is whether the two classifications are related; in particular, are girls from the higher socioeconomic homes physically more mature? Score the Socioeconomic classes A = 4, B = 3, C = 2, D = 1, and find 90% and 95% BCA CIs for the correlation between class and developmental stage.

Concepts 11

Bootstrap Hypothesis Testing

In Concepts 4, we reviewed how statisticians test hypotheses about the value of a population distribution parameter. Hypothesis testing formally depends on two statistical ideas, the null hypothesis and the null sampling distribution. The null hypothesis specifies a value for the population parameter. Frequently, this null value implies that a treatment has had no effect, that two treatments have the same effect, or that the intensity of response is unrelated to the intensity of treatment. The null sampling distribution is the sampling distribution our parameter estimate would have, if the null hypothesis truly described the parameter.

In classical, mathematically based, statistical inference, hypothesis testing and confidence interval (CI) estimation are closely linked. This is because we are required to make one or more assumptions about our population distribution or the statistic we use, assumptions that are sufficient to ensure a near-pivotal relation between the null and true sampling distributions. Bootstrap inference liberates us from the need for such assumptions, which are often unwarranted by our data. However, the lack of pivotality weakens the linkage between the estimation and hypothesis testing aspects of statistical inference. We consider now the implications of that for bootstrap hypothesis testing.

CIs, Null Hypothesis Tests, and p-Values

Consider again the one-sample t-test of the magnitude of the mean, μ, of a population distribution having a normal cumulative distribution function (cdf). The Studentization of the sample mean,

$$t(\bar{x}) = \frac{\bar{x} - \mu}{\widehat{SE}(\bar{x})}$$

has as its sampling distribution that of the t-random variable with $(n - 1)$ degrees of freedom (df) whatever the value of μ.

CI for the Normal Population Mean

We know how to use this pivotal form to estimate a $(1 - 2\alpha)100\%$ CI. We begin with the α and $(1 - \alpha)$ quantiles of the t-distribution with $(n - 1)$ df which can be readily calculated for us by many statistical computing packages. We'll refer to these two quantiles as $F_{t,(n-1)}^{-1}(\alpha)$ and $F_{t,(n-1)}^{-1}(1 - \alpha)$, the subscript reminding us of the distribution to which they belong.

Next, we use the $t(\overline{x})$ formula to estimate the corresponding quantiles of the sampling distribution of $\overline{x}$ by first multiplying our selected quantiles by our estimate of the standard error (SE) of the sample mean and then adding μ. This gives us

$$\left[\widehat{\text{SE}}(\overline{x}) \times F_{t,(n-1)}^{-1}(\alpha) \right] + \mu = F_x^{-1}(\alpha)$$

and

$$\left[\widehat{\text{SE}}(\overline{x}) \times F_{t,(n-1)}^{-1}(1 - \alpha) \right] + \mu = F_x^{-1}(1 - \alpha)$$

By our general rule, to define a $(1 - 2\alpha)100\%$ CI for μ we need the two distances

$$a = F_x^{-1}(1 - \alpha) - \mu$$

and

$$b = \mu - F_x^{-1}(\alpha)$$

Solving for a and b gives

$$a = \left[\widehat{\text{SE}}(\overline{x}) \times F_{t,(n-1)}^{-1}(1 - \alpha) \right]$$

and

$$b = -\left[\widehat{\mathrm{SE}}(\bar{x}) \times F_{t,(n-1)}^{-1}(\alpha)\right]$$

The t-distributions are symmetric as well as centered at zero, so we can use $F_{t,(n-1)}^{-1}(1-\alpha) = -F_{t,(n-1)}^{-1}(\alpha)$ to express the CI more compactly as

$$\bar{x} \pm \left[\widehat{\mathrm{SE}}(\bar{x}) \times F_{t,(n-1)}^{-1}(\alpha)\right]$$

Hypothesis Test for the Normal Mean

The quantities a and b obtained earlier show up in our use of $t(\bar{x})$ to test hypotheses about the value of the population distribution mean, giving rise to the so-called one-sample t-test.

Under the null hypothesis (H_0), $\mu = \mu_0$, the test statistic

$$t_0(\bar{x}) = \frac{\bar{x} - \mu_0}{\widehat{\mathrm{SE}}(\bar{x})}$$

also has as its sampling distribution that of the t-random variable with $(n-1)$ df. Depending on our alternative hypothesis, we have these decision rules:

1. If our alternative hypothesis (H_A) is that $\mu > \mu_0$ and the significance level of our test—the probability of incorrectly rejecting the null hypothesis—is α, then we reject H_0 whenever $t_0 > F_{t,(n-1)}^{-1}(1-\alpha)$, the $(1-\alpha)$ quantile of the t-distribution with $(n-1)$ df.

2. If our H_A is that $\mu < \mu_0$ and the significance level of our test is α, then we reject H_0 whenever $t_0 < F_{t,(n-1)}^{-1}(\alpha)$, the α quantile of the null sampling distribution.

3. If our H_A is that $\mu \neq \mu_0$ and the significance level of our test is 2α, then we reject H_0 whenever $t_0 < F_{t,(n-1)}^{-1}(\alpha)$ or $t_0 > F_{t,(n-1)}^{-1}(1-\alpha)$.

We normally carry out the test this way, using the $t_0(\bar{x})$ statistic. However, we can easily adapt these rules so that they apply directly to $(\bar{x} - \mu_0)$ rather than to t_0. We see from the t_0 formula that all we need do to accomplish this is to multiply the relevant t-distribution quantile by $\widehat{\mathrm{SE}}(\bar{x})$. This gives a new set of decision rules:

1. If our H_A is that $\mu > \mu_0$ and the significance level of our test is α, then we reject H_0 whenever $(\bar{x} - \mu_0) > \left[\widehat{\mathrm{SE}}(\bar{x}) \times F_{t,(n-1)}^{-1}(1-\alpha)\right]$. Or,

recalling the symmetry of the t-distribution, we can say that we would reject H_0 whenever $(\bar{x} - \mu_0) > -\left[\widehat{SE}(\bar{x}) \times F^{-1}_{t,(n-1)}(\alpha)\right]$.

2. If our H_A is that $\mu < \mu_0$ and the significance level of our test is α, then we reject H_0 whenever $(\bar{x} - \mu_0) < \left[\widehat{SE}(\bar{x}) \times F^{-1}_{t,(n-1)}(\alpha)\right]$.

3. If our H_A is that $\mu \neq \mu_0$ and the significance level of our test is 2α, then we reject H_0 whenever either $(\bar{x} - \mu_0) < \left[\widehat{SE}(\bar{x}) \times F^{-1}_{t,(n-1)}(\alpha)\right]$ or $(\bar{x} - \mu_0) > -\left[\widehat{SE}(\bar{x}) \times F^{-1}_{t,(n-1)}(\alpha)\right]$.

We now have decision rules that are closely related to our construction of a CI for μ. We defined the $(1 - 2\alpha)100\%$ CI as $\bar{x} \pm \left[\widehat{SE}(\bar{x}) \times F^{-1}_{t,(n-1)}(\alpha)\right]$. What we've just learned is that the quantity we add to and subtract from $\bar{x}$ to define the limits of this CI is precisely the same quantity against which we compare the distance separating μ_0 from $\bar{x}$ in deciding whether to reject H_0 or not. This leads to a third set of decision rules, which are based on the CI:

1. If our H_A is that $\mu > \mu_0$ and the significance level of our test is α, then we reject H_0 whenever μ_0 falls below the $(1 - 2\alpha)100\%$ CI for μ.

2. If our H_A is that $\mu < \mu_0$ and the significance level of our test is α, then we reject H_0 whenever μ_0 falls above the $(1 - 2\alpha)100\%$ CI for μ.

3. If our H_A is that $\mu \neq \mu_0$ and the significance level of our test is 2α, then we reject H_0 whenever μ_0 falls outside the $(1 - 2\alpha)100\%$ CI for μ.

This third approach to hypothesis testing has considerable appeal because it derives from the computation of the CI estimate rather than requiring a separate computation. There is, however, a subtle difference between the first and third set of decision rules.

1. The t-test approach, in effect, asks if, in the null sampling distribution of the sample mean, our sample $\bar{x}$ is so far away from μ_0 that H_0 is untenable.

2. The CI approach asks if, in the true sampling distribution of the sample mean, μ_0 is so far away from our sample $\bar{x}$ that H_0 is untenable.

The two approaches coincide because of the pivotal nature of $t(\bar{x})$; the result will be the same whether we measure the distance between μ_0 and $\bar{x}$ in the null or in the true sampling distribution. Where we cannot be certain our statistic is pivotal, however, the two approaches might not coincide.

The p-Value of the Test Statistic

We have been describing fixed-level hypothesis testing. First, we fix the level α (or, 2α for a nondirectional test) of our hypothesis test. We then refer our test statistic to its null sampling distribution. If values at least as favorable to the alternative hypothesis as our statistic is make up a proportion of that distribution less than α (or, less than 2α), we reject the null hypothesis.

Classical fixed-level hypothesis testing has fallen out of fashion. When testing statistical hypotheses, statisticians more commonly supplement or even supplant the null hypothesis decision with a report of the p-value of the hypothesis test. This p-value is simply the proportion of the null sampling distribution that is at least as favorable to the alternative hypothesis as the test statistic is.

If the p-value is used as a supplement to a fixed-level null hypothesis decision, a p-value smaller than α (or, smaller than 2α) accompanies a rejection of H_0. Where the researcher does not subscribe to fixed-level hypothesis testing, the p-value often is used as a measure of the amount of support for the alternative hypothesis: the smaller the p-value, the greater the support.

For our Studentization of the sample mean, the α-level null-hypothesis decision could be grounded in either the null sampling distribution (the t-test approach) or the true sampling distribution (the CI approach). The same is true for the computation of the p-value.

By definition, the p-value is obtained from the null sampling distribution. It can also be obtained from a CI construction.

Let's assume that, instead of following an α-level strategy, we were to reject the null hypothesis, $\mu = \mu_0$, at the p-value level attained by our test statistic. That is, we let that p-value become the α level (or, the 2α level) of our hypothesis testing strategy. Given the linkage between the CI and t-test approaches to the null-hypothesis decision, this requires that μ_0 fall at one of the limits of a $(1 - 2p)100\%$ CI, if H_A is directional, or at one of the limits of a $(1 - p)100\%$ CI, if H_A is nondirectional. Thus, an alternate approach to determining the p-value of our test statistic is to determine the size of α that would place μ_0 at the edge of a $(1 - 2\alpha)100\%$ CI.

These two approaches to the p-value computation provide the same result when the test statistic is a pivotal one. In that instance, the CI approach would be needlessly clumsy.

Bootstrap-*t* Hypothesis Testing

In Concepts 9, we described the use of an explicit pivot to obtain bootstrap-*t* CIs for location parameters, θ, other than the normal population mean. We can use that same approach to test a null hypothesis about such a θ, that $\theta = \theta_0$.

We take as our test statistic

$$t_0(t) = \frac{t - \theta_0}{\widehat{SE}(t)}$$

the Studentization of the plug-in location estimator, under the null hypothesis. The location estimator t is obtained from the real-world sample. The SE estimate can be the closed form estimate, also computed from x, or it can be a bootstrap SE estimate.

Next, from a random sequence of B bootstrap samples, we form two bootstrap sampling distributions,

1. $t_b^*, b = 1, \ldots, B$

2. $t_0(t_b^*) = \frac{t_b^* - t}{\widehat{SE}(t_b^*)}, b = 1, \ldots, B$

The standard deviation of the distribution of the t_b^*s provides the denominator of $t_0(t)$ where we do not have or have not used a closed-form SE estimate.

If the Studentization provides a pivot, then the bootstrap sampling distribution of $t_0(t_b^*)$ is an appropriate null distribution for $t(t)$. It is appropriate because we have correctly described the value of θ in the population distribution from which bootstrap samples are drawn, $\widehat{\theta} = t$. That is, we can use the sampling distribution of $t_0(t^*)$ in the same way that we use the distribution of the *t*-random variable with $(n - 1)$ df. Because the bootstrap sampling distribution of $t_0(t_b^*)$ might not be symmetric, the α and $(1 - \alpha)$ quantiles should be evaluated explicitly.

Fixed-α Hypothesis Testing

If your H_A is directional, then determine the appropriate α or $(1 - \alpha)$ quantiles of the sampling distribution of $t_0(t_b^*)$, either $F_{t(t^*)}^{-1}(\alpha)$ or $F_{t(t^*)}^{-1}(1 - \alpha)$.

1. If your H_A is that $\theta > \theta_0$, then reject H_0 if $t_0(t) \geq F_{t(t^*)}^{-1}(1 - \alpha)$.

2. If your H_A is that $\theta < \theta_0$, then reject H_0 if $t_0(t) \leq F^{-1}_{t(t^*)}(\alpha)$.

If your H_A is nondirectional, then determine both the $\alpha/2$ and $(1 - \alpha/2)$ quantiles, $F^{-1}_{t(t^*)}(\alpha/2)$ and $F^{-1}_{t(t^*)}(1 - \alpha/2)$, and reject H_0 if $t_0(t) \leq F^{-1}_{t(t^*)}(\alpha)$ or if $t_0(t) \geq F^{-1}_{t(t^*)}(1 - \alpha)$.

Determining a p-Value for the Test

If your H_A is directional, then determine $\pi = F_{t(t^*)}[t_0(t)]$, the proportion of the $t_0(t_b^*)$s that are smaller than or equal to $t_0(t)$.

1. If your H_A is that $\theta > \theta_0$, then p $= (1 - \pi)$.

2. If your H_A is that $\theta < \theta_0$, then p $= \pi$.

If your H_A is nondirectional, then determine both π and $\pi^- = F_{t(t^*)}[-t_0(t)]$, the proportion of the $t_0(t_b^*)$s that are smaller than or equal to $[-1 \times t_0(t)]$.

1. If $\pi \leq 0.50$, then p $= \pi + (1 - \pi^-)$.

2. If $\pi > 0.50$, then p $= \pi^- + (1 - \pi)$.

One-Time *t*-Tables

In effect, the bootstrap-*t* approach provides us with a one-time nonparametric null sampling distribution for our location test statistic, rather than ask us to rely on the parametric *t*-distribution.

Bootstrap Hypothesis Testing Alternatives

The one-sample *t*-test is an example of an assumption-bound, parametric hypothesis test. The Studentized pivot, $t(\bar{x})$, has a known sampling distribution only if we can make the appropriate parametric assumption about the population distribution sampled. We have shown how to generalize the Studentized pivot to facilitate hypothesis testing for location parameters that do not satisfy these parametric requirements.

When we cannot explicitly pivot our estimator t, how can bootstrap inference be adapted to null hypothesis decision making and p-value computation? In the absence of a pivot, our bootstrap estimates of the null and true sampling distributions of a test statistic can differ, not only in location but in other details as well. As a result, using a CI, grounded in the true sampling distribution, or computing a p-value, from an explicitly-formed null sampling distribution, can lead to different hypothesis testing results.

The Null Bootstrap-World Population

In Concepts 7 and 8, we learned how to obtain a bootstrap sampling distribution for a statistic, s^*, one that we could use to estimate properties of the real-world sampling distribution of the corresponding statistic, s. In the context of hypothesis testing, we would like to be able to develop, in the same way, a second bootstrap sampling distribution that we could use to estimate properties of the null sampling distribution of s. How do Concepts 7 and 8 suggest we do that?

Instead of beginning with X, the real-world population distribution, we would start with X_0, the null-world population distribution. If our null hypothesis is correct, this is the population distribution from which the real-world sample, x, was obtained. Note that one or more attributes of this distribution are determined by our null hypothesis (for example, that $\theta = \theta_0$).

Keeping these null hypothesis attributes in mind, we would create $\widehat{X}_0$, our best estimate of the unknown X_0. By repeatedly and randomly sampling from this null bootstrap-world distribution, we could construct a series of bootstrap samples,

$$x_{0,1}^*, x_{0,2}^*, \ldots, x_{0,b}^*, \ldots, x_{0,B}^*$$

Collecting the test statistics computed in each of these bootstrap samples,

$$s_{0,1}^*, s_{0,2}^*, \ldots, s_{0,b}^*, \ldots, s_{0,B}^*$$

would give us a bootstrap estimate of the null sampling distribution of the test statistic, s. This estimate would provide what we need for hypothesis testing or for p-value determination.

This approach depends on creating $\widehat{X}_0$, an estimate of the null-world population distribution. Creating $\widehat{X}_0$ can be difficult, as an example will illustrate.

Assume that we have collected samples of scores on the same attribute from two different populations of cases:

$$x : (8, 14, 12, 8, 5, 20, 13) \text{ and } y : (14, 19, 12, 10, 8, 20)$$

x is from a population distribution, X, with mean μ_X, and y is from a population distribution, Y, with mean μ_Y. We want to test the null hypothesis that $\mu_X = \mu_Y$ against the alternative hypothesis that $\mu_X < \mu_Y$. How can we estimate the null-world population distributions, X_0 and Y_0?

Usually we want nonparametric estimates insofar as our knowledge about those two distributions will be limited to the following:

    1. x is a random sample from X_0

    2. y is a random sample from Y_0

    3. X_0 and Y_0 have a common mean, μ_0.

We could not form $\widehat{X}_0$ or $\widehat{Y}_0$ from this information alone. We would need to make additional assumptions about those two distributions. For example, population distributions satisfying all three conditions result if we assume that X_0 and Y_0 not only have the same mean but are identical. A nonparametric estimate of this common population distribution would be given by one or more copies of the combined set of 13 scores:

$$(8, 14, 12, 8, 5, 20, 13, 14, 19, 12, 10, 8, 20)$$

The problem with this solution is that it says more about X_0 and Y_0 than that they have the same mean; it says they have the same distribution. We might not want to go that far.

This difficulty is not specific to our example. In general, there are no good guidelines for obtaining from x a nonparametric estimate of X_0. Because we cannot create a nonparametric estimate of the null-world population distribution solely on the basis of the typical null hypothesis, we cannot develop a nonparametric bootstrap estimate of the null sampling distribution.

The CI Approach to Hypothesis Testing

In contrast, we have no difficulty in estimating the real-world population distribution, X, nonparametrically from x, or X and Y from x and y. This means that we can develop an estimate of a statistic's real-world sampling distribution although we cannot do the same for its null-world sampling distribution. As a result, we shall focus in subsequent units on the CI approach to hypothesis testing and p-value estimation. In the final section of this unit, we address the use of our bootstrap-t and BCA CIs in this context.

CI Hypothesis Testing

We have associated two tasks with hypothesis testing—deciding to reject or not reject the null hypothesis, and assigning a p-value to our test statistic.

Fixed-α Hypothesis Testing

If we need only to know whether or not to reject the null hypothesis, H_0: $\theta = \theta_0$, at a predetermined level α, then we can easily adapt the CI approach. We obtain the BCA interval at the appropriate $(1 - 2\alpha)100\%$ level. Then, we inspect that interval:

1. If the alternative hypothesis, H_A, is that $\theta \neq \theta_0$, then reject H_0 if θ_0 lies outside the interval.

2. If H_A is that $\theta < \theta_0$, then reject H_0 if θ_0 lies above the interval.

3. If H_A is that $\theta > \theta_0$, then reject H_0 if θ_0 lies below the interval.

4. Otherwise, do not reject H_0.

If you want to fix the significance level at, say, 5%, then the appropriate confidence levels will be 90% ($\alpha = 0.05$) for the directional H_As and 95% ($\alpha = 0.025$) for the nondirectional one.

p-Value Determination

Can we use the BCA quantiles to define p-values? And, if we can, should we? Our approach in this section is somewhat speculative. We clearly do not have a null sampling distribution to consult. What we do have is the possibility that our BCA CIs really are based on implicit pivots. To the extent that they are, we might be able to use those CIs for hypothesis testing as well, to determine the credibility of the statement that $\theta = \theta_0$ when our scientific expectation (alternate hypothesis) is, say, that $\theta > \theta_0$.

How can we proceed? First, we estimate a BCA equal-tails exclusion CI for θ that places θ_0 just below the lower edge of the interval. We then find the $(1 - 2\alpha)100\%$ confidence level associated with that interval. This is also the $(1 - \alpha)100\%$ lower-bound confidence level—the level of confidence we would have, roughly speaking, that the true value of θ exceeds θ_0.

One possibility, perhaps the preferred one, would be simply to interpret a high level of confidence as strong evidence against the null hypothesis. But, where p-values are required, you could go one step further and interpret α itself as a p-value. This interpretation would be correct if your test statistic were an exact pivot. The interpretation is speculative where pivotality is in doubt.

The conversion of α to a p-value is not always as straightforward as in our example; it will depend on whether θ_0 falls at the lower or upper limit and on the form of the alternative hypothesis. The considerations are described in the following section.

p-Values from the Null-limited BCA CI

We can easily adapt the Concepts 10 algorithm for obtaining a BCA CI to the finding of a p-value. The first eleven steps are carried out without change:

1. Compute the plug-in estimate, t, from the sample distribution, x.

2. Compute the n omitted-case estimates, $t_{(-i)}$.

3. Compute the average of the omitted-case estimates, $t_{(.)}$.

4. Compute n influence statistics, $(U_i/n) = [(n - 1/n)][t_{(.)} - t_{(-i)}]$.

5. Compute the acceleration estimate, $\widehat{a} = \dfrac{\sum\limits_{i=1}^{n}(U_i/n)^3}{6\left[\sum\limits_{i=1}^{n}(U_i/n)^2\right]^{3/2}}$.

6. Form $\widehat{X}$ from one or more copies of x.

7. Set $\widehat{\theta}$ equal to the t of step 1.

8. Draw a bootstrap sample, x_b^*, from $\widehat{X}$.

9. Compute the estimate, t_b^*, from the bootstrap sample.

10. Repeat steps 8 and 9 a total of B times, forming the bootstrap sampling distribution of t_b^*.

11. Compute $\widehat{\pi}$, the proportion of the bootstrap sampling distribution smaller than $\widehat{\theta}$.

We must modify the subsequent flow to find α from a known CI limit, rather than finding CI limits from a known α. The new steps also depend on whether θ_0 is at the lower or upper limit of the null-limited CI.

12a. If $\theta_0 < t$, compute $q_{lo} = (1/B) \times \#[t_b^* < \theta_0]$, the proportion of the sampling distribution of t_b^* that is smaller than θ_0.

12b. If $\theta_0 > t$, compute $q_{up} = (1/B) \times \#[t_b^* > \theta_0]$, the proportion of the sampling distribution of t_b^* that is larger than θ_0.

13a. If $\theta_0 < t$, use the tabled distribution of the standard normal distribution to determine $z_{[\hat{\pi}]}$ and $z_{[q_{lo}]}$. These are the z-scores that cut off the lower $\hat{\pi}$ and q_{lo} proportions of the standard normal distribution.

13b. If $\theta_0 > t$, use the tabled distribution of the standard normal distribution to determine $z_{[\hat{\pi}]}$ and $z_{[1-q_{up}]}$. These are the z-scores that cut off the lower $\hat{\pi}$ and $(1 - q_{up})$ proportions of the standard normal distribution.

14a. If $\theta_0 < t$, compute

$$z_{[1-\alpha]} = z_{[\hat{\pi}]} + \frac{\left(z_{[\hat{\pi}]} - z_{[q_{lo}]}\right)}{1 - \hat{a}\left(z_{[\hat{\pi}]} - z_{[q_{lo}]}\right)}$$

14b. If $\theta_0 > t$, compute

$$z_{[\alpha]} = z_{[\hat{\pi}]} - \frac{\left(z_{[1-q_{up}]} - z_{[\hat{\pi}]}\right)}{1 + \hat{a}\left(z_{[1-q_{up}]} - z_{[\hat{\pi}]}\right)}$$

15a. If $\theta_0 < t$, use the tabled distribution of the standard normal distribution to find the proportion of the distribution falling above the z-score $z_{[1-\alpha]}$. This is α.

15b. If $\theta_0 > t$, use the tabled distribution of the standard normal distribution to find the proportion of the distribution falling below the z-score $z_{[\alpha]}$. This is α.

The step 14 equations come from an algebraic reworking of the similar equations in the Concepts 10 BCA CI algorithm. There, for example, we knew $z_{[1-\alpha]}$ and wanted to solve for $\hat{z}_{lo}$; here we know $z_{[q_{lo}]}$ and need to solve for $z_{[1-\alpha]}$. Steps 13 and 15 can be automated if your computing package can perform the standard normal distribution lookups required.

The value of α you've obtained from your null-limited CI may be the p-value for your test statistic. Or, α might require conversion. The conversion of α to a p-value depends on your alternative hypothesis and on whether your test statistic points toward or away from that alternative. The rules, however, are straightforward.

    1. If your H_A is nondirectional, $\theta \neq \theta_0$, then the p-value for your test statistic is $(2 \times \alpha)$.

2. If your H_A is that $\theta < \theta_0$ and your t is also less than θ_0, then the p-value of your test statistic is α.

3. If your H_A is that $\theta < \theta_0$ but your t actually is greater than θ_0, then the p-value of your test statistic is $(1 - \alpha)$.

4. If your H_A is that $\theta > \theta_0$ and your t is also greater than θ_0, then the p-value of your test statistic is α.

5. If your H_A is that $\theta > \theta_0$ but your t actually is less than θ_0, then the p-value of your test statistic is $(1 - \alpha)$.

If the p-value of your test statistic is very close to 0.50, you might have some problems with these computations. Make certain that the computed value for α is less than 0.50.

Confidence Intervals or p-Values?

I prefer CIs over p-values when I am testing population hypotheses. The level of confidence for the interval provides information equivalent to the p-value, and the width and actual limits of the interval give additional information about the accuracy of our estimate.

Applications 11

Bootstrap p-Values

Concepts 11 describes two approaches to null hypothesis testing using bootstrap sampling. The first requires a plug-in location estimator with closed-form standard error (SE) estimate. We then can take advantage of a Studentized approximation to a pivotal transformation, allowing us to form a t-table that can be used to judge significance. The second approach requires only that we estimate a confidence interval (CI) for the population parameter specified by our null hypothesis and is more generally applicable to bootstrap hypothesis testing. We illustrate both approaches in these Applications.

Computing a Bootstrap-t p-Value

Table 11.1 is taken from Robertson (1991) and gives the reading ages in months of two random samples of primary school children, one sample of boys and one of girls. Higher age scores indicate better reading performances. We do not know the sizes of the populations sampled, but will assume them to be large enough to use our large population approach to population distribution estimation in the analysis that follows.

Table 11.1:   *Reading Ages for Two Samples of Primary School Students*

Boys' Reading Ages (in months)
96 72 78 93 90 96 75 69 84 78 81 84 72 93 84 72 93   84 111 81 93 75 78

Girls' Reading Ages (in months)
90 72 81 78 90 96 93 96 72 75 84 96 72 75 84 72 69   84 117 120 126 117 141

We take as our null hypothesis that, in the two populations sampled, the mean reading age scores are the same. Our alternative hypothesis is that the mean for the girl's population distribution is higher than that for the boy's population. That is, $\theta = \mu(\text{Reading Age}|\text{Girls}) - \mu(\text{Reading Age}|\text{Boys})$, H_0: $\theta = 0$, and H_A: $\theta > 0$.

We'll use as our test statistic the plug-in estimator of θ, the difference between the two sample means,

$$t = \bar{x}(\text{Reading Age}|\text{Girls}) - \bar{x}(\text{Reading Age}|\text{Boys})$$

This estimator has a closed-form SE estimate,

$$\widehat{SE}(\text{Sample Mean Diff}) = \sqrt{\frac{\hat{\sigma}^2(\text{Pop'n 1})}{n_1} + \frac{\hat{\sigma}^2(\text{Pop'n 2})}{n_2}}$$

useful to us because we can estimate the population variances from the same sample distributions that we use to estimate the population means. Thus, we can Studentize our difference in means under the null hypothesis,

$$t_0(t) = \frac{t - \theta_0}{\widehat{SE}(t)}$$

and compare its value against the bootstrap sampling distribution of

$$t_0(t_b^*) = \frac{t_b^* - t}{\widehat{SE}(t_b^*)}, b = 1, \ldots, B$$

the $t_0(t_b^*)$s computed from a random series of B bootstrap samples.

Figure 11.1 gives the text of a Resampling Stats script that finds a p-value for the mean difference in just this way.

The difference in sample means is computed from the original data (MNDIFF and TSTAT) and converted into the numerator of $t_0(t)$ by subtracting from it the null-hypothesized population mean difference (TNUM).

For a sequence of $B = 2000$ bootstrap samples (BSAMP1 and BSAMP2) drawn from the two estimated reading age population distributions (POPN1 and POPN2), the values of t_b^* (BMNDIFF) and of $t_0(t_b^*)$ (TSTAR) are computed and saved as TDIST and TSDIST, respectively.

```
maxsize TDIST 2000 TSDIST 2000
copy (96 72 78 93 90 96 75 69 84 78 81 84 72 93 84) SAMP1
concat SAMP1 (72 93 84 111 81 93 75 78) SAMP1
copy (90 72 81 78 90 96 93 96 72 75 84 96 72 75 84) SAMP2
concat SAMP2 (72 69 84 117 120 126 117 141) SAMP2
copy 2000 MC
copy 0 THETA0
copy SAMP1 POPN1
copy SAMP2 POPN2
'
size SAMP1 N1
size SAMP2 N2
mean SAMP1 MN1
mean SAMP2 MN2
subtract MN2 MN1 MNDIFF
print MN1 MN2 MNDIFF
copy MNDIFF TSTAT
'
subtract TSTAT THETA0 TNUM
repeat MC
  sample N1 POPN1 BSAMP1
  sample N2 POPN2 BSAMP2
  mean BSAMP1 BMN1
  mean BSAMP2 BMN2
  subtract BMN2 BMN1 BMNDIFF
  score BMNDIFF TDIST
  subtract BMNDIFF TSTAT BTNUM
  variance BSAMP1 VAR1
  variance BSAMP2 VAR2
  divide VAR1 N1 MNV1
  divide VAR2 N2 MNV2
  add MNV1 MNV2 MNDV
  sqrt MNDV MNDSE
  divide BTNUM MNDSE TSTAR
  score TSTAR TSDIST
end
variance TDIST SEOUT
sqrt SEOUT SEOUT
divide TNUM SEOUT TOUT
count TSDIST >= TOUT TAIL
divide TAIL MC PVAL
print TOUT PVAL
```

Figure 11.1: *Bootstrap-t Mean Difference Hypothesis Test (Resampling Stats)*

The standard deviation (SD) of the 2,000 elements of TDIST provides the SE estimate, SEOUT, needed as the denominator of $t_0(t)$ or TOUT. Based on our H_A, we expect TOUT to be large, and our p-value is the proportion of the contents of TSDIST, the $t_0(t_b^*)$s, that are as large or larger than TOUT.

Figure 11.2 shows the output of one run of this script.

```
MN1      =          84
MN2      =      91.304
MNDIFF   =      7.3043
TOUT     =      1.6307
PVAL     =      0.0415
```

Figure 11.2: *Output of Resampling Stats Bootstrap-t p-Value Script*

The sample mean difference, $t = 7.30$, yields $t_0(t) = 1.63$. Fewer than 5% of the bootstrap sampling distribution of $t_0(t_b^*)$ are as large as this. Evidence enough to reject the null hypothesis?

Fixed-α CIs and Hypothesis Testing

Where we cannot use a pivotal transformation to estimate a null sampling distribution, we can use CI estimates as a basis for hypothesis testing. We can do this in two different ways:

1. We can estimate a $(1 - 2\alpha)100\%$ CI for a fixed value of α and then reject H_0 if θ_0, the null hypothesized value of the parameter θ, falls above, below, or outside that interval, depending on H_A.

2. If we choose to not adopt a fixed-α approach to hypothesis testing, we can estimate a p-value for our test statistic by determining the confidence level of a CI estimate with θ_0 as one of its limits.

In this section we give an example of the first of these two approaches. The second approach is taken up in the section following.

In Applications 10, we estimated a CI for the variance of a population distribution of scores on a spatial ability test. In the study (Efron and Tibshirani, 1993) the random sample of 26 children actually earned scores on two tests, Spatial Ability Tests A and B. Both sets of test scores for the sample are given in Table 11.2.

We would not expect the children's scores on the two tests to be the same. But do children, in general, score higher on one of the tests, or are the differences from one test to the other, say from A to B, as likely to be negative as positive?

Table 11.2:   *Scores for 26 Children on Spatial Ability Tests A and B*

Case:	1	2	3	4	5	6	7	8	9	10	11	12	13
A:	48	36	20	29	42	42	20	42	22	41	45	14	6
B:	42	33	16	39	38	36	15	33	20	43	34	22	7

Case:	14	15	16	17	18	19	20	21	22	23	24	25	26
A:	0	33	28	34	4	32	24	47	41	24	26	30	41
B:	15	34	29	41	13	38	25	27	41	28	14	28	40

To make these ideas specific, we can formulate our hypotheses, null and alternative, in terms of an appropriate parameter of the population distribution of the test difference scores,

$$d_i = A_i - B_i, \ i = 1, \ldots, n$$

We'll take as the population parameter of interest the 20% trimmed mean, $\mu_{t,0.20}(d)$. We choose the trimmed mean as a population distribution summary for the reasons presented in Applications 9. Our null hypothesis is that $\mu_{t,0.20}(d) = 0$, neither test consistently produces higher scores. Our alternative hypothesis is that $\mu_{t,0.20}(d) \neq 0$. We'd like to reject the null hypothesis if one of the tests produces higher scores, but we cannot specify which one this might be. Our test statistic will be the plug-in estimator of $\mu_{t,0.20}(d)$, the 20% trimmed mean of the difference scores in the sample,

$$t = \bar{x}_{t,0.20}(d)$$

Taking the significance level of our nondirectional test to be 0.05, we can reject the null hypothesis if the $(1 - 2\alpha)100\% = 95\%$ equal tails exclusion CI does not include θ_0, the null-hypothesized value $\mu_{t,0.20}(d) = 0$.

We use a bootstrap-*t* estimate of the 95% CI because, from Applications 9, we know a closed form SE estimate for the trimmed mean location parameter. In Figure 11.3, we use the user-written procedures `tmean_mn` and `tmean_se` introduced in Applications 9 (Figure 9.2) in conjunction with the SC procedure `iboot_t` to estimate a 95% CI.

```
> dvec= TAB'[1][]-TAB'[2][]
> matrix(DM,26,1)
> setcol(DM,dvec,1)
> proc tmean_mn(){}
> proc tmean_se(){}
> vector(gs,1)
> gs:= 26

> iboot_t(tmean_mn,tmean_se,DM,gs,1,2500,FBOOT,95)

_____theta:
  0.6875
_____means of  2500  bootstrap estimates:
  0.70935
_____SDs of  2500  bootstrap estimates:
  1.4915522
      Parameter: 1   95 % CI:   -2.7410822 to   3.552154
      Total iboot_t() running time: : 19.17
```

Figure 11.3: *Hypothesis Testing via CI Inclusion of Hypothesized Value (SC)*

The matrix `TAB`, entered earlier, has two columns, the first with test A scores and the second with test B scores for the 26 cases in the random sample. From this matrix we compute a second matrix, `DM`, that contains a single column of difference scores. The two user-written procedures are declared and the sample size vector, `gs`, is created; this allows us to call `iboot_t`.

As we see, the resulting 95% CI extends some direction in either way from the plug-in estimate, $t = 0.6875$, $[-2.74, 3.55]$, and this interval certainly includes the null hypothesized value of 0 for the 20% trimmed mean of the difference scores. We have no basis, then, to reject the null hypothesis.

Computing a BCA CI p-Value

Our last example is one in which we find the p-value for a hypothesis test by determining the level, α, of that $(1 - 2\alpha)100\%$ CI that has a null-hypothesized value as its lower or upper limit. This approach requires that our test statistic, t, be a plug-in estimator of a population parameter, θ, and that our null hypothesis specify a value, θ_0, for that parameter.

We continue with the sample data of Table 11.2, scores on spatial ability tests A and B for a random sample of 26 children. Now we want to know whether the population test score variances, $\sigma^2(A)$ and $\sigma^2(B)$, are the same (the null hypothesis) or are significantly different (the alternative hypothesis). We'll associate these hypotheses with the population parameter,

$$\theta = \frac{\sigma^2(A)}{\sigma^2(B)}$$

the ratio of the two variances. Our null-hypothesized value for this parameter will be $\theta_0 = 1$. The plug-in estimator of this parameter is the ratio of the two sample variances,

$$t = \frac{S^2(A)}{S^2(B)}$$

We want to use this statistic to test our null hypothesis. In particular, we want to assign a p-value to this statistic, expressing the strength of support for the alternative hypothesis. We'll do this by first estimating a BCA CI for θ that has θ_0 as one of its limits. We next determine the confidence level, α, of that interval. Because our alternative hypothesis is a nondirectional one—the variance ratio can be either greater or smaller than 1.0—the p-value of our statistic will be $p = 2\alpha$.

The S-Plus `bootstrap` function is ideally suited to this task because it allows us quickly to find α by repetitive applications of the `limits.bca` method to the bootstrap sampling distribution—the "replicates"—produced by `bootstrap`. In this way, we can avoid the computations outlined in Concepts 11 for finding a BCA α-level. Figure 11.4 traces the S-Plus interaction.

The matrix `AB` carries the test A scores in column 1 and the test B scores in column 2. The sample variances for the two tests are 171.534 and 109.4098. The ratio of these two, our test statistic, is 1.567813. A user-written function, `varrat`, is constructed for use by the `bootstrap` function. We verify that, when the `varrat` function is applied to the data in `AB`, the function returns the correct value.

The call to the `bootstrap` function specifies that we want 2,000 bootstrap samples drawn.

```
> var(AB[,1],unbiased=F)
 [1] 171.534
> var(AB[,2],unbiased=F)
 [1] 109.4098
> var(AB[,1],unbiased=F)/var(AB[,2],unbiased=F)
 [1] 1.567813
> varrat<- function(X){var(X[,1],unbiased=F)/var(X[,2],unbiased=F)}
> varrat(AB)
 [1] 1.567813
> temp<- bootstrap(AB,varrat,B=2000)
> limits.bca(temp, prob=c(0.02,0.04,0.06,0.08,0.10))
           2%        4%        6%        8%        10%
varrat 0.9592125  1.038551   1.08994 1.1149103  1.1777131
> limits.bca(temp, prob=c(0.02,0.025,0.03,0.035,0.04))
           2%        2.5%       3%        3.5%       4%
varrat 0.9592125  0.9782779  1.00681 1.024593   1.038531
>
```

Figure 11.4: *S-Plus Derivation of Null-Limited BCA α Level*

Because our t is greater than θ_0, we will look for a BCA interval having $\theta_0 = 1$ as its lower limit. Our first invocation of the `limits.bca` method uses an arbitrary set of trial values for α. The return suggests that α lies between 0.02 and 0.04. A second invocation of `limits.bca` narrows this to approximately 0.03. Thus, the approximate p-value for our test of the equality of population variances is $p = 2 \times 0.03 = 0.06$. Attempting any greater precision than this seems unwarranted given the approximate nature of our BCA intervals.

Exercises

Table 11.3 is taken from Hand et al. (1994, Table 54) and lists the expenditures in Hong Kong dollars for each of four household categories for two random samples, 20 unmarried men and 20 unmarried women.

1. Do the two populations have different spending patterns? In particular, test the hypothesis that the mean proportion of household expenditure going

toward housing (category a) is the same in the two populations against the alternative that the two means differ. Use the bootstrap-*t* approach to estimate a p-value for the test.

2. Carry out a second spending pattern comparison of your choice, this time basing your conclusions on a BCA interval of appropriate width.

Table 11.3:   *Household Expenditures of Unmarried Males and Females*

Male Sample				Female Sample			
a	b	c	d	a	b	c	d
497	591	153	291	820	114	183	154
839	942	302	365	184	74	6	20
798	1308	668	584	921	66	1686	455
892	842	287	395	488	80	103	115
1585	781	2476	1740	721	83	176	104
755	764	428	438	614	55	441	193
388	655	153	233	801	56	357	214
617	879	757	719	396	59	61	80
248	438	22	65	864	65	1618	352
1641	440	6471	2063	845	64	1935	414
1180	1243	768	813	404	97	33	47
619	684	99	204	781	47	1906	452
253	422	15	48	457	103	136	108
661	739	71	188	1029	71	244	189
1981	869	1489	1032	1047	90	653	298
1746	746	2662	1594	552	91	185	158
1865	915	5184	1767	718	104	583	304
238	522	29	75	495	114	65	74
1199	1095	261	344	382	77	230	147
1524	964	1739	1410	1090	59	313	177

a: Housing, including fuel & electricity   b: Food, including alcohol & tobacco
c: Other goods, including clothing   d: Services, including vehicles & transport

Concepts 12

Randomized Treatment Assignment

Concepts 2 through 11 dealt almost exclusively with population inference, inference about some characteristic of the score distributions for a population of cases. Inferences of that kind depend on our having scores for one or more random samples of cases. Not many studies are based on random samples of cases, a point made in Concepts 6. We turn now to a second, important class of studies, those in which a nonrandom set of cases are randomly allocated among two or more treatment groups.

Two Functions of Randomization

Randomly assigning cases among alternate treatment groups serves two distinct purposes in experimental design and analysis.

1: Control of Competing Explanations

An important design function of randomization is to provide a measure of control against competing explanations for any differences in response to treatment among the treatment groups. We would like to conclude that these differences reflect differences in the treatments provided the groups rather than any pretreatment differences among the groups. A consequence of the random grouping of cases is that any pretreatment difference among groups will be a purely chance one, not anything built into the study design.

2: A Basis for Statistical Inference

Randomly assigning cases also provides a basis for statistical inference. In random sampling designs, our statistical basis was the sampling distribution, a distribution reflecting what other values our estimate or test statistic would take if our random sample had been a different one. In random assignment designs, our statistical basis for inference can be another distribution, a distribution reflecting what other values our statistic would take if the random assignment of cases had grouped our cases differently, and if a certain hypothesis about differences in response to the treatments were correct. We'll refer to this second kind of inference as causal inference. We'll infer, where we can, that differences in treatment cause differences in response to treatment.

Because the analysis should differ, we make a distinction between two classes of random treatment assignment designs:

> 1. The cases to be randomized are a random sample from a well-defined case population.

> 2. The cases to be randomized are not a random sample from a well-defined case population.

We'll refer to the cases of the first kind as sampled and those of the second kind as available. The latter name reflects a common experimental practice of randomizing cases from among those available to the experimenter. The cases could be patient or student volunteers, a stock of laboratory animals, a divisible agricultural tract, or a set of classrooms—none of them sampled from a larger population.

Randomization of Sampled Cases

When a random sample of cases is subsequently split randomly among several treatment conditions, the result is to create random samples from several prospective populations, one for each treatment condition. As a result, we can estimate or test hypotheses about parameters of the distributions of attribute scores for those populations. That is, we can make population inferences. The approach we'll use for these analyses is the bootstrap one described in Concepts 7 through 11, where we might have thought of our cases as sampled from natural populations.

The randomization of the randomly sampled cases, however, permits us to interpret certain population inferences as having a causal basis. For example, we might test the null hypothesis that the mean response to treatment A is identical to that to treatment B, $\mu_A = \mu_B$, against the alternate, $\mu_A < \mu_B$. If this test results in our rejecting the null hypothesis, we should be able to infer that the higher scores associated with treatment B are caused by that treatment, rather than by some confounding influence.

Randomization of Available Cases

If the cases in our random-assignment study are not a random sample from one or more populations, we cannot make population inferences. That is, estimates of population parameters, either point estimates or confidence intervals (CIs), are not interpretable, nor are those indices associated with the sampling distribution, bias, the standard error (SE), or the RMS error. Importantly, we cannot test hypotheses about population parameters in the absence of a population.

Although population inference is closed to us, causal inference is not. We might conclude, for example, that the superiority of the response of cases assigned to treatment B compared with that of those assigned to treatment A was caused by treatment B and was not the result of our random assignment of cases. This causal inference, from a statistical point of view, must be a local one. Treatment superiority might have been demonstrated among these cases, but we cannot generalize, statistically, to any larger case population. Although we might have a good scientific reason to conclude that other patients, other fertilized fields, or other fourth-grade classrooms would behave as did those available to our study, we have no statistical basis for such generalization.

Statistical Basis for Local Causal Inference

The statistical basis for local causal inference, like that for population inference, lies in considering "what if" our random process had an outcome different to that it actually produced. In population inference, we ask what if we'd obtained a different random sample of cases. In local causal inference, we ask what if our available cases had been randomized differently among the treatments. Actually, there is a bit more to it than this.

Consider a small, hypothetical experiment. Eight cases, labeled I, II,...,VIII, are randomly allocated, four apiece, to two treatment groups. The result, say, is this:

Group A – Cases: II, IV, V, VIII

Group B – Cases: I, III, VI, VII

The cases in group A then receive treatment A and the cases in group B receive treatment B. The effectiveness of treatment is then assessed by measuring an attribute, Z, on each case. These are the attribute Z scores:

Treatment A – II: 12, IV: 6, V: 19, VIII: 13

Treatment B – I: 21, III: 11, VI: 8, VII: 24

We now assess the relative effectiveness of the two treatments with a statistic, s, that is the difference between the medians of the attribute Z scores for treatments B and A:

$$s = \text{Median}(Z_i | \text{Treatment B}) - \text{Median}(Z_i | \text{Treatment A})$$

For the outcome of our experiment, we have

$$s = 16 - 12.5 = 3.5$$

a result that, on the face of it, suggests superiority for treatment B.

But, suppose the random division of our eight cases had resulted in

Group A – Cases: II, III, VI, VIII

Group B – Cases: I, IV, V, VII

What would have been the value of our treatment comparison statistic, s, for this randomization? To compute a new s, we'd need to know the Z scores for all cases. The actual experiment we carried out provides some, but not all, of the case scores for this alternative randomization:

Treatment A – II: 12, III: ?, VI: ?, VIII: 13

Treatment B – I: 21, IV: ?, V: ?, VII: 24

We don't know how cases III, IV, V and VI would have responded to the other treatment, the one they did not receive in the actual study. As a result, we can't say what value s would have taken, given this alternative random assignment of cases to treatments.

That is, we cannot create alternative real-world values of s, values that statistic actually would take if the random assignment had grouped the cases in different ways. But we can create alternative null-world values of s, values that statistic would take if the random assignment had grouped the cases in different ways and if the cases' attribute Z scores followed a particular null treatment-effect hypothesis.

Treatment-Effect Hypothesis

The most common treatment-effect hypothesis is that two or more treatments would have exactly the same effect on a given case. In the context of our two-treatment design, the null treatment-effect hypothesis is that any one of our cases would have earned the same score on attribute Z under either treatment assignment. This treatment-effect hypothesis allows us to replace the "?"s with attribute Z scores for our hypothetical rerandomization of the eight cases:

Treatment A – II: 12, III: 11, VI: 8, VIII: 13

Treatment B – I: 21, IV: 6, V: 19, VII: 24

And the treatment comparison statistic would take the value

$$s^H = 20 - 11.5 = 8.5$$

—larger, for this alternative randomization, than that actually observed.

The treatment-effect hypothesis allows us to fill in the scores for each possible rerandomization of the cases. Note that the treatment-effect hypothesis is specific to the individual case. It does not specify that the average or median score would be the same, but that the individual case's score would be the same.

The Null Reference Distribution

We label the difference-in-medians statistic s^H when it is computed based on the treatment-effect hypothesis for a particular randomization. The null reference distribution for our s statistic is the collection of the s^Hs for all possible rerandomizations of the cases and their scores. Note that s would have to be one of the elements in this distribution; that is, because the real-world randomization arranges the scores in accordance with the null treatment-effect hypothesis, s is one of the s^Hs.

The null reference distribution is the distribution of our test statistic under the null treatment-effect hypothesis, that our two treatments have the same effect on any particular case. This distribution reflects the variability in the statistic because of the chance assignment of cases to treatments. Our s^H will be large and positive for that randomization that puts the four cases that would earn the highest scores on attribute Z in treatment group B and the four that would earn the lowest scores in treatment group A. But s^H will be large and negative for the randomization that puts the four cases that would earn the highest scores in treatment group A and the four who would earn the lowest scores in treatment group B. For our eight cases, s^H will range from $(9.5 - 20) = -10.5$ to $(20 - 9.5) = 10.5$.

Testing the Treatment-Effect Hypothesis

We test the null treatment-effect hypothesis by referring the value of our statistic, s, to its null reference distribution. If there is only a very small chance that our statistic would take so extreme a value, as a result solely of the randomization of cases, then we reject the null treatment-effect hypothesis.

What we count as "so extreme" will depend, as it did in testing population parameter hypotheses, on our alternative treatment-effect hypothesis.

The possible alternative hypotheses will be clearer if we restate the null hypothesis: For any particular case, treatments A and B will yield the same attribute Z scores.

Then, the three alternative hypotheses are the following:

1. For some cases, treatment A will produce higher attribute Z scores than will treatment B.

2. For some cases, treatment B will produce higher attribute Z scores than will treatment A.

3. For some cases, the better treatment will produce higher attribute Z scores than will the poorer treatment.

Large negative values of our statistic,

$$s = \text{Median}(Z_i | \text{Treatment B}) - \text{Median}(Z_i | \text{Treatment A})$$

are consistent with the superiority of treatment A, the first alternative hypothesis, and unlikely under the null. Large positive values, unlikely under the null, would support the superiority of treatment B. And, the nondirectional alternative—one treatment is better than the other for at least some cases— would be supported by either very large positive or very large negative values of s.

What is unlikely under the null hypothesis? Fixed α-level hypothesis testing might require that values of s^H at least as extreme as our s, and in the direction of the alternative hypothesis, make up only 5% or even 1% of the null reference distribution. That is, the p-value of our s must be no larger than 0.05 or 0.01. Those proportions come either from the upper end, the lower end, or both of the null reference distribution, depending on our alternative to the null-treatment effect hypothesis.

Exact and Monte Carlo p-Values

If our null reference distribution contains the s^H from each of the possible rerandomizations of cases, then the p-value we compute by counting values at least as extreme as s is called an exact p-value. Equivalently, the hypothesis test is said to be an exact one.

The number of possible rerandomizations, however, might be quite large. As in developing the bootstrap sampling distribution, we might choose to approximate the reference distribution by computing s^H for a large number of randomly chosen rerandomizations rather than attempting to enumerate all possibilities.

For our small example, eight cases to be randomly divided into two groups of four, the number of possible rerandomizations is only

$$\frac{8!}{4! \times 4!} = \frac{8 \times 7 \times 6 \times 5}{4 \times 3 \times 2 \times 1} = 210$$

They could all be generated in a reasonable time on a personal computer using a fairly simple algorithm. Thus, an exact test would be reasonable.

If the test is based on the cdf of s^H for a sequence of R randomly chosen rerandomizations of cases, the test and p-value are said to be exact to Monte Carlo accuracy. The degree of accuracy, the closeness of the result to the exact one, depends on R, the number of rerandomizations included in the reference set.

How large should R be?

Let's assume that, based on complete enumeration of the possible rerandomizations of some data set, the proportion of the s^Hs that make up the null reference distribution for the test statistic s that are $\geq s$ is p. Now assume we approximate this null reference distribution with R values of s^H, the original s, and the values of s^H from $(R-1)$ randomly chosen rerandomizations.

Figure 12.1 is based on the work of Edgington (1995) and shows—for selected values of R and for two values of p (0.05 and 0.01)—the 99% probability limits for the Monte Carlo accuracy. For example, when the exact p-value is 0.05 and we approximate the null reference distribution with $R = 5000$ randomly chosen rerandomizations, then there is a 99% chance that we will observe a p-value between 0.0422 and 0.0581.

If you want to compute similar 99% limits for large values of R and your choice of p, you can use this formula:

$$p \pm 2.58 \sqrt{\frac{p(1-p)}{R}}$$

This formula describes an aspect to the sampling distribution of the Monte Carlo approximated p-value. Because, for large values of R, that sampling distribution has a normal-cdf, the middle 99% of its elements lie within 2.58 SE's of the exact p-value.

p	R	99% Probability Limits Lower	Upper
0.05	500	0.0268	0.0770
	1000	0.0332	0.0687
	2000	0.0379	0.0630
	3000	0.0401	0.0606
	4000	0.0413	0.0591
	5000	0.0422	0.0581
0.01	500	0.0005	0.0234
	1000	0.0029	0.0191
	2000	0.0048	0.0162
	3000	0.0056	0.0150
	4000	0.0061	0.0143
	5000	0.0066	0.0138
	6000	0.0069	0.0135
	8000	0.0073	0.0130
	10000	0.0075	0.0127

Figure 12.1: *Sampling Limits (99%) for p-Values from Approximated Distributions*

Although the choice of size for the null reference set is yours, values of R in the range of 1,000 to 2,000 are appropriate for a test of the null treatment-effect hypothesis at the 5% level. For tests at the 1% level, your choice of R should be 5,000 or larger.

Exact or Monte Carlo p-Values?

If you have access to statistical software that computes an exact p-value for your choice of treatment comparison statistic, it makes sense to use it. This will not be true always. You should develop a Monte Carlo approximation to the exact p-value for the treatment comparison statistic you want to use rather than abandon that statistic for the sake of an exact result. Your desktop computer makes it easy to obtain a null reference distribution of 1,000, 2,000, or even 5,000 elements.

Population Hypotheses Revisited

We have outlined here a rationale for using what is elsewhere referred to as a permutation or randomization hypothesis test. Available cases are randomly allocated between two treatment groups, their responses to treatment assessed, and a treatment comparison statistic computed. We then take as our null treatment-effect hypothesis that the response to treatment of any particular case would be the same if randomized to either treatment. Under this null hypothesis, we can rerandomize the case assignments and recompute the treatment comparison statistic. Doing this for all possible rerandomizations or for numerous, randomly chosen rerandomizations provides us with a null reference distribution for our treatment comparison statistic. If our treatment comparison statistic is far enough out in the appropriate tail of this distribution, we can reject the null hypothesis and conclude that the observed difference in treatment response is the result of the treatments, rather than of the random assignment of cases.

We call inferences drawn under these circumstances local causal inferences. These inferences indicate whether the treatments caused the responses made by these available cases. These inferences are local inferences, limited to the cases participating in the study. We distinguish local from population inference because the latter requires random samples of cases.

A second rationale for our rerandomization test procedure does support population inference. Consider this example, using the same data we used earlier. Our eight cases now are sampled randomly from a case population. Subsequently, the sampled cases are randomly divided into two treatment groups of four cases. Then, their responses to treatment (A or B) are assessed with the following results:

Responses to treatment A: 12, 6, 19, 13

Responses to treatment B: 21, 11, 8, 24

Because of the random division of the random sample of cases, these two sets of scores are sample distributions ($n_A = n_B = 4$), treatment response scores for random samples from two prospective populations.

We can compute from the sample distributions a plug-in estimate of the difference in population distribution medians

$$t = \text{Median}(\text{Score}|B) - \text{Median}(\text{Score}|A) = 16 - 12.5 = 3.5$$

Is the difference large enough to conclude that the median of the treatment B population distribution is greater than the median of the treatment A population distribution?

In Concepts 11, we discussed the ambiguities of translating a null hypothesis into a null population distribution from which we could resample to develop a null sampling distribution for some test statistic. Because of these ambiguities, we chose to approach population hypothesis testing by estimating a bootstrap confidence interval (CI) for an appropriate population parameter. Here that would be the difference in population medians.

We can grasp the nettle, however, and describe the null population distributions in sufficient detail that we can, through resampling, generate a null sampling distribution for our test statistic. In particular, let us examine the consequences of taking as our null hypothesis that the two population distributions are identical. This goes well beyond postulating that they have the same medians.

Under this null hypothesis, the scores (12, 6, 19, 13) and (21, 11, 8, 24) were sampled from identical population distributions. If we are told that the eight scores (12, 6, 19, 13, 21, 11, 8, 24) were sampled, four apiece, from two identical population distributions, we can only conclude that any division of the eight scores into two sets of four produces a pair of samples from the two populations that is just as likely to have occurred, no more or no less, as the particular two samples we obtained.

That is, each of the

$$\frac{8!}{4! \times 4!} = \frac{8 \times 7 \times 6 \times 5}{4 \times 3 \times 2 \times 1} = 210$$

permutations of the eight scores into two sets of four are equally likely under the null hypothesis. Computing the test statistic for each of these permutations gives us a null reference distribution for our test statistic. We ask what proportion of this distribution is as large or larger than our test statistic. If this proportion, the p-value for the statistic, is small, we can reject the null hypothesis in favor of the alternative—here, that the median of the treatment B population distribution is greater than the median of the treatment A population distribution.

The mechanics of the population permutation test and of the local rerandomization test are identical. The hypotheses tested are different. Where we have random samples, we can test hypotheses about population parameters. Where we have randomization of available cases among treatment groups, we can test local inferences about the causal effect of treatments.

Because excess assumptions are usually needed to formulate one or more null population distributions, we shall make little use of the population permutation test. We return to the test in Applications 23, however, when our random samples are from categorical or multinomial population distributions.

Applications 12

Monte Carlo Reference Distributions

In Concepts 12, we introduced the analysis of studies in which available cases are randomly allocated among two or more treatments. Here we look at how our three statistical packages can be used to analyze an example of the simplest of these random allocation designs, one group of available cases randomly allocated between two treatment levels.

Serum Albumen in Diabetic Mice

The data in Table 12.1 are taken from Hand et al. (1994, Data Set 304) and report the amounts of a certain type of serum albumen produced by mice characterized as alloxan diabetic. The mice were randomly divided between a control group and a group receiving insulin. Insulin administration is postulated to reduce albumen production.

Table 12.1: Serum Albumen Levels Among Alloxan Diabetic Mice

Group	Serum Albumen Levels
Control	391, 46, 469, 86, 174, 133, 13, 499, 168, 62, 127, 276, 176, 146, 108, 276, 50, 73
Insulin	82, 100, 98, 150, 243, 68, 228, 131, 73, 18, 20, 100, 72, 133, 465, 40, 46, 34, 44

In the context of the random allocation design, our null treatment-effect hypothesis is that a given mouse would produce the same amount of albumen whether allocated to the control or insulin treatment group. Our alternative, scientific hypothesis is that at least some of the mice would produce less albumen under insulin treatment than under control treatment.

We need a test statistic that allows us to aggregate responses over the mice assigned to the two treatments and that will be sensitive to the difference between null and alternative hypotheses. A natural choice is a statistic that compares the albumen production of the control-randomized mice with that of the insulin-randomized mice, taking small values under the null hypothesis and large values under the alternative. The statistic we shall use for the following examples is the difference in the 20% trimmed means of the albumen levels for the two treatment groups,

$$s = \overline{x}_{t,0.20}(\text{Albumen}|\text{Control}) - \overline{x}_{t,0.20}(\text{Albumen}|\text{Insulin})$$

where $(\text{Albumen}|\text{Control})$ represents the albumen level under Control treatment.

Our s will be almost zero if mice in the two treatment groups produce about the same amounts of albumen; s will be large and positive if those mice randomized to the insulin treatment group produce much smaller amounts of albumen.

Why do we use trimmed means to summarize the responses to treatment in each group? We want the summaries to describe typical levels of response for the two groups. Had we used untrimmed means as summaries, either summary might assume a quite large or quite small value, solely on the basis of the level of response, atypical for the group, of a single animal.

For the two groups of experimental mice, the value of the test statistic works out to be

$$s = 150.41667 - 87.461538 = 62.955$$

Is this difference large enough that we should believe the insulin treatment results in reduced albumen production, or would differences this large or larger be relatively common solely as a result of the random allocation of mice whose individual albumen productions are uninfluenced by whether they are assigned to insulin or control treatments?

To answer this question, we compare the value of s with a null reference distribution. This is the distribution of the values of our statistic when computed in all possible random allocations of the 37 mice—18 to control treatment, 19 to insulin treatment—where we adopt the null hypothesis that any particular mouse would produce the same amount of albumen under either treatment regime. Our s is just one of those values. Any one of the others we'll denote s^H, a hypothetical value of the statistic, computed under the null hypothesis.

For even moderately sized treatment groups, the reference distribution contains a very large number of s^Hs—there are more than 17 billion different ways in which 37 mice can be divided into two groups of 18 and 19—so we frequently approximate it by computing s^H only for some large number of randomly chosen rerandomizations of the cases and their null-hypothesized scores.

Resampling Stats Analysis

We turn now to how we might use the first of our statistical packages. We begin with Resampling Stats because its script orientation will make it easier to see how the rerandomization of cases to treatments occurs. In the other two packages, the details of this are hidden from us.

```
'A Monte Carlo permutation test: Difference in Trimmed Means
'   cases randomly divided into two treatments.
'
COPY (391 46 469 86 174 133 13 499 168 62 127) SAMP1
CONCAT SAMP1 (276 176 146 108 276 50 73) SAMP1
COPY (82 100 98 150 243 68 228 131 73 18 20 100) SAMP2
CONCAT SAMP2 (72 133 465 40 46 34 44) SAMP2
COPY 0.2 GAMA
COPY 10000 MC
'
SIZE SAMP1 N1
SIZE SAMP2 N2
MULTIPLY GAMA N1 X
ROUND X TRIM1
IF TRIM1 > X
   SUBTRACT TRIM1 1 TRIM1
END
MULTIPLY GAMA N2 X
ROUND X TRIM2
IF TRIM2 > X
   SUBTRACT TRIM2 1 TRIM2
END
ADD TRIM1 1 LO1
SUBTRACT N1 TRIM1 HI1
ADD TRIM2 1 LO2
SUBTRACT N2 TRIM2 HI2
'
```

Figure 12.2: *Part 1, Resampling Stats Trimmed Mean Comparison*

Figures 12.2 through 12.4 contain the full computational script resulting in the computation of a p-value for our treatment comparison statistic, s.

This p-value is the proportion of the null reference distribution that is at least as large and positive as our s.

Much of the script will be familiar from earlier examples. Figure 12.2 shows the input of the scores for the two groups, the declaration of the γ trimming proportion, the declaration of the size of the Monte Carlo approximation, and, finally, the computation of the number of high and low scores to trim from each of the two treatment distributions.

```
SORT SAMP1 SAMP1
TAKE SAMP1 LO1,HI1 TSAMP1
MEAN TSAMP1 MN1
SORT SAMP2 SAMP2
TAKE SAMP2 LO2,HI2 TSAMP2
MEAN TSAMP2 MN2
SUBTRACT MN1 MN2 TRMNDIFF
PRINT MN1 MN2 TRMNDIFF
COPY 1 TAIL
SUBTRACT MC 1 MCM1
'
```

Figure 12.3: *Part 2, Resampling Stats Trimmed Mean Comparison*

Figure 12.3 describes the computation of the difference in trimmed means test statistic, TRMNDIFF, and the initialization of the TAIL and MC counts.

Part 3 of the script, listed in Figure 12.4, generates the null reference distribution and computes from it the p-value for our test statistic. For each rerandomization, the full set of 37 mice, together with their albumen production scores, are shuffled, then the first 18 (1 through N1) are allocated to the control treatment and the remaining 19 (START2 through STOP2) are assigned to the insulin treatment. The test statistic, s^H or SNULL, is computed from the rerandomized cases and compared with s.

Each time that s^H is at least as large as s, the TAIL count is increased by one. After all MC values of s^H (including s) have been evaluated, the script computes the proportion of these that are at least as large as s. This proportion, the p-value for our test statistic, is displayed as PROP. If the p-value is small enough, then the chance determination of an s as large as ours under the null hypothesis is so unlikely that we will conclude that insulin treatment is responsible for lower albumen production.

```
ADD N1 1 START2
ADD N1 N2 STOP2
CONCAT SAMP1 SAMP2 BOTHSAMP
REPEAT MCM1
  SHUFFLE BOTHSAMP BOTHSAMP
  TAKE BOTHSAMP 1,N1 HSAMP1
  TAKE BOTHSAMP START2,STOP2 HSAMP2
  SORT HSAMP1 HSAMP1
  TAKE HSAMP1 LO1,HI1 HSAMP1
  MEAN HSAMP1 MN1
  SORT HSAMP2 HSAMP2
  TAKE HSAMP2 LO2,HI2 HSAMP2
  MEAN HSAMP2 MN2
  SUBTRACT MN1 MN2 SNULL
  '
  IF SNULL => TRMNDIFF       'For the unidirectional
    ADD TAIL 1 TAIL          'alternative: SAMP1 > SAMP2
  END
  '
END
DIVIDE TAIL MC PROP
PRINT PROP
```

Figure 12.4: *Part 3, Resampling Stats Trimmed Mean Comparison*

One run of the script of Figures 12.2 through 12.4 with MC set at 10,000 produced the display of Figure 12.5.

```
MN1       =       150.42
MN2       =       87.462
TRMNDIFF  =       62.955
PROP      =        0.0403
Successful execution. (52.7 seconds)
```

Figure 12.5: *Resampling Stats p-value for the Difference in Trimmed Means*

The p-value is estimated at 0.04. In many instances a p-value this small provides ample support for rejecting the null hypothesis and concluding that there is a treatment effect.

SC Analysis

The same analysis is addressed in Figure 12.6 using SC. The upper panel shows the text of the user-written function usrtrm. This computes the difference in 20% trimmed means for two vectors of scores. In the lower panel, we call this function with the two vectors of albumen scores, for the control mice (con) and for the insulin mice (insu). The result is the *s* described earlier.

```
func usrtrm()
{
 local(a,b)
 a= tmean($1$,floor(0.2*sizeof($1$))/sizeof($1$))
 b= tmean($2$,floor(0.2*sizeof($2$))/sizeof($2$))
 return (a-b)
}


> usrtrm(con,insul)
      62.955128

> perm2___(usrtrm,con,insul,10000)
Using f():
         / 0.9625 / 0.0376 /
Using abs(f()):
         / 0.9253 / 0.0748 /

> perm2___(usrtrm,con,insul,10000)
Using f():
         / 0.9574 / 0.0427 /
Using abs(f()):
         / 0.9182 / 0.0819 /
```

Figure 12.6: *SC Determination of p-Value for Difference in Trimmed Means*

We use the SC function perm2___(f,x,y,MC) to do the work of creating a null reference distribution and computing a p-value.

This function takes either three or four arguments:

f   The name of a user-written function comparing two score vectors

x   The first of two score vectors

y   The second score vector

MC   Optional: If provided, this specifies the number of s^Hs to include in the Monte Carlo approximation; if not provided, all permutations of the two score vectors are considered

This function returns four proportions, two based on the reference distribution for s^H and two based on the reference distribution for $|s^H|$, the absolute value of s^H. The latter can be useful when the alternative hypothesis is nondirectional. In each pair, the two are the proportions of the reference distribution that are less than or equal to s and that are greater than or equal to s.

Figure 12.6 includes two calls to `perm2___`, and each specifies that 10,000 rerandomizations are to be used. The two p-values, 0.038 and 0.043, bracket the 0.40 we obtained in the earlier Resampling Stats computation. Together, these p-values indicate the degree of accuracy to be expected with 10,000 rerandomizations.

S-Plus Analysis

Finally, let's look at an S-Plus approach to this same problem. Figure 12.7 contains the log of an interactive computing session.

```
> albumen<- c(cont,insu)

> albdiff<- function(X){mean(X[1:18],0.2)-mean(X[19:37],0.2)}

> albdiff(albumen)
[1]   62.95513

> refdis<-bootstrap(albumen,albdiff,sampler=samp.permute)

> limits.emp(refdis,prob=c(0.90,0.95,0.975,0.99,0.995))
               90%         95%        97.5%         99%        99.5%
albdiff 43.88141    60.03462    73.37997    84.72558    89.82215
```

Figure 12.7: *S-Plus Evaluation of the Trimmed Means Comparison*

The albumen production scores for the control and insulin mice are present as the vectors `cont` and `insu`. The first command in Figure 12.7 assembles the two sets of scores in a single vector, `albumen`, with the insulin mice scores following those for the control group. Next we write a function—similar to the usrtrm function that we wrote for SC—that will find the difference between the trimmed means of the first 18 scores in a vector and the 19 that follow those, albdiff.

Applying albdiff to the scores in albumen gives the now familiar value of our treatment comparison statistic, $s = 62.95513$.

We modify our earlier use of the S-Plus bootstrap function to develop a reference distribution for our s. The first two arguments are still the names of a data set and of a function that will compute a statistic from those data. Our third argument is a new one. It specifies that an alternative sampler (samp.permute) be used for the resampling. This permutation sampler randomly reallocates cases and their scores.

In the Figure 12.7 bootstrap call, we did not specify a value, B, for the number of resamples to be created, accepting the default number of 1,000. The resulting 1,000 values of s^H are saved as part of a S-Plus object named by us refdis. The final command in Figure 12.7 computes specified quantiles of this null reference distribution. We use the form limits.emp because we want the raw or empirical quantiles of the reference distribution. In finding BCA confidence interval limits, we used the form limits.bca because we wanted the bootstrap sampling distribution quantiles to be adjusted for bias and acceleration before being reported.

Our use of the limits.emp method with refdis does not produce a p-value for our test. However, we do learn that our test statistic, $s = 62.955$, falls in the upper 5% of the reference distribution, above the 0.95 quantile of 60.035. There is less than a 5% chance of having gotten a difference in trimmed means at least this favorable to insulin treatment simply as a function of the random allocation of treatment-indifferent cases. Insulin made a difference.

Exercises

1. Table 12.2 shows two small data sets. For each, work out the complete reference distribution for $s = \text{Mean}(x) - \text{Mean}(y)$. For the equal-sized groups, the reference distribution should be symmetric. What is true for the non-equal groups?

Table 12.2:   *Two Small Data Sets for Rerandomization*

Data Set A		Data Set B	
x	y	x	y
6	7	6	8
8	5	5	4
4	1	7	
		3	
		2	

2. Efron and Tibshirani (1993) report the survival time in days of sixteen mice following surgery. The mice were randomly divided into two groups. One group of seven mice received a treatment expected to prolong their survival:

```
94, 197, 16, 38, 99, 141, 23
```

A second group of 9 mice underwent surgery without the treatment:

```
52, 104, 146, 10, 51, 30, 40, 27, 46
```

Develop a Monte Carlo null treatment-effect reference distribution for the test statistic

$s = \text{Mean(Survival Time}|\text{Treatment}) - \text{Mean(Survival Time}|\text{No Treatment})$

Summarize the characteristics of this distribution.

Concepts 13

Strategies for Randomizing Cases

In Concepts 12, we outlined an approach to testing hypotheses about the differential impacts of two or more treatments, each delivered to a different group of cases where those groups had been composed by randomly dividing a set of available cases. Here, we describe several strategies that can be used to form such randomized groups.

Independent Randomization of Cases

A first strategy is to make an independent, random group assignment for each case. For example, as each successive patient volunteers for a study, that patient is assigned to either the experimental or control group on the basis of a coin toss, "Heads" to the experimental group, "Tails" to the control.

Fixed Total Number of Cases

This independent randomization strategy can be followed where the total number of cases, n, is fixed by the researcher. Each of the n cases then is assigned, sequentially and at random, to one of K treatment groups. The resulting partitioning of the cases is one of the K^n possible divisions of the n cases among the K groups. Some of these possible divisions, however, would place zero cases in one or more of the groups, precluding the computation of one or more treatment comparisons. As a result, no exact p-value can be obtained for any treatment comparison statistic.

If the expected number of cases in any treatment group, n/K, is no smaller than about eight, then these zero group-size outcomes are very unlikely and a Monte Carlo statistical test, based on a random sequence of rerandomizations of the n cases, will be appropriate. If n/K is quite small, however, one or more of the $(R-1)$ randomly chosen rerandomizations might yield a treatment group with zero cases.

Skip over rerandomizations that place no cases in one or more groups as you build the reference set. Actually, you might also have to ignore rerandomizations that place either one or two cases in a treatment group. The principles for skipping over certain rerandomizations are these:

1. If that randomization had occurred initially, the study would have been modified, perhaps by continuing case recruitment.

2. It must be possible to calculate the treatment comparison statistic for each rerandomization. This might require more than two cases in each treatment group.

The first principle is important to the logic of rerandomization analysis; any rerandomization must be subject to the same rules that controlled the original randomization. That is, the rerandomization must have an outcome that could have been the outcome of the original randomization. The second principle is a pragmatic one.

A basic algorithm for independently randomizing, or rerandomizing, n cases among K treatment groups is given in Figure 13.1.

---

1. Create a case identification vector, C, containing the integers from 1 through n.

2. Create a treatment group identification vector, T, containing the integers from 1 through K.

3. Set the case count equal to zero; $i = 0$.

4. Increase i by one.

5. Shuffle the T vector.

6. Place the initial element of T as the i-th element of a treatment assignment vector, A.

7. If i is equal to n, quit. Otherwise, repeat steps 4 through 6.

8. When finished, the n element A vector contains the treatment group assignments of the corresponding n cases identified by the vector C.

Figure 13.1: *Independent Randomization of n Cases Among K Treatment Groups*

Fixed Minimum Treatment Group Size

Rather than fix the total number of cases, a researcher might assign sequential cases at random to one of the K treatments until each treatment group contains at least n_{min} cases. The researcher then stops recruiting additional cases. The motivation for this strategy, of course, is to ensure against the occurrence of a case randomization that leaves one treatment group empty or severely small. The risk is that a large number of cases might need to be assigned, and possibly treated, to ensure that the smallest treatment group reaches the minimum size.

We can rerandomize, of course, only those n cases recruited into the original study. To be faithful to the original design we need to cut off recruitment when the smallest group contains n_{min} cases. Figure 13.2 extends the algorithm of Figure 13.1 to include this feature.

---

1. Create a case identification vector, C, containing the integers from 1 through n. Case identifications should correspond to the original recruitment sequence of cases.

2. Create a treatment group identification vector, T, containing the integers from 1 through K.

3. Create a treatment group count vector, G, with K elements, all set equal to zero.

4. Set n_{min} to the desired smallest group size.

5. Set the case count to zero; $i = 0$.

6. Increase i by one.

7. Shuffle the T vector.

8. Store T_1, the first element of T, as the i-th element of a treatment assignment vector, A.

9. Increase $G_{[T_1]}$, the count of cases assigned to this group, by one.

10. If the smallest element in G is equal to n_{min}, quit. Otherwise, go to step 11.

11. If i is equal to n, quit. Otherwise repeat steps 6 through 10.

12. When you have finished, the i-element A vector contains the treatment group assignments of the cases identified by the first i entries to the vector C.

Figure 13.2: *Rerandomization to Ensure Smallest Group Does Not Exceed n_{min}*

Our rerandomization of cases for this design should respect the order in which the original cases were recruited. One or more of the cases would not have been included if the minimum group sizes had been met before the recruitment of the n-th case. Thus, the case identification vector created at Step 1 of the algorithm should identify cases by their order of entry into the study.

Although the algorithm of Figure 13.2 ensures that the smallest rerandomized group will not contain more than n_{min} cases, we have no assurance—given that the total sample size is fixed at n—that each rerandomized group will contain at least n_{min} cases. Where one or more groups is too small, a particular rerandomization could be rejected to be replaced by a subsequent one.

Alternatively, though this is less faithful to the original randomization, the rerandomizations could all be constrained to yield the same treatment group sizes as resulted from the original randomization. This would ensure a minimum size for each rerandomized treatment group and would reuse all cases in each rerandomization.

Treatment Comparisons

When $K = 2$, the treatment comparison is fairly obvious. When we have more than two treatments, however, we must let the researcher tell us what treatment comparisons are of interest. Occasionally, we will want a comparison involving all K treatments. More often, a treatment comparison will involve fewer than K treatments, perhaps only two of them.

Our original randomization of cases was among K treatments or treatment groups. When we create a null reference distribution for a comparison among some number, g, of treatment groups, $g < K$, we rerandomize subject to the following conditions:

1. We rerandomize only the cases originally randomized to those g groups.

2. We rerandomize those selected cases only among the same g groups.

A simple example establishes why this is true. We originally randomize our cases among treatments A, B, and C. One treatment comparison we want to make is between treatments A and B and we compute our s test statistic to make that comparison.

Our null treatment-effect hypothesis for this comparison is that a case would make the same response to treatment A as to treatment B. Nothing is said in this hypothesis about the response to treatment C. This has two implications:

1. The responses of those cases assigned to treatment C tell us nothing about how they would have responded if they had been assigned to either treatments A or B.

2. The response by a case to either of treatments A or B tells us nothing of what that case's response to treatment C would have been.

Thus, in rerandomizing subject to this particular null hypothesis, we cannot assign a treatment C case (and its response) to either of treatments A or B and we cannot assign a treatment A or B case (and its response) to treatment C.

We avoid these errors by rerandomizing only treatment A and B cases and by limiting their rerandomizations to treatments A or B.

Whenever we rerandomize, we must be faithful both to how the original randomization was carried out and to our current null treatment-effect hypothesis.

In using either of the algorithms described to develop a reference set of rerandomizations, we would need to modify the following: C, the case identification vector; T, the treatment identification vector; and K, the number of treatments—to reflect the treatments and, hence, the cases subject to the null treatment-effect hypothesis.

Independent Randomization Limitations

Independently randomizing each case cannot ensure that the same number of cases is randomized to each treatment group. Unequal treatment group sizes can give rise to asymmetric null reference distributions for our treatment comparison statistics. As a result, some treatment-effect hypotheses will be tested more efficiently than others. Whenever possible, you should use an initial randomization scheme that ensures equal treatment group sizes.

Completely Randomized Designs

The random division of the n cases among the K treatment groups is said to result in a completely randomized design if two conditions are met:

1. The number of cases to be assigned to each treatment group is determined in advance: $n_1 + n_2 + \ldots + n_K = n$.

2. A single random process is used to divide the n available cases among the K treatment groups.

Relative to the first condition, we typically want the same number of cases, n_k, in each treatment group, and we choose $n = K \times n_k$. The second condition distinguishes the completely randomized design from one in which the set of n cases is first split into two or more smaller sets, called blocks, each of which is independently divided at random among the K treatment groups. This latter randomization strategy gives rise to what is known as a randomized block design. In that context, the completely randomized design is a single block design.

The number of different ways in which our n cases can be split into K groups of sizes $n_1, n_2, \ldots, n_K$ is given by

$$\frac{n!}{n_1! \times n_2! \times \cdots \times n_K!}$$

Our initial randomization picks one of these divisions at random.

For a particular null treatment-effect hypothesis the size of the null reference set can be considerably smaller than this because we are concerned only with the random divisions of those cases originally assigned to certain treatment groups, and among those same groups. For instance, if our treatment hypothesis is that treatments 1 and 2 have the same impact on a case, then the reference distribution for our comparison statistic consists of the value of that statistic computed over the

$$\frac{(n_1 + n_2)!}{n_1! \times n_2!}$$

possible divisions of $(n_1 + n_2)$ cases (and their scores) between the two treatments to which they were originally randomized.

A straightforward algorithm for rerandomizing the cases initially randomized to K treatment groups among those same treatments is given in Figure 13.3.

Steps 1 through 6 create the necessary C and T vectors. The rerandomization is carried out at step 7 so that at the final step in the algorithm, the integer stored as T_i identifies the treatment group to which the case identified in C_i has been rerandomized.

Solely for purposes of rerandomization, the algorithm need not know to which treatment group any case was initially randomized. However, the algorithm could be incorporated into a larger program in which the treatment contrast statistic, s, is to be computed for the original randomization. For that purpose, the elements of the vector C are arranged initially so that the T vector describes the original randomization of the cases.

Repeating steps 7 and 8 $(R-1)$ times provides a random sequence of $(R-1)$ rerandomizations of the n cases among the K treatments.

---

1. Create an n-element vector C containing the case identifications of the cases ordered as originally randomized among treatments 1 through K; $n = n_1 + n_2 + \ldots + n_K$.

2. Set the K elements of the vector N equal to the sequence of group sizes: $N_1 = n_1, \ldots, N_K = n_K$.

3. Initialize the group indicator, i, to 1; $i = 1$.

4. Fill the next N_i elements of the T vector with i.

5. Increase i by one.

6. If i exceeds K go to step 7; otherwise, repeat steps 4 and 5.

7. Shuffle C.

8. The elements of the vector T contain the randomized group assignments of the corresponding cases identified in vector C.

Figure 13.3:   *Rerandomizing Cases in a Completely Randomized Design*

Randomized Blocks Designs

In the randomized blocks design, the researcher first divides the available cases into blocks and then randomly allocates each block of cases separately among the treatments. The preliminary blocking of cases is on an attribute of the cases that could influence their responses to treatment. Randomizing each block separately helps ensure that the blocking attribute is not confounded with treatment assignment.

Fifty student volunteers for a psychology experiment are to be randomly divided into five treatment groups. Thirty of the volunteers are women, and 20 are men. If the 50 students are randomized as a single block, 10 to each treatment group, the women might be overrepresented in certain groups and underrepresented in others. If we are worried that women and men might respond differently to treatment, this would be problem; we might not be able to decide whether treatment group differences in response represented differences in treatments or differences in the sex balance of the groups.

If the men and women are randomized separately, however, we guard against such confounding. The 30 women can be randomly divided among the five treatment groups, 6 women to each group. And the 20 men can be randomly divided, 4 men to each treatment group. As a result, each treatment group would contain exactly 6 randomly chosen women and 4 randomly chosen men; the sex balance is the same for all treatment groups.

Some considerations when you set the number of cases for randomized blocks designs are the following:

1. If there are to be the same number of cases, n_k, in each of the K treatment groups, then the total number of cases must be an integer multiple of K, $n = K \times n_k$ $(n_1 = n_2 = \ldots = n_k = \ldots = n_K)$.

2. If there are to be the same number of cases from the j-th block of cases in each of the K treatment groups, then n_j, the number of cases in the j-th block, also must be an integer multiple of K, $n_j = K \times n_{kj}$, where n_{kj} is the number of cases from the j-th block randomized to the k-th treatment $(n_{1j} = n_{2j} = \ldots = n_{kj} = \ldots = n_{Kj})$.

3. If each of J blocks of cases is to contribute the same number of cases to a treatment group, then the number of cases in each block, n_j, must be the same $(n_1 = n_2 = \ldots = n_j = \ldots = n_J)$.

The first two of these considerations describe desirable balances; the first leads to symmetry in the efficiency with which we can compare treatments, and the second ensures a lack of confounding between the blocking attribute and treatments. Relative to the third consideration, it is important not to force an ideal block size by compromising the definition of a block. Blocks of cases should be as homogeneous as possible relative to the attribute used to form blocks. All equally homogeneous levels of a blocking attribute might not attract the same numbers of cases. Some blocks could contain K cases, others $2K$, and yet others $3K$, or $4K$.

An algorithm for rerandomizing cases in a randomized blocks design is described in Figure 13.4. This algorithm is essentially that of Figure 13.3, repeated for each of $j = 1, \ldots, J$ blocks of cases. This repetition, however, illustrates the point that the rerandomization must follow the randomization. Each block of cases is rerandomized separately from every other block, just as the blocks were separately randomized originally. Steps 10 through 14 would be repeated for each of the desired $(R - 1)$ rerandomizations.

---

1. Initialize the block indicator, j, to 1; $j = 1$.

2. Create an n_j-element vector Cj containing the case identifications of the cases in the j-th block ordering them as originally randomized among treatments 1 through K, $n_j = n_{1j} + n_{2j} + \ldots + n_{Kj}$.

3. Set the K elements of the vector nj equal to the sequence of group sizes: $nj_1 = n_{1j}, \ldots, nj_K = n_{Kj}$.

4. Initialize the group indicator, i, to 1; $i = 1$.

5. Fill the next nj_i elements of the Tj vector with i.

6. Increase i by one.

7 If i exceeds K go to step 8; otherwise, repeat steps 5 and 6.

8. Increase j by one.

9. If j exceeds J go to step 10; otherwise, repeat steps 2–8.

10. Initialize the block indicator, j, to 1; $j = 1$.

11. Shuffle Cj.

12. Increase j by one.

13. If j exceeds J go to step 14; otherwise, repeat steps 11 and 12.

14. The elements of the vector Tj contain the randomized group assignments of the corresponding cases identified in vector Cj.

Figure 13.4: *Rerandomizing Cases in a Randomized Blocks Design*

Restricted Randomizations

In each of the randomization strategies described thus far, we have assumed that the same range of treatments is available to all cases. In some cases, this may not be true. For example, in a multicenter study, not all treatments might be available at all centers. Cases will be randomized among those treatments that are available.

Such a restricted randomization scheme can be thought of as another example of a randomized blocks design. The cases are divided into blocks for the original randomization by the range of treatments available. These blocks and their associated restrictions on treatment assignment need to be present at each rerandomization as well.

Constraints on Rerandomization

Two sources constrain how cases can be rerandomized when you develop a reference set of rerandomizations:

1. The null treatment-effect hypothesis specifies which treatment responses are exchangeable. Only cases originally randomized to a treatment group affected by the null hypothesis will be rerandomized, and then only among those treatment groups affected by the hypothesis.

2. Follow the original randomization strategy in the rerandomizations. For example, if cases were randomized originally by blocks, this same block structure must be adhered to in rerandomization.

Applications 13

Implementing Case Rerandomization

Concepts 13 presents algorithms for several case randomization strategies. The most important of these, because of the experimental control they bring with them, are those leading to completely randomized (CR) and randomized blocks (RB) designs. Here, we see how these have been or can be implemented by our three statistical computing packages. We also look at the independent and restricted randomizations of cases. Because CR and RB designs are so important, these designs are treated here first.

Completely Randomized Designs

The hallmark of the CR case randomization is that the available cases are randomly allocated, as a single group, among treatment levels, the number to be assigned to each level having been determined in advance. For the case rerandomizations needed to develop a null reference distribution for a particular treatment-effect hypothesis, this same strategy should be followed, limited to those cases impacted by the treatment-effect hypothesis.

The diabetic mice illustrations of Applications 12 show the CR strategy as implemented in Resampling Stats, SC, and S-Plus for $K = 2$ treatment levels. In CR designs with more than two treatment levels, our science-based treatment-effect hypotheses almost always contrast one level with another or one subset of levels with a second subset. That is, our science almost always predicts that one treatment should be superior to another or that one set of treatments should be superior to a second set. For hypotheses of the second kind, we aggregate some, if not all, of the treatment levels into two sets of treatments and then compare the two sets of case responses. For example, having randomized patients among four levels of treatment—control, drug A, drug B, and drug C—the researcher might want to test the hypothesis that receiving a drug produces greater symptom relief than receiving the control treatment does. To test this scientific or alternative hypothesis, she would join the three drug levels into a single set of treatments to be compared against the control treatment.

The Applications 12 implementations apply directly to hypotheses about sets of treatments; all we need do is to divide the cases involved (and their response scores) into two groups corresponding to the sets to which they were originally randomized. The rerandomization between those sets of treatment levels, rather than among treatment levels, is relevant to the development of a reference distribution for our treatment set comparison statistic.

Rarely, you might want to test a treatment-effect hypothesis that explicitly compares three or more sets of treatments. These hypotheses should be rare because they are less clearly focussed relative to both scientific motivation and practical implications than are two-set comparisons. The classical multiset comparison is the omnibus null hypothesis, that all K treatment levels would have the same effect on a given case. We illustrate this omnibus test with a three level CR example from Loftus and Loftus (1988) described in Figure 13.5, the first part of a Resampling Stats script.

```
'12 students randomly allocated, 4 apiece, among 3 conditions.
'After watching a filmed simulated auto accident Ss read an
'account of the accident:
'    g0: non-attributed article with no misinformation
'    g1: NY Daily News attributed article with misinformation
'    g2: NY Times attributed article with misinformation
'Score: number of errors in test of memory for film contents.
COPY (4 7 1 1) g0
COPY (4 9 10 5) g1
COPY (9 13 15 7) g2
'
```

Figure 13.5: *Part 1, Omnibus Test of Treatments (Resampling Stats)*

The null omnibus treatment-effect hypothesis is that a student experimental subject would make the same number of errors no matter which one of the three reports was read. The alternative is that at least some students will make more errors after reading a particular report or reports than they would after reading one of the other reports. The alternative specifies that the reports differ in promoting errors but does not specify which specific report or reports promote greater errors.

Under the null hypothesis, we should expect approximately equal error scores for the three treatments. Under the alternative, they should differ. We aggregate the scores within treatment level for making the comparison, here using the treatment means. Figure 13.6 shows the computation of our treatment comparison statistic, the variance of the three means. The variance will take a value of zero if the three means are identical and larger positive values as the means diverge in value.

```
MEAN g0 g0m
MEAN g1 g1m
MEAN g2 g2m
CONCAT g0m g1m g2m g012m
VARIANCE divn g012m var
COPY 1 tail
```

Figure 13.6: *Part 2, Omnibus Test of Treatments (Resampling Stats)*

In Figure 13.7, we compare the variance of the three means against a null reference distribution approximated from 1,000 randomly chosen random reassignments of the 12 students, four apiece, to the three treatments. The null hypothesis prescribes that the students' error scores accompany them on any rerandomization.

```
CONCAT g0 g1 g2 g012
REPEAT 999
  SHUFFLE g012 g012x
  TAKE g012x 1,4 g0x
  TAKE g012x 5,8 g1x
  TAKE g012x 9,12 g2x
  MEAN g0x g0xm
  MEAN g1x g1xm
  MEAN g2x g2xm
  CONCAT g0xm g1xm g2xm g012xm
  VARIANCE g012xm varx
  IF varx >= var
    ADD tail 1 tail
  END
END
DIVIDE tail 1000 prop
PRINT var
PRINT prop
```

Figure 13.7: *Part 3, Omnibus Test of Treatments (Resampling Stats)*

All three treatments are involved in this hypothesis. As a consequence, we must rerandomize all those cases originally randomized to the three treatment groups.

A run of the script of Figures 13.5 through 13.7 produced a null reference distribution in which the variance of treatment means of 10.014 was equaled or exceeded by almost 3% of the 1,000 values of s^H, p $= 0.028$. Perhaps this is strong enough evidence that the three treatments would not produce, in the typical subject employed for this study, the same number of errors.

An S-Plus implementation of this omnibus test can be built around the bootstrap function. Figure 13.8 shows how.

```
>grp1<- c(4,7,1,1)
>grp2<- c(4,9,10,5)
>grp3<- c(9,13,15,7)
>allgrp<- c(grp1,grp2,grp3)
>grpvar<- function(X){var(c(mean(X[1:4]),mean(X[5:8]),
+mean(X[9:12]))),unbiased=F)}
>grpvar(allgrp)
[1]   10.01389
>omtest<- bootstrap(allgrp,grpvar,sampler=samp.permute)
>length(omtest$replicates[omtest$replicates>=10.01389])
[1]   32
> 33/1001
[1]   0.03296703
```

Figure 13.8: *S-Plus Evaluation of an Omnibus Treatment Hypothesis*

In this interaction we composed a function, grpvar, to be applied by the bootstrap function to each of 1,000 rerandomizations of the original data set, allgrp. The length function is used to learn that only 32 of the 1,000 values of s^H, the replicates created by our bootstrap function, are at least as large as s.

The ratio $32/1000$, however, is not quite the correct p-value. The bootstrap function does not include the original randomization among the 1,000 replicates, as should be done under the null hypothesis. When we add the original randomization to the 1,000 rerandomizations, we also add s to the reference distribution and, thus, increase by one the number of elements in the reference distribution that equal or exceed s. We now have 33 of the 1001 elements that are at least as large as s; this gives us p $= 0.033$ for this second

Monte Carlo approximation to the reference distribution. Resampling Stats provided the first approximation.

```
proc varmns(){
  local(k,end,un,beg,dvec,mns,i)
  k= sizeof($2$)
  end= $2$
  vector(un,1)
  fill(un,1)
  beg= vcat(un,end)
  cum(beg)
  cum(end)
  vector(dvec,rows($1$))
  getcol($1$,dvec,1)
  vector(mns,k)
  i=0
  repeat(k){
    i++
    limit(dvec,beg[i],end[i])
    mns[i]=mean(dvec)
    delimit(dvec)
  }
  $3$[1]=var(mns)
}
```

Figure 13.9:   *A Procedure in SC for Computing an Omnibus Test Statistic*

Figure 13.9 gives the text of a SC procedure for computing the variance among K treatment means. And in Figure 13.10, this procedure is used by shuffl_k to estimate a p-value for our omnibus comparison of the three CR treatment groups.

The S-Plus function bootstrap with its permutation sampler can compare three or more combinations of treatment levels as easily as it can compare two. The SC function perm2__ of Applications 12, however, estimates p-values specifically for a comparison between two levels or combinations of levels. More generality is possible with the procedure shuffl_k. This procedure has seven arguments:

shuffl_k(p,m,M,gsiz,ssiz,R,RD)

p   A user-written procedure to compute m statistics from a data matrix that is divided by rows either into treatment groups (CR design) or into treatment groups within blocks (RB design)

m   The number of statistics returned by the user-written procedure

M   A data matrix, cases as rows, blocked by treatments (CR) or

> by treatment within case blocks (RB)

gsiz   A vector of sizes of treatment groups (CR) or of treatment groups within case blocks (RB), sum(gsiz)=rows(M)

ssiz   A vector of rerandomization sizes, a single sample size (CR) or J block sizes (RB), sum(ssiz)=sum(gsiz)

R   Number of rerandomizations of data to be added to the original randomization used to approximate reference distributions, for example, 999

RD   Name of a R × m matrix to hold the m reference distributions

```
> vector(errors,12)
> read(errors)
4 7 1 1 4 9 10 5 9 13 15 7
     12
> matrix(MD,12,1)
> setcol(MD,errors,1)
> vector(gps,3)
> read(gps)
4 4 4
     3
> vector(nn,1)
> read(nn)
12
     1
> proc varmns(){}
> shuffl_k(varmns,1,MD,gps,nn,999,RD)
     Full Sample Statistics (theta):     10.013889
     P-values for Statistic Number:   1
        Percent LE theta[ 1 ]:   97.2
        Percent GE theta[ 1 ]:   2.9
        Percent LE abs(theta[ 1 ]):   97.2
        Percent GE abs(theta[ 1 ]):   2.9

     Total shuffl_k() running time:   2.58
```

Figure 13.10: *SC's p-Value for the Three Treatment Omnibus Test*

The user-written procedure, like those needed by the SC bootstrap procedures of earlier applications, takes three arguments, p(M,gsiz,out), where M and gsiz are as defined for shuffl_k and out is an m element vector, to receive the statistics computed by p.

This third Monte Carlo approximation to the p-value of 0.029 (Percent GE theta[1]: 2.9) falls within the range established by the two earlier approximations, 0.028 and 0.033.

Randomized Blocks Designs

In the RB experiment, the cases are first blocked on a shared characteristic. Then each block of cases is randomly distributed among treatment groups. The number of cases from a block to be assigned to each level of treatment is fixed prior to the random allocation. In rerandomizing cases to develop a reference distribution, the block structure of cases is maintained.

Table 13.1:   *Recognition Accuracy Scores for Varying Display Times*

	Display Time			
Reading Speed	1 sec	2 sec	4 sec	8 sec
Block 1	13,16	14,15	17,18	17,18
Block 2	10, 8	11, 9	14,12	16,17
Block 3	8, 8	9,10	12,11	16,15
Block 4	8, 7	8, 8	11,11	13,16

The data in Table 13.1 are taken from Lunneborg (1994) and are recognition scores (maximum of 20) for 32 volunteer university student subjects in an imaginary experiment on the influence of varying the exposure time of a visual display of verbal material. In the experiment, test materials were displayed for either 1, 2, 4, or 8 seconds. The experimenter believed that performance on the task could be influenced not only by exposure time but also by differences in the speed with which the volunteers normally read. To control for this, cases were blocked into four groups of eight on the basis of an assessment of their speed in reading a typical newspaper article. Block 1 consisted of the eight fastest readers, block 4 contained the slowest eight, and blocks 2 and 3 contained those of intermediate speed. For each of the blocks, the eight students were randomly allocated among the four display times, two to each.

We'll consider just one hypothesis about the outcome of this study, a hypothesis that involves all cases.

Substantively, longer display times ought to be associated with higher recognition accuracy scores. Specifically, scores should be higher at 4 or 8 seconds exposure than at 1 or 2 seconds exposure. The null treatment-effect hypothesis, however, is that the recognition accuracy score of a participant would have been the same whatever display time was assigned.

```
copy (13 16 14 15 17 18 17 18) BL1
copy (10 8 11 9 14 12 16 17) BL2
copy (8 8 9 10 12 11 16 15) BL3
copy (8 7 8 8 11 11 13 16) BL4
'
take BL1 1,4 LO1     'Cases 1-4 received short exposures
take BL1 5,8 HI1     'Cases 5-8 received long exposures
take BL2 1,4 LO2     'See Table 13.1
take BL2 5,8 HI2
take BL3 1,4 LO3
take BL3 5,8 HI3
take BL4 1,4 LO4
take BL4 5,8 HI4
concat LO1 LO2 LO3 LO4 LO
concat HI1 HI2 HI3 HI4 HI
mean LO MLO
mean HI MHI
subtract MHI MLO MDIFF
print MLO MHI MDIFF
copy 1 TAIL             'MDIFF is >= MDIFF
'
repeat 999
  shuffle BL1 BL1       'Rerandomizng cases by block
  shuffle BL2 BL2
  shuffle BL3 BL3
  shuffle BL4 BL4
  take BL1 1,4 LO1
  take BL1 5,8 HI1
  take BL2 1,4 LO2
  take BL2 5,8 HI2
  take BL3 1,4 LO3
  take BL3 5,8 HI3
  take BL4 1,4 LO4
  take BL4 5,8 HI4
  concat LO1 LO2 LO3 LO4 LO
  concat HI1 HI2 HI3 HI4 HI
  mean LO MLO
  mean HI MHI
  subtract MHI MLO HMDIFF
  if HMDIFF >= MDIFF
    add TAIL 1 TAIL
  end
end
divide TAIL 1000 PROP
print PROP
```

Figure 13.11: *Rerandomizing Blocked Cases with Resampling Stats*

Figure 13.11 is a Resampling Stats script for this problem. The test statistic, s, is simply the difference in the mean recognition accuracy scores at 4 or 8 seconds, on the one hand, and at 1 or 2 seconds on the other. Note that in the rerandomization loop each block of cases is randomized separately. When executed, the script reports a mean difference of $s = 4.25$. A value this large or larger was found only once in a reference set based on the original and 999 rerandomizations of the cases. None of the 999 values of s^H were as large as s, given the contrived results of this study.

The shuffl_k procedure in SC accommodates block rerandomization with its randomization group vector, ssiz. Figures 13.12 and 13.13 illustrate this.

The first figure displays a user-written procedure, precacc, to compute the difference in means between two sets of scores, here the recognition accuracy scores for long exposure times (4 and 8 seconds) and short exposure times (1 and 2 seconds). In keeping with the requirements of shuffl_k, these scores are in the first column of a matrix (1), arranged by treatment conditions within blocks. Here, short exposure scores are followed by long exposure scores, within reading speed blocks. The vector 2 gives the sizes of the consecutive treatment within blocks groups of cases.

After establishing where groups of cases begin and end among the rows of the input matrix (1), the precacc procedure moves their scores from the first column of that matrix into a vector, vd. Alternate partitions of this vector are then aggregated into the vectors loscor (contains the odd-numbered partitions) and hiscor (contains the even-numbered partitions).

Finally, the difference between the means of these two vectors is stored as the first (and only) element of a vector (3), supplied as the third precacc argument.

Though written for the present example, the procedure precacc can be generalized. This procedure could be used to compare two treatments where the treatment response scores have been arranged within blocks; those for the treatment expected to produce lower scores are followed by those for the treatment expected to produce higher scores.

The difference in means between the two treatments could be replaced with a different treatment comparison statistic. When you develop user-written procedures, think about how they might be modified for use in other data analyses.

We see the procedure precacc at work in the SC interaction captured in Figure 13.13.

```
proc precacc(){
  local(k,end,un,beg,vd,i,loscor,hiscor)
  k= sizeof($2$)
  end= $2$
  vector(un,1)
  fill(un,1)
  beg= vcat(un,end)
  cum(beg)
  cum(end)
  vector(vd,rows($1$))
  getcol($1$,vd,1)
  i=0
  repeat(k){
    i++
    limit(vd,beg[i],end[i])
    if(i==1){
      loscor= vd
    }else{
      if(i==2){
        hiscor= vd
      }else{
        if((i/2)!=int(i/2)){
          loscor= vcat(loscor,vd)
        }else{
          hiscor= vcat(hiscor,vd)
        }
      }
    }
    delimit(vd)
  }
  $3$[1]= mean(hiscor)-mean(loscor)
}
```

Figure 13.12: *SC Procedure to Form Two Sets from Blocked Scores*

The 32 recognition accuracy scores are read into the vector `recacc` in an order determined by `shuffl_k` and `precacc`. The first procedure requires that the scores be arranged by treatments within blocks. The second procedure requires that, within blocks, the scores for the lower-scoring treatment precede the scores for the higher-scoring treatment.

The vector `ssiz` gives the number of scores in each successive rerandomization group or block. In our example, eight cases are in each of the four blocks. The vector `gsiz` describes how the block sizes are divided between treatments. For the example, each block of 8 cases consists of 4 short-display-time cases followed by 4 long-display-time cases. `shuffl_k` uses `ssiz` to rerandomize cases within blocks among treatments and passes `gsiz` to `precacc` for each calculation of s^H from a rerandomization of the cases.

```
> vector(recacc,32)
> read(recacc)
13 16 14 15 17 18 17 18
10 8 11 9 14 12 16 17
8 8 9 10 12 11 16 15
8 7 8 8 11 11 13 16
   32
> vector(gsiz,8,ssiz,4)
> fill(gsiz,4)
> fill(ssiz,8)
> matrix(MDAT,32,1)
> setcol(MDAT,recacc,1)
> proc precacc(){}
> shuffl_k(precacc,1,MDAT,gsiz,ssiz,999,RD)
    Full Sample Statistics (theta):   4.5

P-values for Statistic Number:  1
    Percent LE theta[ 1 ]:   100
    Percent GE theta[ 1 ]:   0.1
    Percent LE abs(theta[ 1 ]):   100
    Percent GE abs(theta[ 1 ]):   0.1
```

Figure 13.13: *SC Evaluation of RB Reference Distribution*

In preparation for the call to `shuffl_k`, the scores are moved into the first column of a matrix, `MDAT`, and the user-written `precacc` is declared.

`shuffl_k` reports, as did our Resampling Stats script, that none of 999 values of s^H were as large as s.

Finally, the S-Plus `bootstrap` function can accommodate RB rerandomizations if we supply, in addition to the data vector, a second vector identifying the block to which each case belongs. Figure 13.14 shows the use of a blocking vector for the present example.

The rerandomization group size vector used by SC's `shuffl_k` requires that we organize the cases and scores by blocks. Using, instead, a block-identification vector, S-Plus's `bootstrap` function allows us to organize cases in a way that facilitates the computation of s. Here, we have grouped the recognition accuracy scores in the vector `recacc` so that the 16 1-second and

2-second display time scores are followed by the 16 4-second and 8-second scores. This makes it easy to create a user-written function `mndif` that will find the difference in means between long and short display times.

```
> sec1<- c(13,16,10,8,8,8,8,7)
> sec2<- c(14,15,11,9,9,10,8,8)
> sec4<- c(17,18,14,12,12,11,11,11)
> sec8<- c(17,18,16,17,16,15,13,16)
> blk<- c(1,1,2,2,3,3,4,4)
> recacc<- c(sec1,sec2,sec4,sec8)
> blk<- c(blk,blk,blk,blk)
> mndif<- function(X){mean(X[17:32])-mean(X[1:16])}
> mndif(recacc)
[1] 4.5
> hilo<-bootstrap(recacc,mndif,group=blk,sampler=samp.permute)
> length(hilo$replicates[hilo$replicates>=4.5])
[1] 0
> max(hilo$replicates)
[1] 3.125
```

Figure 13.14: *S-Plus RB Rerandomization Analysis*

The vector `blk` is created to contain the block identification of each of the 32 cases. Passing this information to the permutation sampler, as the additional argument `group=blk` in the `bootstrap` function call, ensures that the rerandomization of cases takes place separately for each of the 4 blocks of cases.

Again, not one of the 1,000 values of s^H is as large as $s = 4.5$. In fact, the largest value of s^H in this run was 3.125.

Independent Randomization of Cases

An infrequent experimental design calls for each available case to be randomized independently. In the analysis of data obtained this way, we must rerandomize in the same independent way.

Table 13.2 taken from Agresti (1990) summarizes the outcome of a study comparing radiation therapy with surgery in treating cancer of the larynx.

Patients were assigned, individually and at random (we are assuming), to either surgery or radiation and after treatment their cancers were judged to have been controlled or not controlled. Each patient was as likely to be assigned to one treatment as another, but the independent random assignments resulted in an imbalance; 23 patients were assigned to surgery and 18 to radiation.

Table 13.2:   *Outcomes of a Randomized Treatment Study*

	Cancer	
Treatment	Controlled	Not Controlled
Surgery	21	2
Radiation	15	3

The null treatment-effect hypothesis is that if a patient's cancer was controlled under one treatment it would have been controlled under the other, and if a patient's cancer was not controlled under one treatment it would have been not controlled under the other as well.

Our alternative hypothesis, the basis for conducting the study, is that cancer of the larynx is more likely to be controlled with surgery than with radiation. We'll take as our test statistic the odds ratio in favor of surgery. The obtained odds of success for surgery, the ratio of controlled to not controlled cancers, was $(21/2) = 10.5$. The obtained odds of success for radiation was $(15/3) = 5$. The odds ratio in favor of surgery is the ratio of these two odds, that for surgery in the numerator,

$$ s = \frac{10.5}{5} = 2.1 $$

The odds of success following surgery were about twice those following radiation.

The question for the statistician is whether an odds ratio of this magnitude or larger could be a chance outcome, the result of the randomization of patients whose cancers would respond identically to either treatment.

Figures 13.15 and 13.16 show the text of a Resampling Stats script for comparing our odds ratio s with an approximation to the null reference distribution. This latter is based on 1,000 randomly chosen, independent rerandomizations of the patients between the two treatments.

Figure 13.15 shows the input of data and reference distribution size followed by the computation of s, the odds ratio favoring surgery. This s enters into the null reference distribution and, consequently, is counted among those that are at least as large as s.

```
copy 21 SS                  'Observed Surgical Successes
copy 2 SF                   'Surgical Failures
copy 15 RS                  'Radiation Successes and
copy 3 RF                   'Radiation Failures
copy 1000 REFSIZE           'Size of Null Reference Set
'
multiply SS RF NUM          'Numerator of Odds Ratio
multiply RS SF DEN          'Denominator
divide NUM DEN ORAT         'Odds Ratio for Observed Data
print ORAT
'  °
copy 1 TAIL                 '1st Reference Set Value Counts
subtract REFSIZE 1 MONTE    'Needed to Complete
'
```

Figure 13.15: *Part 1, Independent Rerandomization in Resampling Stats*

Figure 13.16 details the computation of a random sequence of s^H values, each then compared against s. The script uses the image of a coin to be flipped to determine the assignment of each patient. If the coin lands heads up (TOP takes the value 1), the patient is assigned to surgery, if tails the patient is assigned to radiation. In each rerandomization, the patients with controlled cancers are rerandomized first, followed by the rerandomization of patients with not controlled cancers.

Because the data set contains so few failures, there is the chance that all patients with not controlled cancers will be assigned to one treatment. If none of the not controlled cancers are assigned to surgery, the denominator of s^H will be zero. We cannot compute s^H in this situation, but we know its value ought to be very large. In particular, we can count it as an s^H that is at least as large as s.

One run of the script of Figures 13.15 and 13.16 found 23.1% of the s^Hs to be as large or larger than $s = 2.1$. A p-value of 0.231 offers little support for the superiority of surgery over radiation.

```
copy (0 1) COIN                    'A coin to flip
add SS RS SUC                      'No of Successes in Sample
add SF RF FAL                      'No of Failures in Sample
'
repeat MONTE                       'Execute this loop MONTE times
  copy 0 SS                        'Begin each with 0 in these
  copy 0 SF                        'categories
  repeat SUC                       'Go through the Successes
    shuffle COIN COIN              'Flip the coin
    take COIN 1 TOP                '50/50: TOP will be 0 or 1
    add SS TOP SS                  'TOP=1: Add Patient to Surgery
  end
  subtract SUC SS RS               'Other Successes to Radiation
  repeat FAL                       'Now, go through the Failures
    shuffle COIN COIN
    take COIN 1 TOP                '50/50: TOP will be 0 or 1
    add SF TOP SF                  'TOP=1: Add Patient to Surgery
  end
  subtract FAL SF RF               'Other Failures to Radiation
  if SF >0
    multiply SS RF NUM             'Compute odds ratio for this
    multiply RS SF DEN             'rerandomization of cases
    divide NUM DEN ORATX
    if ORATX >= ORAT               'Is this s(H) large?
      add TAIL 1 TAIL
    end
  end
  if SF =0                         'Cannot compute odds ratio
    add TAIL 1 TAIL                's(H) would have been large
  end
end
divide TAIL REFSIZE PROP           'Propor in ref set >= ORAT
print PROP                         'A directional p-value
```

Figure 13.16: Part 2, Independent Rerandomization in Resampling Stats

Restricted Rerandomization

For one reason or another, not all treatment options may be open to all the available cases. In this situation, we should block cases on the range of options available to them and use this block structure in randomization and rerandomization. Edgington (1995) provides an example that has been modified somewhat to provide the data of Table 13.3.

Four small and four large animals are randomized among drug dosages in this example. However, the small animals cannot receive dosages above 10. As a result, they were randomly assigned, two apiece, to the 0 and 10 levels. The four large animals were randomly assigned, one apiece, to levels 0, 10, 20, and 30.

Table 13.3:   *Testing for a Positive Linear Relation, Dosage and Response*

Cases	Drug Dosage			
	0	10	20	30
Small	7	8		
	10	11		
Large	12	16	14	18

The substantive hypothesis is that the size of the drug response should be positively correlated with the drug dosage. However, researchers anticipated that the size of the response will be systematically bigger for the large animals. To account for this possibility, the large animal responses should be adjusted downward before the responses are correlated with the dosages. Given that both groups were assessed at dosages of 0 and 10, we could adjust the large animal response to have the same average as the small animals at dosage 0, at dosage 10, or for the average of those two dosages. An argument for the latter is that the adjustment is based on the responses of more animals and, hence, might be more stable. This is the adjustment that we shall use.

On the way to computing s, we first find the average of the average response of the small animals to dosages of 0 and 10,

$$\frac{\left(\frac{7+10}{2}\right) + \left(\frac{8+11}{2}\right)}{2} = \frac{8.5 + 9.5}{2} = 9$$

and the average of the average response of large animals to those two dosages,

$$\frac{12 + 16}{2} = 14$$

The difference between these two, $14 - 9 = 5$, is the amount we will subtract from all large animal responses to adjust them to a common level with the small animal responses.

Table 13.4:   *Drug Dosages and Adjusted Responses*

Dosage:	0	0	10	10	0	10	20	30
Adj Resp:	7	10	8	11	7	11	9	13

Our test statistic, s, is the correlation between the drug dosages and the response scores, as given in Table 13.4. The final four response scores are the adjusted scores for the large animals. This correlation is $s = 0.6875$.

The null treatment-effect hypothesis is that a given animal's response would be the same at any of the dosage levels to which it might have been assigned. Following the study design, we can rerandomize the large and small animals among the dosage levels. The null hypothesis requires that we let their Table 13.3 response scores accompany them. To build a reference distribution for s we'll compute s^H in this rerandomized data set by the same rule we used to calculate s, adjusting the scores for the large animals before computing a correlation.

Figure 13.17 shows how we could use S-Plus to test our hypothesis of a linear correlation between drug dosage and (animal size adjusted) response scores.

The work is done here by our user-written function `adjcor`, which computes the correlation between `dosage` and `response` where the last four scores on the latter, for the large animals, are first adjusted. Including the qualifier `group=sizes` in the `bootstrap` function call ensures that the size blocks (1 = small and 2 = large) of `response` scores are rerandomized separately.

In the random series of 1,000 s^Hs formed in Figure 13.17, only 92 were as large or larger than our s. Adding s into the reference distribution, the estimated p-value for our test is $(93/1001) = 0.093$. The obtained correlation is not strong enough, in this small data set, to establish a linear relation firmly.

```
> dosage<-c(0,0,10,10,0,10,20,30)
> response<-c(7,10,8,11,12,16,14,18)
> sizes<- c(1,1,1,1,2,2,2,2)
> adjcor<- function(X){cor(dosage,c(X[1:4],X[5:8]-
+(mean(X[5:6])-mean(X[1:4]))))}
> adjcor(response)
[1] 0.6875
> rrefdis<-
+bootstrap(response,adjcor,group=sizes,sampler=samp.permute)
>length(rrefdis$replicates[rrefdis$replicates>=0.6875])
[1] 92
>93/1001
[1] 0.09290709

> choose(4,2)
[1] 6
> factorial(4)
[1] 24
> choose(4,2)*factorial(4)
[1] 144
```

Figure 13.17: *Testing for a Linear Dosage-Response Relation (S-Plus)*

We note, relative to the final lines in Figure 13.17, that there are only $4!/(2! \times 2!) = 6$ different ways in which the four small animals could be randomized, 2 apiece, into two dosage levels and $4! = 24$ ways in which the four large animals could be randomized, 1 apiece, into four dosage levels. As these randomizations take place independently, there are $6 \times 24 = 144$ different ways in which the eight animals might be assigned to different dosage levels. Our obtained randomization is one of those. With some effort, we could determine systematically how many of the 144 produce a value of s^H as large or larger than s and thus have an exact p-value. Our approximation of the p-value suggests that the number of arrangements is about 14.

Exercises

1. Reanalyze the larynx cancer data in Table 13.2 using the assumption that patients were randomized not independently but, rather, as a single block, 19 to one treatment and 23 to the other. How different are the results?

2. The data in Table 13.5 are from a study of the effect of caffeine on a finger-tapping task and are taken from Draper and Smith (1981). Thirty male college

student volunteers were trained at the task and then divided at random into three groups of 10 subjects. The groups received different amounts of caffeine (0, 100, or 200 mg). Two hours later, each subject performed the finger-tapping task and the tapping rate, the number of taps per minute, was recorded.

Table 13.5:   *Finger Tapping Rates and Caffeine Levels of Treatment*

Tapping Rates	
Treatment:	
0 mg Caffeine	242, 245, 244, 248, 247, 248, 242, 244, 246, 242
100 mg Caffeine	248, 246, 245, 247, 248, 250, 247, 246, 243, 244
200 mg Caffeine	246, 248, 250, 252, 248, 250, 246, 248, 245, 250

The study follows the completely randomized design; the full complement of 30 subjects was randomized among the three treatment levels as a single block. What is interesting in this study?

a. Did those administered caffeine really tap faster than those not administered caffeine?

b. Did those administered 200 mg caffeine really tap faster than those administered only 100 mg caffeine?

Both hypotheses are directional. Evaluate them by comparing each of two test statistics against null treatment-effect reference distributions.

3. Table 13.6 is adapted from Mead, Curnow, and Hasted (1993) although all three factors have been redefined. Four methods of teaching high school U.S. History (Methods A, B, C, and D) are to be compared. To control for potential differences among the participating schools, the schools were recruited from eight geographical regions. Within each region, four high

schools were chosen. This block of four schools was then randomly allocated, one apiece to the four teaching methods.

Table 13.6:   *Percent Historical Figures Identified by History Teaching Method*

Region:	1	2	3	4	5	6	7	8
Schools:								
1	A	C	B	D	D	B	C	A
	45.4	42.4	58.8	32.2	32.2	34.6	38.1	30.8
2	B	D	A	C	B	D	A	C
	33.4	38.6	24.8	36.4	47.6	44.0	27.2	44.9
3	C	A	D	B	A	C	B	D
	45.6	37.8	46.8	28.2	32.0	42.4	40.8	50.8
4	D	B	C	A	C	A	D	B
	42.7	41.6	45.8	30.4	34.0	39.0	35.8	39.3

Each cell in the table identifies both the method of instruction and the resulting score for a particular school. The score is the average percentage of historical figures correctly identified by students at that school.

Methods B, C, and D are new methods, thought to be superior to the older method A. Compare each of the new methods with the older one. Choose a test statistic for each comparison. Then, develop p-values for the three null hypotheses, basing each on rerandomizing the appropriate schools and accounting for the randomized blocks design.

Concepts 14

Random Treatment Sequences

In Concepts 12 and 13, we described a rerandomization approach to statistical inference in study designs where a set of available cases is randomly divided among two or more treatment groups. Each group then receives a distinct treatment, and we want to compare the effectiveness of one treatment with that of another. Because each treatment is delivered to a different group of cases, these treatment comparisons are said to be between cases comparisons.

Now we discuss designs that allow for treatment comparisons to be made within cases. The hallmark of these designs is that each case receives not one but all K treatments.

Between- and Within-Cases Designs

The research designs of Concepts 12 and 13 evaluate treatments by comparing the responses of different groups of cases, those assigned to different treatments. Thus, two potential sources are reflected in those treatment group differences: differential treatment effects and treatment-unrelated response differences among cases. By randomly assigning cases to treatment groups, perhaps after first blocking the cases, we hope to even out the differences among cases over the treatment groups. Despite the randomization, however, we are making comparisons between different cases. Variability among cases can make it difficult to detect a relatively modest treatment effect.

If all cases are observed at each treatment level, then we can observe treatment differences for individual cases. These differences cannot be influenced by differences between cases. Hence the attractiveness of the within-cases design. It cannot be a universal remedy, though, particularly where our cases are human or animal subjects. To employ the within-cases approach we must satisfy ourselves that

1. It is possible physically, sensible scientifically, and acceptable ethically, to expose cases to the proposed series of treatments and, in particular, that

2. The cases involved in the experiment will not respond differently to one treatment because of their exposure to another treatment.

Not all treatment comparison studies can pass this test. A case can be irrevocably changed by treatment. Even where it is feasible to administer more than one treatment to a case, it is important that the study design ensures sufficient separation between treatments that the effect of one does not carry over to another.

Randomizing the Sequence of Treatments

In the between-cases design, we randomly divided cases into treatment groups. This served both a design and an analysis purpose. On the design side, the random division militated against cases with certain characteristics piling up in one particular treatment group. On the analysis side, when our study involved available rather than sampled cases, the random division provided a statistical basis for causal inference, permitting us to conclude, or not, that an observed difference in response to treatment was caused by treatment differences and was not the chance result of the random grouping of treatment-indifferent cases.

For very much the same reasons, in the within-cases design, we randomly divide the cases into treatment sequence groups.

Treatment Position

A key idea in the within-cases design is the position of a particular treatment in the sequence of treatments received by a case. Most commonly, this sequence is either temporal or spatial.

Medical and psychological researchers frequently use a temporal sequence, with the treatments administered to the case one after another in time. Studies in which patients are moved from one drug to another are known as crossover studies.

A two-treatment, two-position crossover study is illustrated in Figure 14.1. Some cases are assigned to treatment sequence AB, receiving treatment A in period 1 (i.e., the first position) and then crossing over to receive treatment B in period 2. Other cases, assigned to sequence BA, begin on treatment B and crossover to treatment A.

The spatial sequence, where the several treatments are administered to different locations on the case, is more common in agricultural experiments. In such experiments, the case is a plot of land that is further subdivided to accommodate several distinct treatments, resulting in what is called a split-plot design.

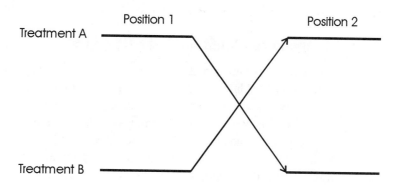

Figure 14.1: *A Two-Treatment, Two-Position Crossover Study*

Avoiding Position Treatment Confounding

Because every case receives all treatments, we don't have the risk of confounding cases with treatments. There is the possibility, however, of confounding treatment with position. If drug A is always administered first and drug B second, then we can never discover whether the apparent superiority of drug A is due to its chemistry or, simply, to its being administered first.

We could, of course, counterbalance the design, ensuring that half the patients in the study receive drug sequence AB and the other half receive sequence BA. Drug A would no longer have a position advantage. If we make the division into two treatment sequences at random, we also militate against the healthier cases being assigned, other than by chance, to one sequence and the less-healthy cases to the other.

Random division of cases among treatment sequences, then, provides protection against a confounding of treatment sequence with any pretreatment case characteristics.

Analysis of Sampled Case Designs

If the cases in our within-cases design are a random sample from a population of cases, then we will be interested in estimating population characteristics. For example, "What is the median difference in symptom-relief time between drugs A and B in this population of asthmatics?" We'd turn to our bootstrap inference techniques for answers to such population questions.

Analysis of Available Case Designs

If our cases are not a random sample but, rather, a set of available cases, then there is no population and, hence, no population characteristic to estimate. We were able to use the random division of available cases among treatment groups in the between-cases design as a basis for local causal inference, for inferring that among this set of cases, one treatment was superior to another. In the within-cases design we can use the random division of available cases among treatment sequences to this same end. That is the topic of the next section.

Causal Inference for Within-Cases Designs

In the between-cases design, we were able to test hypotheses about differences in treatment effectiveness by referring our treatment comparison statistic to a null reference distribution. This reference distribution was composed of values of the comparison statistic computed for all or for a very large number of rerandomizations of the cases and their response scores, rerandomizations that followed the original scheme for randomization and a null treatment-effect hypothesis. Most often, the between-cases null hypothesis stipulates that a case would have made the same response to some other treatment as it made to the one to which it was randomized. Under the null hypothesis, the response is a characteristic of the case and not of a particular treatment.

Null Treatment-Effect Hypothesis

We follow very much the same logic in testing treatment-effect hypotheses in the within-cases design. However, it is no longer sensible for us to postulate that a case would respond the same to two or more treatments. Each case responds to all treatments, so we know whether a case's responses are the same or different.

When we come to compose null treatment-effect hypotheses for within-cases designs, the position of a treatment, temporally or spatially, plays an important role. Our null hypothesis is that the case would have made the same response to two or more treatments, if those treatments were administered in the same position of the treatment sequence. Under the null hypothesis, then, the observed response is a characteristic of the position of that response and not of the treatment delivered at that position.

Here is a simple example. Case I has been assigned to treatment sequence AB and responds to the two treatments with the two scores, $(8, 15)$. The case's response to treatment A is 8, and to treatment B it is 15. The null hypothesis, that treatments A and B would evoke the same response if administered in the same position, means that the observed 8 and 15 are treatment-neutral; they are position responses. That is, 8 is this case's period 1 response to either of treatments A or B and 15 is its period 2 response to either of these treatments. Under the null hypothesis, this case, if assigned to treatment sequence BA also would have responded $(8, 15)$.

In a later section we extend this null treatment hypothesis to studies with more than two treatments.

Treatment Comparison Statistic

First, we'll need to develop a general method for obtaining a treatment comparison statistic:

1. Our n available cases are randomly divided into a number, G, of treatment sequence groups.

2. Each case then receives all K treatments in the assigned sequence.

3. Our substantive hypothesis is that a case's responses to treatments $a_1, \ldots, a_\ell$ will be greater than (different from) that case's responses to treatments $b_1, \ldots, b_m$.

4. For each case, we form a treatment comparison score, c_i, typically the difference between the average response to treatments $a_1, \ldots, a_\ell$ and the average response to treatments $b_1, \ldots, b_m$.

5. Our treatment comparison statistic, s, is an aggregate of the n treatment comparison scores, c_i, and this might be the sum, the mean, the median, or a trimmed mean of the c_is.

Our substantive hypothesis is about the responses of a case to each of two sets of treatments and each case provides a treatment comparison. These case-level comparisons then must be aggregated in some way to provide an overall comparison statistic.

Developing a Reference Distribution

Is our s large or extreme enough that we think it unlikely to be solely the result of our randomly dividing cases among G different treatment sequences?

To answer that question, we'll need to refer our s to a null reference distribution, just as we did in between-cases studies. This distribution consists of the values of our treatment-comparison statistic when computed for each of the permissible rerandomizations of our n cases. As in between-cases studies, the permissible rerandomizations are determined by the original randomization strategy and by the current null hypothesis. Any rerandomization must be consistent with:

1. The original mechanism for the random division of our n cases among G treatment sequences

2. The null treatment-effect hypothesis that the observed differences in response to the $\ell + m$ treatments are positional differences only and not treatment differences

We demonstrate with an example how these two factors together restrict the way in which a case and its response scores can be rerandomized. In this example:

1. Each case receives $K = 3$ treatments, A, B, and C.

2. The n available cases are randomly divided, in equal numbers, among $G = 6$ treatment sequences: ABC, ACB, BAC, BCA, CAB, and CBA.

3. We want to test the substantive hypothesis that treatment A should produce higher response scores than treatment C.

We compute the treatment comparison score for the i-th case, c_i, by subtracting that case's treatment C response from its treatment A response. To illustrate, we'll assume that the i-th case, randomly assigned to treatment sequence BCA, responds to the three treatments with scores of

$$(12, 15, 10)$$

The treatment comparison score for this case is $c_i = 10 - 15 = -5$.

The null treatment-effect hypothesis associated with our substantive hypothesis is that the i-th case's responses to treatments A and C would have been the same if the two treatments had been administered to that case in the same position. Because the two treatments were administered to the i-th case in different positions, however, any observed difference in treatment responses is, under this null hypothesis, solely a difference in positional responses for that case.

What restrictions does this null hypothesis place on the rerandomization of the i-th case? The six initially possible treatment sequences give us these rerandomizations of the i-th case and its sequence of responses:

$$(A = 12, B = 15, C = 10) \quad (A = 12, C = 15, B = 10)$$

$$(B = 12, A = 15, C = 10) \quad (B = 12, C = 15, A = 10)$$

$$(C = 12, A = 15, B = 10) \quad (C = 12, B = 15, A = 10)$$

Of these six, however, those on the top and bottom rows are not permissible rerandomizations under the null hypothesis. These four rerandomizations would assign the treatment B response of 12 either to treatment A (top row) or to treatment C (bottom row). Although the treatment A and C scores are interchangeable under the null hypothesis, neither may be replaced by a treatment B score.

On rerandomization, our i-th case can be assigned only to treatment sequences BAC or BCA. A similar restriction on rerandomization holds true for any other case in this study; only two of the six treatment sequences are permissible rerandomizations under our null hypothesis.

In general, treatment responses involved in the current null hypothesis cannot be exchanged, on rerandomization, with treatment responses that are not involved. As a result, any rerandomization must assign the two sets of responses to the two sets of positions initially occupied by the corresponding sets of treatments. Preserving this separation of the two sets of responses into their respective initial sets of positions induces a symmetry between sets of initial treatment sequences and their permissible rerandomizations.

For example, we've established that, for our hypothesis, any case initially randomized to sequence BCA can be rerandomized only to sequences BAC or BCA. That same argument would also establish that a case initially assigned to sequence BAC also can be rerandomized only to sequences BAC or BCA. Putting these two findings together, we can say that cases initially assigned either to sequences BCA or BAC can be rerandomized only to one of these same sequences. These two sequences form one of the symmetric partitions of the initial set of six treatment sequences. There are two other partitions: ABC and CBA form one, and CAB and ACB form the other.

For some other treatment comparison hypothesis, say between treatments A and B, the six sequences would again form three symmetric partitions, but these would involve different pairings of the sequences.

The restrictions placed on rerandomization by our null treatment-effect hypotheses must be considered when we design our within-cases study. Otherwise, we might not be able to test those hypotheses. Take the example of our three-treatment study. Let's assume that we randomly had divided cases among just three treatment sequences, say, ABC, CAB, and BCA, rather than among all six possible sequences. We would no longer be able to rerandomize the assignment of any case if our treatment hypothesis involved, as it did earlier, only two of the three treatments. We have already seen, as an example, that a case initially assigned to sequence BCA could not be rerandomized to either sequence ABC or CAB under the null hypothesis that treatments A and C are equally effective.

Figure 14.2 describes how we might evaluate a proposed within-cases study design to ensure that we will be able to test each treatment comparison hypothesis of interest to us.

---

1. Assign integers from 1 through G to the proposed initial treatment sequences.

2. Set the treatment sequence counter, g, to one; $g = 1$.

3. For the g-th treatment sequence and the current null treatment-effect hypothesis, count the number of initial treatment sequences that are permissible rerandomization sequences. If this number is less than two, the current hypothesis is not testable; there are too few initial treatment sequences.

4. Increase g by one. If g exceeds G go to step 5. Otherwise repeat steps 3 and 4.

5. Repeat steps 2 through 4 for each proposed null treatment-effect hypothesis to be tested.

Figure 14.2: *Checking the Design of a Within-Cases Study*

As a general principle, we will be able to test the greatest number of treatment comparison hypotheses if we can design the study so that G, the number of treatment sequences, is as large as possible.

If each case is to receive K treatments, then the number of possible sequences in which the treatments can be administered is $K!$. For $K = 2, 3, 4, 5$ these maximum values for G are $2, 6, 24, 120$.

Practical considerations, however, might prevent us from using all possible treatment sequences. We might not have as many as 120 cases available to us. Or certain treatment sequences might be impossible or undesirable. Whenever we reduce G below its maximum, however, we lose the ability to test certain treatment-comparison hypotheses.

Sequence Randomization Strategies

We have the same techniques for dividing our n cases among G treatment sequences as we used in Concepts 13 for dividing n cases among K treatments.

Independent Randomization of Cases

Each case might be initially assigned randomly and independently to one of the G treatment sequences. Figures 13.1 and 13.2 provide algorithms for carrying out these initial assignments. All that is required to make them useful in the present context is to replace references to treatment groups 1 through K with references to treatment sequences 1 through G.

Because every case receives all treatments, balancing the numbers of cases assigned to the different treatment sequences might not be as important as is the corresponding balancing of treatment group assignments in the between-cases designs.

On rerandomization, to test a particular treatment-effect hypothesis, each case is independently randomized again, but the range of treatment sequences open to the case, as noted earlier, is a function of the null treatment-effect hypothesis.

Complete Randomization

Just as in between-cases studies, our cases can be randomly assigned, now to treatment sequences, not one case at a time but as a single block of n cases. The result is a completely randomized within-cases study.

Figure 14.3 describes an algorithm for distributing n cases among G groups. The number of cases to be assigned to each group is established in advance,

$n = n_1 + \ldots + n_G$. Typically, these group sizes will be identical, $n_g = n/G$ for $g = 1, \ldots, G$.

---

1. Create a vector of group sizes, with elements $n1, n2, \ldots, nG$.

2. Create a vector of case identifications, labels from 1 through n.

3. Shuffle the case identification vector.

4. Initialize the start index at 1.

5. Initialize the stop index at 0.

6. Initialize the group indicator at 1, $g = 1$.

7. Select the g-th group size as ng.

8. Increase the stop index by ng.

9. Assign the cases with labels in positions from the start index through the stop index to the g-th group.

10. Increase the start index by ng.

11. Increase the group indicator by 1, $g = g + 1$.

12. If g is greater than G, all assignments have been made. Otherwise, repeat steps 7 through 11.

Figure 14.3:   *Randomly Allocating n Cases Among G Sequences*

On rerandomizing cases subject to a treatment comparison hypothesis, this same algorithm could be used. If that hypothesis involves fewer than all G of the treatment sequences, the algorithm needs revision along the following lines:

1. The algorithm is executed separately for each symmetric partition of the G treatment sequences.

2. The vector of case identifications contains the labels only of those cases in the current partition.

3. The vector of group sizes contains the sizes of the treatment sequence groups belonging to the current partition.

That is, steps 1 and 2 of the Figure 14.3 algorithm would be made specific to each symmetric partition of the treatment-sequence groups for a particular treatment-comparison hypothesis.

Additional treatment-comparison hypotheses are likely to produce different symmetric partitions of the G treatment sequences. The random allocation algorithm, though, is easily adapted to each set of partitions.

Randomized Blocks

Our third major case randomization strategy is to independently distribute each of J blocks of cases among the alternative treatment sequences at random. These blocks, as in Concepts 13, bring together cases having a common level on some attribute researchers believe might affect the response to treatment. For example, patients in a drug trial might be blocked by originating clinic or date of entry into study.

The value of blocking is that cases in each block, sharing the same amount of the blocking attribute, are evenly distributed among treatment sequences. Thus, there can be no concentration, even by chance, of cases from a certain clinic, for example, in a particular sequence. To ensure complete balance, the number of cases in each block should be an integer multiple of G, the number of treatment sequences.

We can use the algorithm of Figure 14.3 to assign cases by block as well as completely:

> 1. For the initial randomization, the algorithm is implemented separately for each block of cases.
>
> 2. For rerandomizations subject to any hypothesis-induced restrictions, the algorithm is applied separately to each symmetric partition within each block of cases.

Applications 14

Rerandomizing Treatment Sequences

Concepts 14 introduces within-cases designs in which individual cases receive two or more treatments. This allows us to directly compare the response of the individual case to different treatments. By randomizing the sequence of treatments, we gain a measure of experimental control, and where our cases are an available set, rather than a random sample, we create a probabilistic basis for the statistical evaluation of comparisons of treatment effectiveness. Here, we see how our statistical packages can be used to analyze some of the more popular within-cases designs.

Analysis of the AB-BA Design

An important within-cases design is the two-treatment, two-period crossover or AB-BA design. Cases are randomly allocated among two treatment sequences, treatment A followed by treatment B, or treatment B followed by treatment A. The study's goal is to determine whether one treatment is better than the other is. Using two sequences helps guard against confusing the effectiveness of a treatment with the effectiveness of a period of treatment. Randomly allocating cases between the two sequences minimizes the likelihood of the more (or, less) responsive cases receiving a particular sequence of treatments. Where the cases are not a random sample but, rather a collection of available cases, we have a basis for evaluating the chance versus causal nature of any observed treatment difference.

Table 14.1 describes the results of one AB-BA study. The data are taken from Senn (1993) and are for a study "comparing the effects of a single inhaled dose of 200 μg salbutamol, a well-established bronchodilator, and 12 μg formeterol, a more recently developed bronchodilator." The cases were children aged 7 to 14 years with moderate to severe asthma. The children were randomized to one of two groups. Those in the for-sal group were administered formeterol on a first clinic visit and salbutamol on a second visit, the two visits separated by a washout period of at least 24 hours. The order of the two medications' administration was reversed for those children in the sal-for group. Eight hours following the administration of either bronchodilator, a measurement was made of the child's peak expiratory flow (PEF) in liters/minute. Larger PEF values signal a better outcome.

Table 14.1: *Peak Expiratory Flow Measured 8 Hours Postmedication*

		PEF			
Order	Pt. No.	Period 1	Period 2	Difference	Sum
for-sal	1	310	270	40	580
	4	310	260	50	570
	6	370	300	70	670
	7	410	390	20	800
	10	250	210	40	460
	11	380	350	30	730
	14	330	365	−35	695
sal-for	2	370	385	−15	755
	3	310	400	−90	710
	5	380	410	−30	790
	9	290	320	−30	610
	12	260	340	−80	600
	13	90	220	−130	310

The patient numbers reported in Table 14.1 reflect the result of the randomization of the children to the two treatment orders. Notice that patient number 8 is missing. We shall return to this point when deciding on a rerandomization strategy for the analysis. In addition to the PEF scores for the two periods, Table 14.1 displays a difference and sum for each patient. These are defined more completely as

$$\text{PEF Period Difference} = \text{Period 1 PEF} - \text{Period 2 PEF}$$

and

$$\text{PEF Period Sum} = \text{Period 1 PEF} + \text{Period 2 PEF}$$

emphasizing their origins in PEF scores for the two periods. These two derived patient scores will be used in the construction of two test statistics.

Differential Effectiveness Measure

The study was motivated by a desire to learn whether formeterol, the newer drug, might provide better bronchodilation than salbutamol, the established drug. Accordingly, an obvious basis for a test statistic reflecting the anticipated superiority of formeterol is the collection of differences,

$$[PEF(formeterol)_i - PEF(salbutamol)_i], \; i = 1, 2, \dots, 13$$

We'll compute such a test statistic from the PEF Period Differences of Table 14.1 as

$$s_{trtw} = \left(\frac{1}{n_{AB} + n_{BA}} \right) \left[\sum_{i=1}^{n_{AB}} PerDiff_i - \sum_{i=1}^{n_{BA}} PerDiff_i \right]$$

where n_{AB} is the number of patients randomized to the AB (for-sal) treatment order and n_{BA} is the number randomized to the BA (sal-for) sequence. A positive PEF Period Difference reflects a formeterol superiority for a patient in the first treatment order whereas a negative PEF Period Difference shows superior bronchodilation under formeterol for a patient in the second order, hence the signs of the PEF Period Differences for the sal-for patients must be reversed before they are pooled with those of the for-sal patients. We accomplish this by subtracting the sal-for group's period differences from those of the for-sal group.

Differential Carryover Effect Detection

s_{trtw} is the average $[PEF(formeterol)_i - PEF(salbutamol)_i]$, considering the responses of all patients to both medications and is a natural statistic to use for the comparison. Despite the provision of a washout period between treatments, however, investigators who use the crossover design worry that the results might be contaminated by a carryover effect. If the first treatment enhances or reduces the effectiveness of the second treatment, we say the first treatment has a carryover effect.

The carryover effects of the two treatments might be the same or they might differ. If the carryover effects are the same, for example, either treatment when administered first enhances the effectiveness of the second treatment by the same amount, then the carryover, though present, might not invalidate the use of the s_{trtw} statistic to assess the superiority of one drug to another. An argument in support of this follows.

If the two treatments have differential carryover effects, however, the design does not provide for this kind of balancing-out. For example, if A when in the first period enhances the effectiveness of B, and prior exposure to B has no influence on the subsequent effectiveness of A, then a statistic such as s_{trtw} would overstate the effectiveness of B relative to A.

Clearly, we would like to be able to detect differential carryover. That is the role of a second test statistic, one based on the period-sum scores. Each case's period-sum score can be thought to be made up of three components or effects. That is, the score can be expressed as

$$\text{Period Sum}_i = \text{Treatment A}_i + (\text{Treatment B}_i + \text{Carryover A}_i)$$

if the treatment order for the i-th case is AB and as

$$\text{Period Sum}_i = \text{Treatment B}_i + (\text{Treatment A}_i + \text{Carryover B}_i)$$

if the i-th case was assigned treatment order BA. The parenthesized terms on the right sides of these equations reflect our belief that the period 2 contribution is a sum of the effect the second treatment would have if it were not influenced by the first treatment and a carryover effect from that first treatment.

The two period sums also can be expressed as

$$\text{Period Sum}_i = (\text{Treatment A}_i + \text{Treatment B}_i) + \text{Carryover A}_i$$

for treatment order AB and as

$$\text{Period Sum}_i = (\text{Treatment B}_i + \text{Treatment A}_i) + \text{Carryover B}_i$$

for treatment order BA.

Because of the random assignment of cases to treatment orders, the average of the (Treatment A_i + Treatment B_i) components should be about the same for the two treatment order groups. If there is no differential carryover effect, the average of the Carryover A_i components for the AB group should be about the same as the average of the Carryover B_i components for the BA group. As a result, in the absence of a differential carryover, the statistic

$$s_{\text{cov}} = \text{abs}\left[\left(\frac{1}{n_{\text{AB}}}\right)\sum_{i=1}^{n_{\text{AB}}}\text{PerSum}_i - \left(\frac{1}{n_{\text{BA}}}\right)\sum_{i=1}^{n_{\text{BA}}}\text{PerSum}_i\right]$$

ought to take a value close to zero. Equivalently, if this absolute value of the difference in mean period-sums is large, we should suspect a differential carryover effect.

An Analysis Strategy

A differential carryover effect casts doubt on the validity of the period 2 response data; they are contaminated by inconsistent lingering effects of the period 1 treatments. Confronted with a suspiciously large s_{cov}, it is good practice to abandon s_{trtw} as the assessment of treatment difference because it is based on both period 1 and period 2 data. Instead, we can make the treatment comparison using only the uncontaminated period 1 data using an alternative statistic,

$$s_{\text{trtb}} = \left(\frac{1}{n_{\text{AB}}}\right)\sum_{i=1}^{n_{\text{AB}}}\text{Period }1_i - \left(\frac{1}{n_{\text{BA}}}\right)\sum_{i=1}^{n_{\text{BA}}}\text{Period }1_i$$

the difference in mean response to the two treatments when administered in the first period of the sequence. s_{trtw} gives us a within-cases assessment of differential treatment effectiveness, whereas s_{trtb} provides only a less-sensitive between-cases evaluation.

The possibility of a differential carryover effect suggests the following flow in the analysis of AB-BA designs:

1. Compute and evaluate the statistic s_{cov}.

2a. If s_{cov} is satisfactorily small, compute and evaluate the statistic s_{trtw}.

2b. If s_{cov} is suspiciously large, compute and evaluate the statistic s_{trtb}.

At each stage, the evaluation of a statistic takes the form of relating the value of that statistic to its null reference distribution. Because all three reference distributions are based on rerandomizing cases between the two treatment orders, it might be easier to form all three reference distributions in a single run, rather to proceed sequentially.

The three statistics have all been defined in terms of arithmetic means. This is standard. You might choose, however, to replace the mean with another indicator of the size of a group of scores, perhaps a trimmed mean. Similarly, the differences in the A and B scores for individual subjects are used to compute s_{trtw}. Some treatments might be better compared, substantively, by aggregating and summarizing ratios rather than differences. Or, the carryover effect might be thought to have a multiplicative influence, rather than an additive one, on the effect of the period 2 treatment, suggesting the use of period-products, rather than period-sums in the definition of s_{cov}. The question, as always in the analysis of scientific data, is what makes best sense in terms of the science?

Example of a Rerandomization Analysis

We should be ready to apply these ideas to the analysis of the data of Table 14.1. Computing s_{trtw}, s_{cov}, and s_{trtb} is straightforward enough. But to develop a reference distribution for any one of these statistics requires that we know enough about how the children were randomized in the actual study to organize the correct rerandomizations. Senn (1993) tells us only that "the children were randomized to one of two sequence groups." How was this accomplished? We don't know. Here are two possibilities:

1. Each child was independently randomized. This just happened to result in seven children in the for-sal sequence and six in the sal-for sequence. It could have been six and seven or eight and five or five and eight or a result even less balanced between the two sequences.

2. A decision was taken at the outset to randomize seven children to for-sal and six to sal-for.

Were we to choose one of these possibilities, we would then rerandomize in exactly that same way. The presence of a patient numbered 14 in the study and the absence of one numbered 8, however, suggests a third randomization scenario:

3. Fourteen children were recruited into the study. They were randomized seven to for-sal and seven to sal-for. Patient number 8, who was randomly assigned to the sal-for sequence, did not complete the study.

Though we cannot know it is the correct choice, we'll use this third possibility in arranging our rerandomizations. These rerandomizations will account for the presence of a patient number 8, a patient whom, we further assume, also would not have completed the study if randomized instead to the for-sal sequence.

Figure 14.4 is the log of an S-Plus session in which we investigated first the possibility of a differential carryover effect and, not having found one, then went on to test for a formeterol advantage over salbutamol.

Note that in the entry of the data the S-Plus missing data indicator (the characters NA) was used to supply data for patient number 8. Subsequently, the analysis assumes that seven patients are randomized to each of the two sequences.

The user-supplied function scov, described in Figure 14.4, calculates the differential carryover statistic,

$$s_{\text{cov}} = \text{abs}\left[\left(\frac{1}{n_{\text{AB}}}\right)\sum_{i=1}^{n_{\text{AB}}}\text{Per Sum}_i - \left(\frac{1}{n_{\text{BA}}}\right)\sum_{i=1}^{n_{\text{BA}}}\text{Per Sum}_i\right]$$

accounting for the noncontributing patient—n_{AB} and n_{BA} are the numbers of patients in the two sequences that completed the study, supplying responses to both treatments.

```
> per1
 [1] 310 310 370 410 250 380 330 370 310 380   NA 290 260   90
> per2
 [1] 270 260 300 390 210 350 365 385 400 410   NA 320 340 220

> scov<-function(X)
+{abs(mean(X[1:7],na.rm=T)-mean(X[8:14],na.rm=T))}

> scov(per1+per2)
[1] 14.40476

> abtest1<-bootstrap(per1+per2,scov,sampler=samp.permute)
> length(abtest1$replicates[abtest1$replicates>=14.40476])
[1] 870
> 871/1001
[1] 0.8701299

> strtw<-function(X){(1/length(na.omit(X)))*
+(sum(X[1:7],na.rm=T)-sum(X[8:14],na.rm=T))}

> strtw(per1-per2)
[1] 45.38462

> abtest2<-bootstrap(per1-per2,strtw,sampler=samp.permute)
> length(abtest2$replicates[abtest2$replicates>=45.38462])
[1] 1
> 2/1001
[1] 0.001998002
```

Figure 14.4: *Assessment of Carryover and Treatment Effects (S-Plus)*

The differential carryover effect statistic takes the value $s_{\text{cov}} = 14.40$ liters/minute for these data. Is that a large value, suggesting a differential carryover effect?

Our null hypothesis for testing s_{cov} is that of no differential carryover. No differential carryover means that, on the average, the period-sum scores are the same size for the AB treatment sequence as they are for the BA sequence. In turn, this means that, under the null hypothesis, we can rerandomize cases and their period-sum scores between the two treatment sequences. We do so in the S-Plus interaction of Figure 14.4 to build a Monte Carlo null reference distribution, as `abtest1$replicates`, for our s_{cov}. Fully 87% of this reference distribution consists of values at least as large as 14.40. Our data provide no evidence, then, of a differential carryover effect.

Had s_{cov} been larger than we could attribute to the chance alignment of patients with treatment sequences, we would have had to abandon the within-cases assessment of the two drugs and made a between-cases comparison, using only the period 1 data. That did not happen, and we were able to continue with the within-cases treatment comparison. The user-supplied function `strtw`, also described in Figure 14.4, computes the within-cases differential treatment effectiveness measure,

$$s_{\text{trtw}} = \left(\frac{1}{n_{\text{AB}} + n_{\text{BA}}} \right) \left[\sum_{i=1}^{n_{\text{BA}}} \text{Per Diff}_i - \sum_{i=1}^{n_{\text{BA}}} \text{Per Diff}_i \right]$$

with the `per1` and `per2` PEF scores arranged so that the period-difference scores (period 1 − period 2) will be aggregated to produce a positive s_{trtw} if there was better bronchial performance under formeterol.

For the study, $s_{\text{trtw}} = 45.38$ liters/minute. Is that large enough to establish, for these patients, a formeterol advantage?

Our null hypothesis for the within-cases treatment comparison is that a case's period-difference score would be the same, whether randomized to treatment order AB or to treatment order BA. Under this null hypothesis we can rerandomize cases and their period-difference scores between the two orders to build a null reference distribution for s_{trtw}. When we did this, in Figure 14.4, only one of 1,000 rerandomizations of the patients produced a value as large as 45.38. We approximate the p-value for our treatment comparison statistic, then, at $p = 2/1001 = 0.002$. The result provides strong evidence that the observed superiority of formeterol is not the result of the random assignment of patients to treatment sequences. Rather, formeterol was more effective than the older drug, salbutamol.

Sequences of $K > 2$ Treatments

With two levels of treatment, only two treatment sequences are possible, and there is a single treatment comparison. With $K > 2$ levels of treatment, there are $K!$ possible treatment sequences. Depending on the focus of our study, we might have to make many treatment comparisons.

If we want to ensure that all the treatment comparisons that might be of interest to us can be made, we must have all $K!$ treatment sequences represented in our design. If we want all our comparisons to be made with the same efficiency, we should randomize the same number of cases to each treatment sequence.

In the following example, volunteer cases were randomized among sequences of three treatments. We'll see how it is helpful to be able to reorganize the resulting data for each treatment comparison of interest. In this design, more cases were randomized to some treatment sequences than to others, so we'll see how this restricts our statistical inferences.

We now consider a study of drowsiness as a side-effect of antihistamines as reported in Hand et al. (1994, Data Set 11) and reproduced as Table 14.2.

Each of three drugs

  A: Meclastine   B: Placebo   C: Promethazine

was administered, on separate days, to nine volunteers. Promethazine, at the time of the study, was an established antihistamine, known to cause drowsiness. Meclastine was a new antihistamine. The study was motivated, we shall assume, by two questions:

  1. Does the new drug meclastine (treatment A) produce less drowsiness than the old drug promethazine (treatment C)?

  2. Does meclastine (treatment A) produce more drowsiness than the placebo (treatment B)?

Drowsiness as a side-effect was assessed, six hours postmedication, by having subjects observe a light source flickering on and off at a changing rate and ascertaining the slowest flicker rate (per minute) at which the light is perceived to have fused into a steady light. This rate is referred to as the flicker-fusion frequency (FFF) for the subject. Smaller FFF values are symptomatic of greater drowsiness.

Table 14.2:   *Subjects Randomized to Rows of Three 3 × 3 Latin Squares*

Flicker-fusion Frequencies, 6 hr. Postmedication			
Subject No.:	Day 1	Day 2	Day 3
1	31.25 (A)	31.25 (C)	33.12 (B)
2	25.87 (C)	26.63 (A)	26.00 (B)
3	23.75 (C)	26.13 (B)	24.87 (A)
4	28.75 (A)	29.63 (B)	29.87 (C)
5	24.50 (C)	28.63 (A)	28.37 (B)
6	31.25 (B)	30.63 (A)	29.37 (C)
7	25.50 (B)	23.87 (C)	24.00 (A)
8	28.50 (B)	27.87 (C)	30.12 (A)
9	25.13 (A)	27.00 (B)	24.63 (C)
Medications — A: meclastine, B: placebo, C: promethazine			

Table 14.2 gives the FFFs for each subject for the three treatments and the treatment order.

The nine subjects were randomized to treatment order in this study by the use of latin squares. The latin square is a $K \times K$ array of K treatment labels that follows these two rules:

1. Each treatment label appears exactly once in each row

2. Each treatment label appears exactly once in each column

As each subject was to receive three treatments, the researcher started with a particular 3×3 latin square. Based on the data reported, we can take it to be this one:

A B C
C A B
B C A

Each row of the square prescribes a treatment order. If only three subjects were to be used, they could be randomized to the three rows. To include nine volunteers in the study, however, nine rows or treatment orders were needed. The nine rows were created by the researcher from the initial latin square in the following way:

1. The columns of the initial latin square were randomly permuted, column 2, followed by column 3, followed by column 1, to give a new latin square with rows identified as R_1, R_2, and R_3.

2. The columns of the initial latin square were randomly permuted a second time, column 3, followed by column 2, followed by column 1. This gives another latin square with rows identified as R_4, R_5, and R_6.

3. The columns of the initial latin square were randomly permuted a third time, column 1, followed by column 2, followed by column 3. This gives a latin square with rows identified as R_7, R_8, and R_9.

The nine rows are displayed below. Each describes a treatment order to be assigned, at random, to a subject in the study.

The columns labeled D_1, D_2, and D_3 refer to the days on which the drugs are to be administered.

The nine volunteers, S_1 to S_9, were randomly paired with the nine preselected treatment orders, R_1 to R_9. The results of this randomization and the FFF data subsequently collected appear as Table 14.2.

	D_1	D_2	D_3
R_1	B	C	A
R_2	A	B	C
R_3	C	A	B
R_4	C	B	A
R_5	B	A	C
R_6	A	C	B
R_7	A	B	C
R_8	C	A	B
R_9	B	C	A

We return now to the first of our two substantive questions,

1. Does the new drug meclastine (treatment A) produce higher FFFs (less drowsiness) than the old drug promethazine (treatment C)?

To answer this question we will compare the obtained   [Treatment A − Treatment C]$_i$ FFF differences, aggregated over subjects, with the rerandomization reference set for this statistic obtained under the null treatment-effect hypothesis that the two treatments are equally influential on FFF. Under that null hypothesis, any observed [Treatment A − Treatment C]$_i$ FFF difference reflects only the difference in the positions these two treatments occupied in the order of administration for the i-th case.

For ease of explanation, we'll aggregate the FFF differences by taking the average of the nine [Treatment A − Treatment C] FFF differences,

$$s_{\mathrm{AC}} = (1/9)\sum_{i=1}^{9}[(\mathrm{FFF}|\mathrm{A})_i - (\mathrm{FFF}|\mathrm{C})_i]$$

Table 14.3:   *FFF Subjects Grouped by Treatment Sequence*

Flicker-fusion Frequencies, 6 hr. Postmedication			
Sequence:	Day 1	Day 2	Day 3
1 (ACB)	31.25	31.25	33.12
2 (CAB)	25.87	26.63	26.00
2 (CAB)	24.50	28.63	28.37
3 (CBA)	23.75	26.13	24.87
4 (ABC)	28.75	29.63	29.87
4 (ABC)	25.13	27.00	24.63
5 (BAC)	31.25	30.63	29.37
6 (BCA)	25.50	23.87	24.00
6 (BCA)	28.50	27.87	30.12
Medications — A: meclastine, B: placebo, C: promethazine			

We could have used a trimmed mean or some other aggregate difference score instead of the mean. To organize the computation of s_{AC} from the experiment and of s_{AC}^{H} from each rerandomization of the cases, as well as to facilitate the rerandomizations, it will be helpful to rearrange the data, grouping the cases by treatment sequence.

Table 14.3 shows the same data as make up Table 14.2 but with the rows organized by treatment sequence. The case numbers have been replaced by a numeric code for the sequence, from 1 to 6, to which that case was randomized. Some sequences (e.g., CAB) appear twice whereas others (e.g., ACB) appear only once. We shall return to this point later.

In Table 14.3, the treatment sequences have been arranged in pairs, ACB and CAB, CBA and ABC, and BAC and BCA. These three pairs of sequences form, in the vocabulary of Concepts 14, the symmetric partitions of the cases, relative to the comparison of treatments A and C. Under the null hypothesis that, if administered at the same position in the sequence, treatments A and C would have the same effect on a case (though treatment B might have a different effect), the cases originally randomized to sequences ACB or CAB can only be rerandomized among those same sequences. A similar restriction holds for the other two partitions.

Each group of three cases, each symmetric partition, can be shuffled separately as part of a rerandomization of the cases under the null hypothesis.

Let's see one way in which SC could handle the required computations and rerandomizations.

Figure 14.5 displays the input of the data from Table 14.3. A matrix, FFF, is created containing the following data:

Column 1:   An integer identifying one of the six possible sequences
              of the three treatments—coded as in Table 14.3

Columns 2–4: Integers indicating the positions of treatments A, B, and C in
              this treatment sequence, for example, treatment sequence 1
              (ACB) is coded as 1 3 2

Columns 5–7: FFF scores for positions (days) 1, 2, and 3 in the
              sequence

Columns 2–4 are then extracted as the vectors apos, bpos, and cpos, and columns 5–7 are copied to a second matrix, M.

The vector gsiz identifies the number of consecutive rows of M associated with each of the six unique treatment sequences. In turn, the vector ssiz shows the number of consecutive rows of M associated with each symmetric partition of the R_is. Here each `ssiz` element includes two consecutive `gsiz` elements.

```
> matrix(FFF,9,7)
> read(FFF)
1 1 3 2 31.25 31.25 33.12
2 2 3 1 25.87 26.63 26.00
2 2 3 1 24.50 28.63 28.37
3 3 2 1 23.75 26.13 24.87
4 1 2 3 28.75 29.63 29.87
4 1 2 3 25.13 27.00 24.63
5 2 1 3 31.25 30.63 29.37
6 3 1 2 25.50 23.87 24.00
6 3 1 2 28.50 27.87 30.12
      63

> vector(apos,9,bpos,9,cpos,9)
> getcol(FFF,apos,2)
> getcol(FFF,bpos,3)
> getcol(FFF,cpos,4)
> mlimit(FFF,1,rows(FFF),5,7)
> M= FFF
> mdelimit(FFF)
> gsiz:= 1,2,1,2,1,2
> ssiz:= 3,3,3
```

Figure 14.5: *Part 1, SC Rerandomization and Hypothesis Tests, FFF Study*

The rows of FFF (and, hence, of M) were arranged keeping our treatment A versus treatment C hypothesis in mind; the symmetric partitions are those appropriate to the rerandomizations needed to compare treatments A and C.

The data have been arranged to facilitate the use of two SC procedures, the standard `shuffl_k` procedure for the formation of a reference set, and a user-written procedure `accontr` to be called by `shuffl_k` for computing the treatment comparison statistic. The text of `accontr` appears in Figure 14.6.

The procedure `accontr` extracts two values from each row of an input matrix 1 and finds the difference between them. The positions of the elements to be extracted from the i-th row and the order in which they enter the subtraction is determined by the i-th rows of the vectors `apos` and `cpos`. If 1 is the matrix M, the effect is to subtract the treatment C FFF score from the treatment A FFF score for each case. The average of these differences is returned as the first element of the input vector 3. The procedure `accontr` will be used by `shuffl_k` to compute s_{AC} and s_{AC}^H.

```
proc accontr(){
  local(MM,n,ac,i)
  MM= $1$
  n=rows(MM)
  ac= 0
  i= 0
  repeat(n){
    i++
    ac= ac + (MM[i][apos[i]]-MM[i][cpos[i]])
  }
  $3$[1]= ac/n
}
```

Figure 14.6: *Part 2, SC Rerandomization and Hypothesis Tests, FFF Study*

In Figure 14.7, we apply `accontr` to M and find $s_{AC} = 1.003$. We then call `shuffl_k` with the arguments `accontr`, M, `gsiz`, and `ssiz` and request a Monte Carlo reference distribution of 1,000 scores, s_{AC} plus s_{AC}^H computed in each of 999 rerandomizations of the cases.

```
> vector(out,1)
> proc accontr(){}
> accontr(M,gsiz,out)
> out
  1.0033333

> shuffl_k(accontr,1,M,gsiz,ssiz,999,RD)
    Full Sample Statistics (theta):   1.0033333

P-values for Statistic Number:  1
    Percent LE theta[ 1 ]:   100
    Percent GE theta[ 1 ]:   6.9
```

Figure 14.7: *Part 3, SC Rerandomization and Hypothesis Tests, FFF Study*

Because our scientific hypothesis is directional, that FFF scores should be higher under treatment A (the new drug, meclastine) than under treatment C (the old drug, promethazine), we want to know the proportion of the reference distribuion that is as large as s_{AC}. Our approximate p-value is p = 0.069.

To test our second, treatment A versus treatment B, hypothesis, we must rearrange the rows of FFF to form the symmetric partitions for that hypothesis. We'll also have to revise accontr to compute a different statistic, s_{BA}.

Figure 14.8 describes the rearrangement of the R_is.

```
> FF2= FFF
> ab:= 1,8,9,4,2,3,7,5,6
       9
> morder(FF2,ab)

> getcol(FF2,apos,2)
> getcol(FF2,bpos,3)
> getcol(FF2,cpos,4)
> mlimit(FF2,1,rows(FF2),5,7)
> M2= FF2
> mdelimit(FF2)
```

Figure 14.8:   *Part 4, SC Rerandomization and Hypothesis Tests, FFF Study*

We want the symmetric partitions now to consist of the treatment sequence pairs (1,6), (3,2), and (5,4) rather than the (1,2), (3,4), and (5,6) appropriate to the first hypothesis. The vector ab describes how the rows of FFF should be reordered to achieve the desired order, row 1 followed by rows 8 and 9, followed by row 4, and so forth. The function morder imposes this row order on the matrix, now copied to FF2.

From FF2 we extract the reordered treatment position vectors, apos, bpos, and cpos and the reordered matrix of FFF scores, M2.

The revision of accontr is minor. To compare the FFF scores under the placebo (treatment B) with those under meclastine (treatment A), all we need do is to change the line

```
ac= ac + (MM[i][apos[i]]-MM[i][cpos[i]])
```

in Figure 14.6 to read

```
ac= ac + (MM[i][bpos[i]]-MM[i][apos[i]])
```

Note that, in this example, our new symmetric partitions are consistent with the way we described the vectors `gsiz` and `ssiz` for the treatment A versus treatment C hypothesis. This will not always be true, and you should check these two vectors for consistency after reordering the rows of a matrix like `FFF`.

In the final part of the SC analysis, Figure 14.9, we call `shuffl_k` with a revised `accontr` and a reordered M2. Our hypothesis test is again directional, the new drug may produce greater drowsiness (lower FFF scores) than the placebo. Our estimated p-value for this test is p = 0.041.

```
> proc accontr(){}
> accontr(M2,gsiz,out)
> out
  0.61
> shuffl_k(accontr,1,M2,gsiz,ssiz,999,RD)
     Full Sample Statistics (theta):   0.61

P-values for Statistic Number:  1
     Percent LE theta[ 1 ]:   100
     Percent GE theta[ 1 ]:   4.1
```

Figure 14.9:   *Part 5, SC Rerandomization and Hypothesis Tests, FFF Study*

It appears, then, the new drug might produce less drowsiness than the old one (p = 0.069). It is somewhat clearer that the new drug produces more drowsiness, though, than does the placebo (p = 0.041).

Exercises

1.   Table 14.4 from Senn (1993) gives data for another response measure collected from asthmatic children treated with the bronchodilators formeterol and salbutamol. The patients were asked to rate the efficacy of each drug on a scale ranging from 0 (good) to 100 (bad). Senn suggests that the limited range of the scale can cause analysis and interpretation difficulties. He suggests transforming the scale score given by the i-th patient in the j-th period as follows,

$$y_{ij} = \log\left(\frac{x_{ij}}{100 - x_{ij}}\right)$$

a. How do you think a rerandomization analysis based on the y_{ij}s might differ from one based on the x_{ij}s?

b. Using either the scale scores as given or the transformed scores, carry out an analysis of the patient ratings following the pattern given for the PEF scores.

Table 14.4: *Patients' Judgments of Efficacy of Two Bronchodilators*

		Rating	
Sequence	Patient	Formeterol	Salbutamol
for-sal	1	63	82
	3	24	84
	5	32	66
	6	33	51
	10	24	75
	12	30	38
sal-for	2	40	68
	4	5	4
	7	21	57
	8	5	23
	9	32	56
	11	4	53

2. In a third bronchodilator study reported in Senn (1993), 24 children were randomized, 12 apiece, between the treatment orders for-sal and sal-for. The children reported which of the two medications they preferred, as shown in Table 14.5. Carry out a rerandomization analysis using as your statistic the odds ratio of favoring formeterol. The odds ratio was defined in Applications 13, in connection with the analysis of the data in Table 13.2. Do the analysis for two definitions of the odds ratio:

a. Ignoring the No Preferences in the computation, though not in the rerandomizations.

b.   Pooling the No Preferences with the Prefer Salbutamol responses in the computation of the odds ratio.

How do the two analyses differ in interpretation?

Table 14.5:   *Patients' Preference for One of Two Bronchodilators*

	Preference		
Sequence	First Period	No Preference	Second Period
for-sal	9	3	0
sal-for	1	5	6

3.   In the Flicker Fusion Frequency study, Table 14.2, cases were assessed under three conditions. Test the null hypothesis that all three treatments have the same influence on a case. An appropriate treatment response statistic for the individual case might be the variance of that case's three responses. Why? Aggregate over the cases by averaging to obtain an overall treatment comparison statistic. What precautions do you need to take in rerandomizing cases? Is the hypothesis test directional or nondirectional? Explain. What do you conclude?

Concepts 15

Between- and Within-Cases Designs

In Concepts 14, we distinguished between-cases designs in which cases are divided into K treatment groups, the cases in any one group receiving a single level of treatment, from within-cases designs in which cases are divided into G treatment sequence groups, each case receiving all K levels of treatment but in a randomly determined sequence. Now we describe some designs that incorporate both within- and between-cases treatment comparisons.

Between/Within Designs

The reader will have noted earlier a reference to "K levels of treatment." The phrase really means the same as "K treatments." For the class of designs we are about to consider, we will need to distinguish between two varieties of treatments. We'll refer to two treatment factors. The treatments contributing to each factor will be referred to as the levels of that factor. We can relate factors and levels to some of the designs considered earlier.

In Concepts 14, we noted that although some treatment factors could be administered at all levels sequentially to the same case, it might not be possible, desirable, or practical to administer more than one level of other treatment factors to any particular case. Treatment factors of the first kind lead to within-cases designs, and treatment factors of the second kind to between-cases designs.

When treatment factors of both kinds are called for, the resulting design sometimes is referred to as a whole plot/split plot design. The terminology derives from agricultural research. The whole of a plot of land (a case, in our vocabulary) receives one level of a first factor (our between-cases factor), while that plot is split into portions to allow exposure to all levels of a second factor (our within-cases factor). Outside agriculture, however, the "plot" terminology can be confusing, and we will refer to these designs simply as between/within designs.

Types of Between/Within Designs

If the within-cases treatment sequence is to be varied, between/within designs require the cases to be divided twice. The cases are first divided into as many groups as there are levels of the between-cases factor. Second, each between-factor group is divided into as many groups as there are within-cases treatment sequences.

Here are two examples:

1. A random sample of 250 university freshmen is split between males and females. The two groups of students are each subsequently divided, at random, into six groups to receive training, in different sequences, on each of three word-processing programs, WS, WP, and MW.

2. A volunteer group of 100 patients is randomly divided into two groups of 50. Patients in one group will take a prescribed drug by intramuscular injection; those in the other group will take the drug in an oral form. The intramuscular group of 50 is further divided, again at random, into two groups of 25, one to receive drug A for one week followed by the administration of drug B for a second week. This sequence is reversed for the second group. The oral group is also subdivided at random into two drug sequence groups.

In the first study, sex at levels male and female is a between-cases factor and word-processing package at levels WS, WP, and MW is a within-cases factor. In the second study, delivery setting at levels intramuscular and oral is a between-cases factor and drug at levels A and B is a within-cases factor.

How these two divisions of the cases are carried out will determine how we assess, statistically, the impact on cases of the two factors and what we can say about the interaction of the factors with one another. The six rows of Figure 15.1 depict the variety of between/within designs and their associated analyses.

Forming Between-Cases Factor Level Groups

The first column of Figure 15.1 identifies three mechanisms for dividing cases among the levels of the between-cases factor:

1. Random Sample: randomly sampling several case populations

2. Random Division: randomly dividing a set of available cases

3. Nonrandom: assigning available cases nonrandomly

The random samples of column 1 can be the result of:

1. Drawing separate random samples from each of K populations. We could take samples from two populations of laboratory mice. The populations, for example, might differ genetically or they might have been bred in two different laboratories.

2. Sorting a single random sample into K groups according to some characteristic of the sampled cases. This is illustrated by the hypothetical word-processing study. We have a random sample of university students divided between males and females. The result is a random sample of male students and a random sample of female students.

3. Randomly dividing the cases constituting a random sample into K treatment groups. A random sample of 50 laboratory animals could be randomly divided into two equal-sized groups, to receive different diets. This would give random samples from two different prospective populations.

Between Cases	Within Cases	Between Infer	Within Infer	B × W Infer
Random Sample	Random Division	Population	Population	Population
Random Sample	Nonrandom	Population	Descriptive	Descriptive
Random Division	Random Division	Causal	Causal	Causal
Random Division	Nonrandom	Causal	Descriptive	Descriptive
Nonrandom	Random Division	Descriptive	Causal	Causal
Nonrandom	Nonrandom	Descriptive	Descriptive	Descriptive

Figure 15.1: *Statistical Inference for Six Types of Between/Within Designs*

The between-factor groups might not be random samples but, rather, a Random Division of available cases or Nonrandom sets of available cases. We illustrate the two possibilities with the drug-delivery example:

1. As the study was described earlier, an available set of 100 volunteers is divided randomly between intramuscular and oral drug-delivery groups.

2. Instead, Dr. X recruits 50 patients for the intramuscular drug-delivery group, and Dr Y recruits another 50 patients for the oral drug-delivery group. Available cases are nonrandomly assigned to the two between-factor levels.

Allocating Cases to Within-Cases Treatment Sequences

Entries in the second column of Figure 15.1 describe how cases are allocated among different sequences of the within-cases factor levels. Here, we distinguish between the Random Division strategies presented in Concepts 13 and 14, and a Nonrandom allocation of cases among treatment sequences. If, in a psychological experiment, the first 50 subjects are shown nonsense syllables before words and the second 50 subjects are shown words before nonsense syllables, then the division into the two treatment-sequences is nonrandom.

Taken together, the first and second columns of Figure 15.1 describe six types of between/within designs.

Treatment Comparisons

The third and fourth columns of Figure 15.1—Between Infer and Within Infer—summarize the kinds of statistical inferences that can be drawn relative to the influence of our between- and within-cases treatment factors. Whether the inference is Population, Causal but local, or Descriptive depends on the manner in which cases were selected for levels of treatment.

Between-Cases Treatment Comparisons

How do we make between-cases treatment comparisons in the between/within design? These comparisons are made as they would be in a strictly between-cases design. Typically, we use the sum of the responses over all levels of the within-cases factor as the between-factor response for a case. For example, in the word-processing study, students' between-factor scores would be the sum of their responses to the three word-processing packages. In summing over the levels of the within-cases factor, we are examining the main effect of the between-cases factor.

Within-Cases Treatment Comparisons

How do we make within-cases treatment comparisons in the between/within design? These comparisons often are subject to the question of whether the within-cases factor interacts with the between-cases factor in determining a case's response to treatment.

What do we mean by an interaction of the two factors? If a within-cases treatment comparison is the same at all levels of the between-cases factor, the two factors do not interact. If the within-cases treatment comparison yields different results at different levels of the between-cases factor, the two factors interact.

For example, in the drug administration study, the researcher would be interested in whether the same difference in response to drugs A and B was observed for the oral and intramuscular groups.

If there is no interaction of the between- and within-cases factors, we can evaluate the main effects of the within-cases factor. That is, we can pool data from all the between-factor groups in making a within-cases treatment comparison. If there is an interaction, however, we must evaluate our within-cases treatment comparisons separately for each level of the between-cases factor. The within-cases treatment effects evaluated in this second way are called simple effects of the within-cases factor.

Interaction of Factors

How do we decide if there is an interaction of the between- and within-cases factors or not? The fifth column of Figure 15.1 shows the kind of inference appropriate for an assessment of factor-interaction in each of the six between/within designs. Note that these are the same as for within-cases treatment comparisons.

Between/Within Resampling Strategies

The three statistical inference possibilities, population, causal, and descriptive, are associated with our three resampling strategies. Where we have random samples but cannot justify the assumptions associated with standard, parametric, or classical inference—normal population score distributions, large sample sizes, homogeneity of population score variances, acceptance of a conventional parameter—our approach to population inference is based on bootstrap resampling, as developed in Concepts 7 through 11.

When our design involves the random allocation of available cases among treatment levels or sequences of treatments, we can rerandomize those allocations, keeping in mind a particular null treatment-comparison hypothesis, to provide a basis for inferring a causal link between treatment and response. This mode of resampling inference has been developed in Concepts 12 through 14.

Concepts 16 will describe a resampling approach suitable when our cases have been neither randomly sampled nor randomly allocated among treatment options. There we outline the use of data from subsamples of a set of available cases as a way of supporting our statistical description of the outcomes of such nonrandom studies.

In Applications 21, you will encounter examples of the variety of between/within designs outlined in Figure 15.1. We will use each of these three resampling approaches for the analyses of those studies.

We conclude this introduction to the between/within study with a description of the analysis of one particular design, the one defined in the third row of Figure 15.1 That design is based on a double Random Division of available cases, first to levels of the between-cases factor and then to sequences of within-cases factor levels.

Doubly Randomized Available Cases

Let's develop further our hypothetical drug-delivery study. Random halves of the intramuscular and oral patient groups receive drug A for an initial period and then crossover to receive drug B in a second period. The other halves of the two delivery-setting groups of patients crossover from drug B to drug A, receiving the same drugs, but in the reverse order.

In practice, n might be 60 or 100, but Figure 15.2 illustrates this double randomization for an abbreviated version of the study, showing the assignments and responses of only $n = 8$ patients.

Half of the patients, cases 2, 5, 6, and 8, are randomly assigned to intramuscular and the other four are assigned to oral drug settings. Within each of these settings, the four patients are further randomized, two to each of the drug sequences AB and BA. The patient randomized to drug sequence AB receives drug A in the first period of the study and receives drug B in the second period. Following the double randomization, cases 5 and 6 have been assigned to intramuscular treatment with drug sequence AB, cases 2 and 8 have been assigned to intramuscular treatment with drug sequence BA, cases 1 and 7 have been assigned to oral treatment with drug sequence AB, and cases 3 and 4 have been assigned to oral treatment with drug sequence BA.

The Scores column of Figure 15.2 gives the responses for each patient to the drugs administered in the first and second periods of the study. The Sum column gives the sums of the two period responses and the Diff column gives the result of subtracting the second period response from the first period response. The individual responses are times to symptom relief, so smaller responses are better.

Patients	Setting	Sequence	Scores	Sum	Diff
C1	I	AB	10,12	22	-2
C2	I	AB	13,8	21	5
C3	I	BA	8,7	15	1
C4	I	BA	6,9	15	-3
C5	O	AB	9,7	16	2
C6	O	AB	11,11	22	0
C7	O	BA	8,6	14	2
C8	O	BA	11,13	24	-2

Figure 15.2: *Randomizing Cases to Levels and to Sequences*

Between-Cases Levels Comparison

The between-cases factor is the setting in which drug administration takes place. Each patient is randomized to drug administration either by intramuscular injection or in oral form. We'll take as our treatment-comparison statistic the result of subtracting the mean response-sum for intramuscular patients from the mean response-sum for oral patients. We sum the two drug responses for each patient because we want to know the main effect of drug-setting, across the two drugs. Our setting comparison statistic should be positive if, as the researcher expects, intramuscular administration provides faster symptom relief. The actual value of the statistic is

$$s_{set} = \frac{16 + 22 + 14 + 24}{4} - \frac{22 + 21 + 15 + 15}{4} = 19 - 18.75 = 0.25$$

giving, at best, scant support for the substantive hypothesis.

Our null treatment-effect hypothesis is that the two settings would have the same effect on any given patient. Thus, a patient would have had the same response-sum if randomized to the other setting. We develop a null reference distribution for s_{set} by computing the statistic for all possible rerandomizations of the eight patients to the two treatment settings. The small example of Figure 15.2 permits only $[8!/(4! \times 4!)] = 70$ randomizations, including the one actually obtained. In evaluating whether a value of 0.25 represents only a chance advantage to intramuscular administration, we want to know the proportion of the 70 s_{set}^{H} values that are as large as or larger than 0.25. This p-value is likely to be rather large in this instance, suggesting that any apparent advantage of the intramuscular delivery-setting might be only a chance result of our randomization strategy; one or more slightly less responsive patients were randomized to the oral group.

Interaction of Between- and Within-Factors

The notion of an interaction in the between/within design takes this form: The difference in effectiveness of the within-cases factor levels is not the same for all between-cases factor levels.

The no-interaction or null hypothesis is that: The difference in effectiveness of the within-cases factor levels is the same at all between-cases factor levels.

An evaluation of the interaction hypothesis takes its most natural form when, as in this example, both the within- and between-cases factors have just two levels. If our substantive hypothesis were that drug B is the more effective (lower response time) of the two, we might formulate an interaction statistic as follows:

$$s_{int} = \text{Mean}(A - B | \text{intramuscular}) - \text{Mean}(A - B | \text{oral})$$

This is the difference between the mean $(A - B)$ scores for the intramuscular and oral groups. If there is no interaction, s_{int} should be close to zero. Large positive or negative values provide evidence of an interaction. We use means here simply to obtain a summary $(A - B)$ score under each of the two treatment settings. Other summarizations could be used.

Based on the period-difference (Diff) column of Figure 15.2, we can compute the interaction test statistic as

$$s_{int} = \frac{-2 + 5 - 1 + 3}{4} - \frac{2 + 0 - 2 + 2}{4} = 1.25 - 0.50 = 0.75$$

suggesting a bigger $(A - B)$ difference, on average, for patients assigned to the intramuscular setting. Notice that in computing s_{int}, the period-differences were multiplied by -1 for patients in the BA sequence to produce $(A - B)$ response-differences.

What are the chances of obtaining a s_{int} statistic as large as 0.75 or larger or as small as -0.75 or smaller, if there is no interaction? For our choice of between/within design, the null interaction hypothesis takes the form of specifying that a patient would have the same $(A - B)$ score in either treatment setting. We can form a null reference distribution for s_{int} by computing the value of that statistic for all possible rerandomizations of the patients and their $(A - B)$ scores between the two treatment settings. This reference distribution will be the same size, 70, as for the between-cases or settings comparison statistic, s_{set}.

Within-Cases Levels Comparison

How we compare the effectiveness of the levels of the within-cases factor depends on the outcome of our test of the interaction hypothesis. If we decide there is no interaction, we have determined that the drug effect is the same at both settings. We can then assess a single drug effect, across setting. We refer to this across-levels assessment as a main effect.

On the other hand, if we decide there is an interaction, then we must assess any drug effect separately, at each level of setting. We refer to these by-levels assessments as simple effects.

Main Effect of the Within-Cases Factor

Our drug main effect might be the mean of the $(A - B)$ response-differences for all n patients,

$$s_{drug} = \frac{-2 + 5 - 1 + 3 + 2 + 0 - 2 + 2}{8} = 0.875$$

a result that suggests that drug B works faster than drug A.

Is a s_{drug} of 0.875 large enough to establish a drug difference, or could such a value be a chance result, the result of randomizing cases with no drug preferences? Our null treatment-effect hypothesis for our within-cases factor comparison is that a case would have the same response to the two drugs, if administered in the same period of the treatment sequence. Any observed difference in responses, then, is a period-difference, not a treatment difference. To test this hypothesis, we can rerandomize our patients among treatment sequences and recompute the statistic. Doing this for all possible rerandomizations provides a null reference distribution for our within-cases factor comparison statistic, s_{drug}.

It is possible under this null treatment-effect hypothesis to reassign a case to a different treatment sequence, but we cannot reassign it to a different between-cases level. When we rerandomize between drug sequences, we must do it separately for the intramuscular and oral cases.

The four cases randomized to each delivery setting were, in turn, randomized, two apiece, to sequences AB and BA. There are, then, $[4!/(2! \times 2!)] = 6$ ways in which the four intramuscular cases could have been randomly allocated among the two treatment sequences. The same is true for the four oral cases. We carry out the two random sequence allocations independently, so the result is a total of $6 \times 6 = 36$ different arrangements of the eight patients among the two treatment sequences.

We could calculate s_{drug}^H for each of these 36 case arrangements and then determine the proportion that are as large or larger than 0.875. Note that for this small set of data, the smallest possible p-value would be 1/36 as the observed value of s_{drug} is one of the 36 scores in the reference distribution.

Simple Effects of the Within-Cases Factor

If we decide that there is an interaction, then we assess the drug effect separately at the two settings.

Using our Figure 15.2 data, we would have two drug comparison statistics,

$$s_{\text{druglintra}} = \frac{-2 + 5 - 1 + 3}{4} = 1.25$$

and

$$s_{\text{drugloral}} = \frac{2 + 0 - 2 + 2}{4} = 0.50$$

We can test each for departure from chance by comparing it to its null reference distribution. The two reference distributions are based on the rerandomization to treatment sequence of either the intramuscular or the oral cases, but not both. For our example, these reference distributions will contain only a very small number of $s_{\text{drugloral}}^H$ or $s_{\text{druglintra}}^H$ scores, $[4!/(2! \times 2!)] = 6$ in each.

Applications 15

Interactions and Simple Effects

The resampling analysis of between/within model designs relies heavily on techniques you've already mastered. Figure 15.1 in Concepts 15 describes six between/within designs. When the between-cases or within-cases comparisons involve random samples from populations, we can develop bootstrap CIs or hypothesis tests to facilitate treatment comparisons. If the between-cases or within-cases treatment assignments of available cases are made on a random basis, rerandomization provides a statistical basis for determining whether treatments differ from one another by more than we should expect by the chance alignment of cases with treatment levels or sequences. In the absence of either random mechanism, between-cases or within-cases treatment comparisons will have only descriptive, rather than population or causal, inference—the topic of Concepts 16.

Simple and Main Effects

In these Applications we will illustrate the distinction introduced in Concepts 15 between the main and simple effects of a factor. The two kinds of effects arise in designs involving two factors—call them factors A and B.

1. If all levels of factor B are paired with all levels of factor A, the two factors are said to be fully crossed with one another.

2. If each level of factor A is associated with a different set of levels of factor B, the levels of factor B are said to be nested within the levels of factor A.

3. Between these two designs fall the partially crossed designs. For these, not all levels of factor B are paired with all levels of factor A.

If a treatment comparison for factor B is made at a particular level of factor A, as it must be in nested studies, that comparison assesses a simple effect of factor B.

On the other hand, if a treatment comparison for factor B is made across all levels of factor A, as might be done in a fully crossed design, or across all possible levels, as might be done in a partially crossed design, that comparison assesses a main effect of factor B.

In the crossed designs, however, it is not always appropriate to assess the main effect of factor B. If the result of a factor B comparison would change from one level of factor A to another, then we clearly should make the factor B comparisons separately at each level of factor A, assessing simple effects. Where a treatment B comparison has a different result at different levels of factor A, the two factors are said to interact.

One of the two factors can be a between-cases factor and the other can be a within-cases factor, as in the between/within models of Concepts 15. Both can be within-cases factors, or both can be between-cases factors. We illustrate main and simple effects with a fully crossed two-factor between-cases example, in which available cases have been randomized among the crossed-levels of the two factors.

Table 15.1:   *Estimated IQ Scores for Target Persons*

Attractiveness of Target			
High		Low	
Sex of Target		Sex of Target	
Male	Female	Male	Female
112	124	120	122
114	124	121	123
116	126	123	125
118	126	124	126

The data in Table 15.1 are taken from Woodward, Bonet, and Brecht (1990) and were obtained in a (alas, fictional) fully-crossed or factorial study. Sixteen college student volunteers were randomly allocated, four apiece, to four treatment conditions. In each condition, the student subject is shown a photograph of a college-age person and asked to estimate the IQ of the pictured person or target. The four treatment conditions result from the crossing of the target's sex (male or female) with the experimenter's judgment of the attractiveness of the target (high or low attractiveness). The two factors, then, were sex of target and attractiveness of target, each at two levels.

We are interested in whether the estimated IQ of a target was influenced by the sex or attractiveness of that person.

Main Effects of Factors

One standard approach to analyzing the two-factor between-cases study is to consider both factors to be at the same level of substantive interest, thus we can evaluate the influence of each, in turn, by the same technique. We'll refer to this approach as evaluating the main effect of each factor. A factor has a main effect if we conclude that it is unlikely that each case would respond in the same way at all levels of that factor. That is, the null treatment-effect hypothesis for the sex of target factor is that any one of our college student subjects would make the same IQ estimate for a target, whether the target is male or female. The null treatment-effect hypothesis for the main effect of target attractiveness is that any one of the subjects would make the same IQ estimate for a target, whether the target is of high or low attractiveness.

Notice here that each treatment-effect hypothesis deals with one of the factors, but is silent relative to the influence of the other factor. Consider, for example, the student subject who, assigned the male high-attractive target, made an IQ estimate of 112. The null treatment-effect hypothesis for sex specifies that this subject would estimate the female high-attractive target also to have an IQ of 112. The hypothesis does not predict, however, how this subject would estimate the IQ for either male low-attractive or female low-attractive targets. Similarly, the treatment-effect hypothesis for attractiveness specifies that this same subject would estimate the male low-attractive target to have an IQ of 112, but says nothing about the estimates for female low-attractive or female high-attractive targets.

The specificity of these treatment-effect hypotheses shapes how we carry out case rerandomization to build the relevant reference distributions. We see this in the Resampling Stats scripts beginning with Figure 15.3.

This initial part of the script inputs the estimated IQ scores, rearranges them by levels of the two factors, and computes the two main effect statistics. The first (SXDIFF) compares the estimates given male and female targets, and the second (ATDIFF) compares the estimates given high- and low-attractive targets.

```
COPY (112 114 116 118) MLHA
COPY (120 121 123 124) MLLA
COPY (124 124 126 126) FEHA
COPY (122 123 125 126) FELA
'
CONCAT MLHA MLLA ML
CONCAT FEHA FELA FE
CONCAT MLHA FEHA HA
CONCAT MLLA FELA LA
MEAN ML MLMN
MEAN FE FEMN
MEAN HA HAMN
MEAN LA LAMN
SUBTRACT MLMN FEMN SXDIFF    'Obtained Sex Difference
SUBTRACT HAMN LAMN ATDIFF    'Obtained Attractiveness Diff
SCORE SXDIFF SD
SCORE ATDIFF AD
'
```

Figure 15.3: *Part 1, Resampling Stats, Main Effects of Sex and Attractiveness*

Part 2 of the script, Figure 15.4, develops Monte Carlo approximations to the two reference distributions.

Note that in the rerandomization of subjects between levels of attractiveness (lines 2 through 9 in Figure 15.4) the rerandomization is carried out separately for subjects initially randomized to the two levels of sex of target (lines 2 and 3). In accordance with the treatment-effect hypothesis for attractiveness, the exchangeability of data is between levels of that particular factor; we are not authorized to exchange judgments of male and female targets. For a similar reason, the rerandomization of subjects between the levels of sex of target (lines 14 through 21) is carried out separately for subjects initially randomized to the two attractiveness levels. The treatment effect hypothesis for sex of target allows us to exchange data between levels of sex but not between levels of attractiveness.

The rerandomization analysis carried out by the scripts of Figures 15.3 and 15.4 computes p-values on the assumption that each of the two hypotheses is nondirectional. That is, in each case we want to know the probability of a difference in mean IQ estimates as large or larger than that observed, in either direction, under the null treatment-effect hypothesis. We find these p-values by comparing the absolute value of each reference set element against the absolute value of the obtained difference.

```
REPEAT 999
  SHUFFLE ML MLX
  SHUFFLE FE FEX
  TAKE MLX 1,4 MLHAX
  TAKE MLX 5,8 MLLAX
  TAKE FEX 1,4 FEHAX
  TAKE FEX 5,8 FELAX
  CONCAT MLHAX FEHAX HAX
  CONCAT MLLAX FELAX LAX
  MEAN HAX HAXMN
  MEAN LAX LAXMN
  SUBTRACT HAXMN LAXMN XAT     'Shuffled Attractiveness Diff
  SCORE XAT AD
  SHUFFLE HA HAX
  SHUFFLE LA LAX
  TAKE HAX 1,4 MLHAX
  TAKE HAX 5,8 FEHAX
  TAKE LAX 1,4 MLLAX
  TAKE LAX 5,8 FELAX
  CONCAT MLHAX MLLAX MLX
  CONCAT FEHAX FELAX FEX
  MEAN MLX MLXMN
  MEAN FEX FEXMN
  SUBTRACT MLXMN FEXMN XSX     'Shuffled Sex Difference
  SCORE XSX SD
END
'
ABS AD ABSA                    'Absolute values for
ABS SD ABSS                    'nondirectional
ABS ATDIFF ABA                 'hypotheses
ABS SXDIFF ABS
COUNT ABSA >= ABA ATWO         'Nondirectional counts
COUNT ABSS >= ABS STWO
DIVIDE ATWO 1000 ATWO          'Nondirectional p-values
DIVIDE STWO 1000 STWO
PRINT ATDIFF ATWO SXDIFF STWO

ATDIFF = -3        ATWO = 0.076      SXDIFF = -6   STWO = 0.006
```

Figure 15.4:   *Part 2, Resampling Stats, Main Effects of Sex and Attractiveness*

For example, the obtained difference for attractiveness, reported at the bottom of Figure 15.4, was

$$\text{Mean(IQ|High Attractiveness)} - \text{Mean(IQ|Low Attractiveness)} = -3$$

indicating higher IQ estimates for low-attractive targets. Of the mean differences computed for 1,000 rerandomizations under the no attractiveness-effect hypothesis, 7.6% were -3 or smaller or $+3$ or larger, $p = 0.076$.

The obtained difference for sex of target was

$$\text{Mean(IQ|Male Target)} - \text{Mean(IQ|Female Target)} = -6$$

and only 6 of 1,000, fewer than 1%, of the null reference set were either -6 or smaller, or $+6$ or larger, $p = 0.006$. There was a clear sex of target effect in this study—female targets were estimated to have higher IQs than male targets were.

Factor Interaction and Simple Effects

The main effect of a particular factor is evaluated across all levels of other factors in the study. Thus, in the evaluation of the main effect for attractiveness, our test statistic compared the IQ estimates for low-attractive targets with those for high-attractive targets, after pooling together the estimates for male and female targets at each level of attractiveness. This pooling might reflect our belief that any attractiveness effect will be the same for male as for female targets, or it might be that we want to ignore any such difference when assessing the effect of attractiveness.

On the other hand, we might have anticipated in planning the study that the attractiveness effect could be different for male than for female targets. For example, we might have expected that there would be no attractiveness effect for female targets, but that the estimated IQ for the low-attractive male target would be higher than that for the high-attractive male target. Or, we might have hypothesized an attractiveness effect for each target sex, but that the effect for male targets would be in the opposite direction as that for female targets, that is, the low-attractive male would be judged to have a higher IQ than the high-attractive male but the low-attractive female would be judged to have a lower IQ than the high-attractive female.

To illustrate the assessment of simple effects in the IQ estimation study, we'll use some specific differential hypotheses. Let's assume we had hypothesized in advance of the study that

  1. Although the low-attractive male target would be judged to have a higher IQ than the high-attractive male target,

  2. The IQ judgments of low- and high-attractive female targets would not differ.

That is, the attractiveness of target factor was postulated to have specifiably different effects at the two levels of sex of target.

```
> fh:= 124,124,126,126
> ml:= 120,121,123,124
> fl:= 122,123,125,126

> mean(ml)-mean(mh)
      7

> perm2(ml,mh)
        / 0.014285714 / 1 /

> 1/70
      0.014285714

> abs(mean(fl)-mean(fh))
      1

> func f() return abs(mean($1$)-mean($2$))
> f(fl,fh)
      1

> perm2__(f,fl,fh)
Using f():
        / 0.71428571 / 0.48571429 /
Using abs(f()):
        / 0.71428571 / 0.48571429 /
```

Figure 15.5:   *Assessing Simple Attractiveness Effects (SC)*

In Figure 15.5, we use the SC `perm2` function, which finds the difference in means between two vectors and then produces a null reference distribution for that statistic, to obtain a p-value for the null attractiveness effect hypothesis for male targets, where the alternative is directional. We can use the more general version of that function, `perm2__`, introduced to evaluate the null attractiveness-effect hypothesis for female targets, where the alternative is nondirectional.

We begin the interactive SC session by inputting IQ estimates for the male high-attractive (`mh`), female high-attractive (`fh`), male low-attractive (`ml`), and female low-attractive (`fl`) targets. Our first scientific hypothesis was that the male low-attractive target would be judged higher than the male high-attractive target. Thus, a natural choice of a test statistic, though not the only one, is the one computed in Figure 15.5,

$$s_{\text{att|male}} = \text{mean(ml)} - \text{mean(mh)} = 7$$

The SC function `perm2(ml,mh)` returns two proportions, `/p1/p2/`. The two vectors (`ml` and `mh`) are the same length, so these proportions are `p1`, the proportion of rerandomizations of the eight IQ judgments between the male low-attractiveness and high-attractiveness levels that yield $s^H_{\text{attlmale}} \geq 7$, and `p2`, the proportion that result in $s^H_{\text{attlmale}} \leq 7$. SC reports the first of these as 0.0143 and the second as 1.0. Given the directionality of our scientific hypothesis, we are interested in the first of these. That is, under the null treatment-effect hypothesis, the probability of obtaining a value of s_{attlmale} greater than or equal to 7 was just p = 0.0143. It would seem unlikely that the difference in mean IQ judgments was a chance result.

In this example, the two proportions are based on a complete enumeration of the difference in mean IQ judgments for all possible rerandomizations between low- and high-attractiveness of those cases initially randomized to male targets. These rerandomizations and recomputations were possible under the null treatment-effect hypothesis that the male-target cases would make the same IQ judgment of low- and high-attractive targets. There were eight subjects to be rerandomized, four to each of the two attractiveness levels, so this complete enumeration consisted of

$$\frac{8!}{4! \times 4!} = \frac{8 \times 7 \times 6 \times 5}{4 \times 3 \times 2 \times 1} = 70$$

values of s^H_{attlmale}. SC reports that only one of the 70 possible randomizations of the IQ estimates produced an $s^H_{\text{attlmale}} \geq 7$, that is, $1/70 = 0.01428$. We know that the randomization actually used in the study contributes an s_{attlmale} of 7, so we've learned that no other rerandomization results in an s^H_{attlmale} as large as 7. A glance at the data in Table 15.1 validates this conclusion. All IQ judgments for the male low-attractive target were higher than any of the judgments for the male high-attractive target; thus, no rearrangement of the eight scores could yield a bigger difference in means.

We anticipated no difference in the IQ judgments of the high- and low-attractive female targets, so our alternative to the second null treatment-effect hypothesis is nondirectional; we would be surprised if the low-attractive female target were judged either more or less intelligent than the high-attractive one. We can accommodate deviations from the null in either direction if we use as our test statistic the absolute value of the difference in mean IQ judgments,

$$s_{\text{attlfem}} = \text{abs}[\text{mean}(\text{fl}) - \text{mean}(\text{fh})]$$

a statistic that will be large and positive for large deviations in either direction. The value of the statistic, as computed in Figure 15.5 by the user-written function f(fl,fh), was $s_{\text{attlfem}} = 1$.

What are the chances of an s_{attlfem} this large or larger if each of the eight subjects who made a judgment about the intelligence of one of the female targets would have made that same judgment for the other female target? We can answer this by repeatedly rerandomizing the eight subjects, together with their IQ judgments, between the two attractiveness levels, computing s^H_{attlfem} for each rerandomization, and then determining the proportion of these that are as large or larger than our s_{attlfem} of 1.

The SC function perm2__(f,fl,fh) is shown doing this work for us in Figure 15.5. This function takes three arguments. As with perm2(ml,mh) you must supply the names of the two data vectors (fl and fh). The function also needs, as a first argument, the name of the SC function to use to compute s from those two vectors. This function might be one already built-in to SC or we might write it for the occasion. Here, we define our own function

```
func f() {return abs(mean($1$) - mean($2$))}
```

This will return the absolute value of the difference in the means of the two vectors whose names are supplied to it, replacing the generic names 1 and 2 that we used in defining the function. Thus, f(fl,fh) returns a value of 1, replicating an earlier computation.

Using our new function, perm2__(f,fl,fh) reports, in turn,

1. 0.714: The proportion of randomizations yielding $s^H \geq s$
2. 0.486: The proportion of randomizations yielding $s^H \leq s$
3. 0.714: The proportion yielding $\text{abs}(s^H) \geq \text{abs}(s)$
4. 0.486: The proportion yielding $\text{abs}(s^H) \leq \text{abs}(s)$

Our s and s^H are absolute values of differences and, thus, can never be negative, so the latter two results are identical to the first two. The finding that for 71% of the randomizations $s^H_{\text{attlfem}} \geq s_{\text{attlfem}}$ suggests that a difference in mean IQ judgments of the magnitude we observed is readily accountable to the chance effect of the randomization of female-target cases between the two levels of attractiveness and does not signify an attractiveness effect for female targets.

Exercise

1. Table 15.2 describes data from a completely randomized 3×4 factorial study. Forty-eight experimental animals were randomized, four apiece, to the 12 treatment levels that resulted from crossing three levels of Anesthetic Drug with four levels of Mode of Administration. The response is Anesthetic Duration, in fractions of an hour. Assess the main effects of each factor with an omnibus statistic of your choice. Focus just on drugs 1 and 2. Compare the two drugs as a main effect. Then make the comparison separately for each mode—as a set of simple effects. These simple effects are based on small numbers of cases, but can you see evidence here of an interaction between drugs and mode of administration?

Table 15.2:   *Duration of Anesthesia as Function of Drug and Application Mode*

Drug	Mode	Anesthetic Duration			
1	1	0.31	0.45	0.46	0.43
	2	0.82	1.1	0.88	0.72
	3	0.43	0.45	0.63	0.76
	4	0.45	0.71	0.66	0.62
2	1	0.36	0.29	0.4	0.23
	2	0.92	0.61	0.49	1.24
	3	0.44	0.35	0.31	0.4
	4	0.56	1.02	0.71	0.38
3	1	0.22	0.21	0.18	0.23
	2	0.3	0.37	0.38	0.29
	3	0.23	0.25	0.24	0.22
	4	0.3	0.36	0.31	0.33

Concepts 16

Subsamples: Stability of Description

In Concepts 7 through 11, we developed bootstrap resampling as an approach to population inference. In Concepts 12 through 15, we developed rerandomization as a basis for causal inference. The bootstrap requires that we have random samples from one or more case populations. Rerandomization requires an original randomization, that we randomly allocate cases among treatment levels or sequences of treatments. If these cases constitute one or more random samples, our causal inference extends to populations, otherwise it is local to a particular set of available cases. Both population and causal inference are driven by the randomness built into the design of a study, random sampling, or random allocation.

Now we turn to the problem of analyzing score distributions for cases that have been neither randomly sampled or allocated. What should we do when there is no randomness in the design of a study?

Nonrandom Studies and Data Sets

Randomness is an important study design element. Whenever possible we should randomly sample our cases from one or more well-defined populations and, if the cases are to be assigned to different treatment levels or sequences, then we should carry out these assignments by one of the randomization schemes covered in earlier Concepts.

There are occasions, however, when one or both random processes are not available to us. Random sampling might prove too expensive or we might not be interested, at a particular stage of our research program, in population inference. Randomization of cases might be unacceptable, politically or morally, or we might have to work with intact groups of available cases.

Here are some examples of nonrandom studies:

1. The American states vary in the proportion of tax monies budgeted for higher education and in the proportion of the adult population employed at the managerial or professional level. How strong is the relationship between the two attributes? You have data on both attributes for all 50 states.

2. Second-grade students at the Cedar Park elementary school have been taught reading by a whole-language method. Those at Spruce Hill elementary have been taught by a phonemic method. We want to compare their reading performances at the end of the school year.

3. Forty-five volunteer patients, with histories of vascular headaches, are switched from a quick-acting, short-duration formulation of their prescribed headache suppressant to a formulation that enters the bloodstream more slowly but over a longer period of time. How does the frequency and severity of headaches differ between the two formulations for the typical patient?

Our first study is a population study; we have attribute scores for all members of the population, not for a random sample. Furthermore, we have not manipulated, randomly or otherwise, the level of, say, spending on higher education.

In the second study, students have not been assigned at random either to elementary school or to reading teaching method. The students were enrolled in their respective schools, we shall suppose, because of where they live. Furthermore, the two groups of students are not random samples from any populations.

The volunteer (available, nonrandom) patients in the third study all follow the same treatment sequence, quick-acting formulation followed by long-acting formulation. We have no basis for evaluating whether any observed difference in effectiveness of the two formulations might have been a chance result. (Exercise 2 in Applications 21 provides an example of how a study of this kind might be conducted to provide a basis for evaluating the two treatments statistically, even when they are always given in the same order.)

Population and causal inference are denied us in each of the three illustrations. Yet, examples of designs such as these are not uncommon. What can we infer from such research?

Local Descriptive Inference

Introductory statistics texts often distinguish between descriptive and inferential statistics. In that context, inference means inference beyond the data set at hand and corresponds to what we here call population inference. Inferential statistics, then, would include parameter estimates, standard errors, confidence intervals, and hypothesis test statistics such as the t-test and the F-ratio that require an associated p-value for their interpretation. Descriptive statistics, on the other hand, are those that simply describe the data set at hand but make no inference about a larger population. The average, median, standard deviation, and interquartile range are cited as descriptive statistics.

In fact, the distinction is better drawn between the intended or permissible use of a statistic than between statistics. We know, for example, that the mean of the distribution of scores for a random sample of cases is an estimate of the mean of the distribution of scores for the population of cases sampled. The mean, for an appropriate study design, can be an inferential as well as a descriptive statistic.

We use inference in a broader sense here: that there is a statistical basis for drawing a certain inference from our data. We can draw population inferences from random samples and causal inferences from studies in which cases are randomly allocated among two or more treatments. We'll use descriptive in a broader sense as well, as we see next.

Not every research design, however, involves the random sampling or random assignment of cases. What can we infer from such studies? Whatever inference we make, it must be local; we have no statistical basis on which to generalize to other cases. Inferences from nonrandom studies also must be descriptive, now in the sense of not attributing causality to any score difference or relationship between variables.

We can describe one or more outcomes of the nonrandom study, and we can choose whatever best summarizes those outcomes for our description. Our description can be as simple as a correlation coefficient or a difference in mean scores or as complex as a prediction or discrimination model.

Is there a sense in which we want our description to be inferential? We certainly want to be able to infer that our chosen description fairly summarizes the data set.

We take the position here that a fair description is one that is stable, that is, one that is relatively uninfluenced by the presence or absence of specific cases.

Description Stability and Case Homogeneity

When we select our cases at random from some population, not only are we assured that they all belong to that population but we also know the sample to be representative, at least randomly so, of that population. Because of their known common source and the unbiasedness of their selection, we can expect the randomly sampled cases to make up a homogeneous set. An available set of cases, on the other hand, cannot be assumed to have the same degree of homogeneity. We do not know the population from which they were selected, nor can we be certain there were no biases in their selection.

We are interested in a particular homogeneity of cases, their homogeneity relative to study description. If cases are inhomogeneous, then the presence or absence of certain cases will alter considerably the description of the study. The description will lack stability.

Subsample Descriptions

Our choice of technique for assessing description stability—or, equivalently, case homogeneity—is to describe not only the full data set but subsamples of the data set as well. A subsample contains some but not all the cases present in the full data set. If we leave out some of the cases, is the description appreciably altered?

How we form these subsamples will depend on what we know about how the full set of available cases was recruited into the study. We distinguish between structured and unstructured data sets.

Structured Data Sets

Data sets can be structured by our knowledge of the way in which cases entered the study. Consider these three examples.

1. Time Structured Data

Mothers of newborns are invited to join a study of infant development. They are recruited into the study from one hospital but over a three-year period. We are concerned that over such a long time the mode of recruitment might not remain constant. This, in turn, might lead to an inhomogeneous set of mothers. If we know the dates at which mothers entered the study, we can investigate the potential for an unstable description of the study by describing not just the entire data set but subsamples that sytematically omit time periods.

2. Site Structured Data

A new treatment for arthritis is being investigated. Six different clinics agree to offer the treatment to patients. Every effort is made to ensure that the treatment is delivered in the same way at each clinic and that the eligibility requirements for patients are the same. Nonetheless, it is important to ensure that the overall description of the investigation does not depend too heavily on the results obtained at just one clinic.

3. Investigator Structured Data

A third, identifiable potential source of structured variability of outcome is the experimenter/therapist/teacher/assistant, where treatment delivery duties are shared. Each of six research assistants (i.e., assistants a, b, c, d, e, and f) is responsible for the surgical preparation of four mice used in a study of subcortical control of eating behavior. Any summarization of the performance of all 24 of the mice should be checked against summaries of six subsamples, each the result of dropping out those mice operated on by a particular assistant—as illustrated in Figure 16.1.

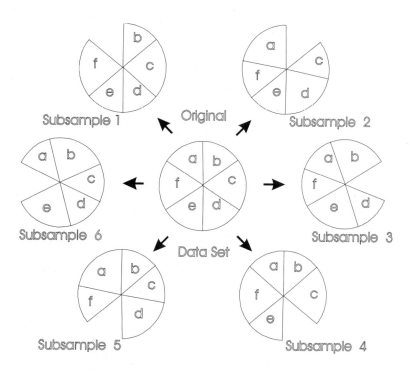

Figure 16.1: *Forming Subsamples from a Structured Data Set*

Note that time of recruitment, clinic site, and research assistant differences are not the targets of investigation in these three studies. Rather, we simply seek assurance that they have not introduced a case inhomogeneity that has biased the overall description of a study's outcome toward one time period, clinic, or assistant.

Figure 16.1 illustrates the construction of subsamples from a structured data set. Our complete data set is made up of six, structured partitions, such as, sites, investigators, or time periods. We systematically drop these partitions, one at a time, to create six subsamples.

For each subsample we compute the same description as we did for the full data set. If the partitions are homogeneous relative to this description, then the descriptions of the subsamples should be in close accord with one another and with the description of the full data set.

If, on the other hand, the description of a particular subsample is markedly at variance with those for the other subsamples, and that for the full data set, then this indicates that the full data set description relies too heavily on the cases omitted from that subsample. The omitted cases are not homogeneous with the others, so far as the description is concerned. Later in this unit we describe what we might do after discovering such an inhomogeneous partition.

The data set structure that we use in partitioning our cases to form structured subsamples is a structuring of the cases by source. If our cases enter the study from different sources—different investigators, different dates, different sites—we should be alert to a compromise of case homogeneity.

Of the three nonrandom studies that introduced this Concepts, only the American states higher-education spending example involves a structured data set. The structure is inherent in the use of political subdivisions as cases. The obvious way to compose subsamples is to drop the states one at a time and recompute the association measure over the remaining 49 cases. In this way, we would evaluate whether the association measure, computed for all 50 states, is strongly influenced by the data for a single state.

Another possibility would be to group the states geographically (and, perhaps, geopolitically) into regions such as New England, North Atlantic Coast, and so on, and form subsamples by dropping the regions one at a time. Comparing the resulting subsample descriptions with the description for all 50 states would tell us whether the overall description depended heavily on the data from a particular region. Note that the states are aggregated into regions solely for the purpose of forming subsamples; the states remain the cases for purposes of computing subsample and full data set descriptions.

Unstructured Data Sets

The other two examples of nonrandom studies—of second-grade reading performance and the treatment of vascular headaches—involve unstructured data sets.

The lack of source structure is clear for the headache study; we have one group of patients recruited at one time, in one place, and receiving the same treatment.

There is what appears to be source structure in the reading study: We have students from two different elementary schools. Alas, the two sources are fully confounded with the two treatments; all second graders at Cedar Park are taught by one method, all those at Spruce Hill by a second method. We cannot form structured subsamples, dropping one school at a time, and still compare the teaching methods. To use structured subsampling for treatment comparisons, we must have source structure at each of the treatment levels being compared.

Our approach to assessing the stability of description of an unstructured data set is to repeat the description on a series of randomly chosen subsamples, all of the same size. The subsample size that we use here gives us half-samples. That is, half of the cases in the full data set, randomly selected, are dropped to form a typical half-sample.

The reason for choosing half-samples is that there will be a larger number of distinct subsamples of this size than of any other size. This can be illustrated for a small data set. If there are 10 cases in the full data set, then there are $ss_\ell = 10!/[\ell! \times (10 - \ell)!]$ different subsamples of size ℓ, where ℓ takes one of the integer values 1 through 9. A little arithmetic gives these results:

ℓ	1	2	3	4	5	6	7	8	9
ss_ℓ	10	45	120	210	252	210	120	45	10

For a set of 10 cases, there are, for example, only 45 subsamples of size 8, whereas there are 252 of size $10/2 = 5$. By studying the stability of our description over half-samples, we do so where there is the greatest diversity of subsample membership.

There will be some variability among half-sample descriptions. For example, the mean of a half-sample containing a large number of small scores will be smaller than the mean of a half-sample containing a large number of large scores. But, if the full data set description is to be regarded as stable, it should be broadly typical of the distribution of half-sample descriptions. The full data set description should be solidly in the middle of the distribution of subsample descriptions. It should not be at or near one of the limits of that distribution, or that would be evidence that the full data set description depends on including certain cases in the data set described.

How Many Subsamples?

In source-structured data sets, the structuring very often dictates the number of subsamples we should employ. When it doesn't, as in time-structured observations, we can partition the series of cases into a number of equal-time spans. Depending on the total number of cases and how they are distributed over time, the number of partitions may be as few as two or three, or as large as ten or twelve. These are removed, one at a time, to construct an equivalent number of subsamples.

Unstructured data sets, on the other hand, do not suggest the appropriate number of subsamples. The number of randomly chosen half-samples that we need will depend largely on the complexity of our description. If our description is a simple, numerical one, for example, a difference in means or a correlation coefficient, then we'll want the distribution of descriptions for a moderately large number of subsamples, say 100. On the other hand, if our description is complex—for example, a factor analysis or the selection of a subset of attributes, where we need to examine each description in some detail—then as few as 5 or 10 subsamples will give us all the information we need.

Employing Subsample Descriptions

Finally, how do we report, interpret, and act on our subsample descriptions? The answer depends largely on whether the subsample descriptions support or cast doubt on the stability of the overall data set description.

Stable Data Set Description

If the subsample descriptions cluster about the full data set description, they can be used to bracket our basically stable overall description:

1. When there are only a few subsample descriptions, we should either report the full range, in the case of simple numeric descriptions, or describe how the subsample descriptions depart from that for the full data set, in the case of more complex descriptions.

2. When there are many subsample descriptions, we can report the range encompassing, say, the middle 50% of the subsample descriptions.

In studies with randomly sampled cases, we use a confidence interval to report our uncertainty about a more precise point estimate. In available case studies, we don't have that mechanism available to us. But we can use the variability among subsample descriptions to do somewhat the same thing. Keep in mind, however, that the confidence interval reflects uncertainty about a population parameter, based on sample-to-sample variability, whereas our subsample description range reflects only our uncertainty about the description of the present data set, based on the variability of the available cases.

Unstable Data Set Description

Where our data set is unstable, depending heavily on certain cases, we might want to do more than simply report that fact. Here are some options:

1. If the data set is structured and omitting one contributing partition leads to a quite different description, then that partition is inhomogeneous relative to the description. What is distinctive about that partition? Does it contain the earliest cases treated? Are the therapists at clinic C less experienced than those at the other sites? Do car drivers in Alaska and Hawaii face different traveling constraints than do those in the other 48 states? If so, you might want to consider limiting your overall description by excluding the aberrant partition. Of course, you must report both descriptions, and your reason for reducing the data set.

2. A highly skewed distribution of half-sample descriptions, with the full data set description displaced from the middle, can be corrected if we employ a more robust description. Some very frequently used descriptive statistics, notably the mean, standard deviation, and product moment correlation, are strongly influenced by an inhomogeneous case and, hence, can provide unstable descriptions. In later units, we illustrate the use of descriptions of location, variability, and association that are more resistant to the presence in a data set of such cases.

3. The subsample variability in description might be the result of a very unreliable measurement or scoring of our attribute. We might be able to rescore our attribute or measure another, closely related attribute and achieve a more stable description of our data set.

4. We might not be able to alter the description or the choice of attribute. In the presence of skewed variability of the subsample descriptions, we might decide that the cases contributing to the present data set are

hopelessly inhomogeneous and elect to repeat the study with newly recruited cases.

We can choose to change our choice of interpretation, description, attribute, or cases in the face of evidence of lack of stability in our original description. We are obligated, of course, to tell the whole story in reporting our results so that our readers can judge for themselves how much importance to attach to our final description.

Subsamples and Randomized Studies

We introduced subsample descriptions in this unit for making local inferences for nonrandom studies. The technique also can play an important role in random assignment studies with available cases. Having established, for example, that an observed difference in response to a pair of treatments is not likely to be the result of our chance assignment of cases to treatments, we might want to say something about how large that difference is. We can use subsample ranges, say the mid-50%, to good effect here to reflect uncertainty about the exact size of the difference. We use this approach in subsequent applications.

Applications 16

Structured and Unstructured Data

Concepts 16 addressed the task of assessing the stability of the results of a nonrandom study. A description will be stable if the available cases are homogeneous relative to that description. If, on the other hand, some cases are inhomogeneous, they will strongly influence the description and detract from its stability. We search for inhomogeneous cases by repeating the description on a series of subsamples of the data set. Each subsample omits some of the available cases. How the subsamples are composed depends on whether the data set is structured by source or not.

Half-Samples of Unstructured Data Sets

If the data set is unstructured, we take as our subsamples a series of randomly chosen half-samples, each containing a randomly selected half of the available cases.

Design-Divided Cases

This first example of a nonrandom design is taken from Manly (1991). The data in Table 16.1 report the "mandible lengths in millimeters for male and female golden jackals (*Canis aureus*) for 10 of each sex in the collection in the British Museum (Natural History)."

Table 16.1: Golden Jackal Mandible Lengths

Males:	120	107	110	116	114	111	113	117	114	112
Females:	110	111	107	108	110	105	107	106	111	111

Manly wanted to know whether the data provide evidence of a difference in the mean lengths for the two sexes. We can assume that the observations are not on random samples of golden jackal mandibles, not even random samples from the full British Museum collection. The data set is from a nonrandom study.

How stable, then, is the observed difference, (Male Mean − Female Mean), in mandible lengths of 4.8 mm? How should the subsamples be drawn? We have no information about any source structure for the two sets of data, so we will form subsamples randomly.

```
copy (120 107 110 116 114 111 113 117 114 112) SET1 'Set 1
copy (110 111 107 108 110 105 107 106 111 111) SET2 'Set 2
copy 1000 REPS              'No. of half-samples wanted
'
mean SET1 M1               'Mean length, Male skulls
mean SET2 M2               'Mean length, Female skulls
subtract M1 M2 DM          'Observed difference in means
'
size SET1 N1               'Size of male sample
divide N1 2 N1H
round N1H N1H              'Half-sample size for males
size SET2 N2               'Size of female sample
divide N2 2 N2H
round N2H N2H              'Half-sample size for females
'
repeat REPS
  shuffle SET1 H1
  take H1 1,N1H H1         'Random half of male lengths
  shuffle SET2 H2
  take H2 1,N2H H2         'Random half of female lengths
  mean H1 MH1
  mean H2 MH2
  subtract MH1 MH2 DMH     'Difference in means for pair of
  score DMH Z              'random half-samples, saved in Z
end
'
percentile Z (25 75) IQR   'Limits of middle 50%
percentile Z (10 90) M80   'Limits of middle 80%
max Z TOP                  'Largest of half-sample mean diffs
min Z BOT                  'Smallest
concat BOT TOP XTRMS       'Extremes: smallest and largest
print DM IQR M80 XTRMS     'Display results

DM      =       4.8
IQR     =       3.8         5.8
M80     =         3         6.6
XTRMS   =       1.4         8.6
```

Figure 16.2: *Half Samples for a Design-Divided Data Set (Resampling Stats)*

The computing script described in Figure 16.2 illustrates half-sample stability analysis for a design-divided set of cases. To compose half-samples of cases that are divided, though not structured by source, we need to know whether the division was by design or not. Did the design of the jackal mandible study call for measuring 20 mandibles, 10 of which proved to be male and 10 female? Or did it call for measuring 10 male mandibles and 10 female mandibles? If we assume the first scenario, the data set was divided not by design; the division into male and female mandibles occurred after the cases had been selected for the study. In the second scenario, however, the division of the cases was by design—exactly 10 male and 10 female mandibles were selected for the study.

If the division of the data set is not by design, we form half-samples from the full data set and then divide each half-sample before computing the description. For design-divided data sets, however, it is important to preserve the design in the half-samples. So, we form our half-samples separately within each division. The Resampling Stats script of Figure 16.2 forms separate half-samples from the male and female divisions of the jackal mandible data set. We've assumed, in doing so, that the sex-division of the 20 mandibles was by design.

The number of distinct half-samples for this design is given by

$$\left(\frac{10!}{5! \times 5!}\right) \times \left(\frac{10!}{5! \times 5!}\right) = 252 \times 252 = 63504$$

The Resampling Stats script samples, with replacement, 1,000 times from this set of half-samples. The full data set difference of 4.8 mm is solidly in the middle of the distribution of half-sample differences in means. The middle 50% of the half-sample differences are between 3.8 mm and 5.8 mm. For none of the 1,000 half-samples was the difference in means negative. The cases are homogeneous in showing longer mandible lengths, of about 5 mm on average, for male jackals.

Division of Cases after Selection

The data set of Table 16.2 illustrates the difficulty of carrying out the most appropriate analysis of a data set when we have incomplete information about how those data were obtained.

The data are the cholesterol levels of the 40 heaviest men in a health-related study of 3,154 middle-aged men as published in Hand et al. (1994, Data Set

47). The men were classified behaviorally as Type A (urgent, aggressive, ambitious) or Type B (relaxed, less-hurried, noncompetitive), and the question of interest is whether, in these heavy middle-aged men, cholesterol level is related to behavior type.

Table 16.2: *Cholesterol Levels of Heavy Middle-aged Men*

Type A Men: Cholesterol Levels

233 291 312 250 246 197 268 224 239 239 254 276 234 181
248 252 202 218 212 325

Type B Men: Cholesterol Levels

344 185 263 246 224 212 188 250 148 169 226 175 242 252
153 183 137 202 194 213

The fragmentary description of the study establishes that 40 men were selected on the basis of their weights. They were then divided between behavioral types. Thus, in our stability analysis, our random half-samples should each be a random sample of 20 of the 40 men. We would then divide each half-sample between types A and B before computing our description. Picking 20 men at random from the original 40 is likely to lead to unequal numbers of the two types in any given half-sample.

That is true because, given the form in which we have the data, we must use the initial division of cases given in Table 16.2 to assign a type A or B label to a half-sample case. How, though, were the 40 men originally divided into behavioral types? That there were the same number in the two categories suggests that they might have been divided on the basis of their scores on some behavioral test; those with scores above the median for the 40 receiving one label, those with scores below the median, the other. If those scores were available to us, we could replicate this design element in the construction of half-sample data sets; we could split each half-sample of 20 men at the median score for that half-sample to form groups of 10 men for each type. Alas, we do not have those scores (if they were used) and must depend upon the division between types made for the original set of cases.

As a result, we will have learned something about the stability of a description when using the particular type division created for the 40 cases, rather than about the stability of a description based on using a median-split definition of the two types.

Subsamples of Source-Structured Cases

The data in Table 16.3 are for the U.S. presidential election of 1844 and are taken from Noreen (1989). For each state then in the Union, two values are presented: Participation (Partic) is the percent of the eligible voters in the state who voted and Margin is the absolute value of the difference in the percentages voting for the two candidates, Polk and Clay.

Table 16.3: *Voter Participation, US Presidential Election of 1844, by Region*

Northeast States		Atlantic States		Southern States		Western States	
Partic	*Margin*	*Partic*	*Margin*	*Partic*	*Margin*	*Partic*	*Margin*
67.5	13	39.8	20	80.3	5	83.6	2
65.6	19	76.1	5	54.5	6	84.9	2
65.7	18	73.6	1	79.1	5	76.3	12
59.3	12	81.6	1	94.0	4	74.7	17
		75.5	2	80.3	8	68.8	26
		85.0	3	89.6	1	79.3	6
				44.7	3		
				82.7	18		
				89.7	13		

We want to know whether a close race is associated with greater voter turnout. We anticipate that the correlation between participation and margin should be negative.

Note that the states are aggregated in the table geographically; in our stability analysis, we seek assurance that any nationwide correlation between participation and margin was not unduly influenced by one region of the country.

Figure 16.3 shows an SC interaction in which the correlation is computed first over all 25 states, and then after dropping out, in turn, the Northeast, Atlantic, Southern, and Western states.

We begin by creating three vectors to carry the participation and margin scores for the states and the states' index numbers, their positions in the first two vectors.

```
> vector(partic,25,marg,25,indx,25)
> read(partic)
67.5 65.6 65.7 59.3 39.8 76.1 73.6 81.6 75.5 85.0 80.3 54.5
79.1 94.0 80.3 89.6 44.7 82.7 89.7 83.6 84.9 76.3 74.7 68.8
79.3
     25
> read(marg)
13 19 18 12 20 5 1 1 2 3 5 6 5 4 8 1 3 18 13 2 2 12 17 26 6
     25
> fill(indx,1)
> cum(indx)
> indx
    1           2           3           4           5
    6           7           8           9          10
   11          12          13          14          15
   16          17          18          19          20
   21          22          23          24          25

> pmcorr(partic,marg)
    -0.37397543           # Full set of States
> limit(indx,1,4)
> pmcorr(nopick(partic,indx),nopick(marg,indx))
    -0.29899165           # Omit Northeast States
> delimit(indx)
> limit(indx,5,10)
> pmcorr(nopick(partic,indx),nopick(marg,indx))
    -0.22995819           # Omit Atlantic States
> delimit(indx)
> limit(indx,11,19)
> pmcorr(nopick(partic,indx),nopick(marg,indx))
    -0.67690738           # Omit Southern States
> delimit(indx)
> limit(indx,20,25)
> pmcorr(nopick(partic,indx),nopick(marg,indx))
    -0.36750876           # Omit Western States
> delimit(indx)
> bundle("preselec",partic,marg,indx)
[ done ]
```

Figure 16.3: *Correlations Over Structured Subsamples (SC)*

The indx vector is used to eliminate, through the SC nopick function, systematic groups of states from the correlation. The participation-margin correlation is decidedly negative ($r = -0.37$) overall, and it varies from -0.68 to -0.23 as any regional grouping of states is excluded. Although no regional group can be held responsible for the overall description of the relationship of voter participation and margin of victory, the data for the Southern states tend to depress the relationship.

Exercises

1. The following study could have been carried out as a pilot for a larger intervention study. The question in this small-scale study is whether delinquent youths who are also visually impaired are less likely to wear corrective lenses (spectacles) than are visually impaired nondelinquent youths. A positive response might provide part of the motivation for a well-controlled intervention study—"Would increased eye-testing and the provision of free corrective lenses reduce the number of youths becoming delinquent?"

The present study provides no basis for answering that question. It is based on two very small and nonrandom groups, one of delinquent youths who failed an eye test and the other of nondelinquent youths also failing that test. The data are reported in Hand et al. (1994) as Data Set 20.

Table 16.4:   *Spectacle Wearing and Delinquency*

	Juvenile Delinquents	Nondelinquents
Spectacles	1	5
Non-spectacles	8	2
Total	9	7

The odds ratio provides a convenient descriptive statistic that allows us to compare spectacle wearing in the two groups. The odds of spectacle wearing among the juvenile delinquents are just 1 to 8 or $1/8 = 0.125$. Among nondelinquents, though, the odds of spectacle wearing are 5 to 2 or $5/2 = 2.5$. An odds ratio can be formed from these two odds; setting the odds of spectacle wearing among the nondelinquents as the numerator in the ratio and the odds for the delinquents as the denominator gives

$$\text{Odds Ratio(Spectacles)} = \frac{5 \times 8}{2 \times 1} = 20$$

The odds of a nondelinquent but visually impaired youth wearing spectacles are 20 times those for the visually impaired delinquent. How stable is that description of these observations?

2. This second example of a pilot study is one of a number of small-sample studies carried out to explore the properties of a new method of measuring body composition. This pilot study focuses on the relation of age to body fat percentage, as assessed by the new method. A nonrandom group of 18 normal adults, aged between 23 and 61 years, were assessed. The data shown in Table 16.5 are those reported in Hand et al. (1994) as Data Set 17.

Table 16.5: *Sex, Age, and Body Fat Percentage*

Age	23	23	27	27	39	41	45	49
Fat%	9.5	27.9	7.8	17.8	31.4	25.9	27.4	25.2
Sex	M	F	M	M	F	F	M	F
Age	50	53	53	54	56	57	58	58
Fat%	31.1	34.7	42.0	29.1	32.5	30.3	33.0	33.8
Sex	F	F	F	F	F	F	F	F
Age	60	61						
Fat%	41.1	34.5						
Sex	F	F						

Compute the correlation between Age and Fat% over the complete set of 18 cases. What is the magnitude of this correlation? Now, note that there is only a small number of males in the data set. What happens to the correlation when these cases are removed? Did the minority of males exert a strong influence on the correlation? If you conclude they did, go on to assess the stability of the correlation for female cases.

3. Populations can be small. That certainly is the case here. Table 16.6 is taken from Hand et al. (1994, Data Set 122) and provides 1978 information about each of the ten Canadian provinces. Reported are the rates of Cirrhosis mortality (deaths per 100,000 population), Alcohol consumption (liters per person per year) and Pork consumption (kilograms per person per year). We have a trivariate (three values) observation on each of the provinces. Together they constitute a population distribution; they are the collection of all observations of this kind on all Canadian provinces. We want to know whether pork consumption plays a role in the development of cirrhosis of the liver.

Table 16.6:   *Pork and Alcohol Consumption and Cirrhosis*

Province	Cirrhosis	Alcohol	Pork
Prince Edward I.	6.5	11.00	5.8
Newfoundland	10.2	10.68	6.8
Nova Scotia	10.6	10.32	3.6
Saskatchewan	13.4	10.14	4.3
New Brunswick	14.5	9.23	4.4
Alberta	16.4	13.05	5.7
Manitoba	16.6	10.68	6.9
Ontario	18.2	11.50	7.2
Quebec	19.0	10.46	14.9
British Columbia	27.5	12.82	8.4

As excessive alcohol consumption is known, at the level of the individual, to be a risk factor for cirrhosis, aggregate alcohol data were gathered for the provinces as well as aggregate cirrhosis and pork figures. Fit the linear model,

$$\text{Pred(Cirrhosis)}_i = \beta_0 + \beta_1 \text{ Alcohol}_i + \beta_2 \text{ Pork}_i$$

predicting cirrhosis mortality as a linear function of alcohol and pork consumption. The descriptive statistic of interest is β_2. This coefficient models, at the province level, the increase in cirrhosis mortality associated with a one-unit increase in pork consumption when alcohol consumption is held constant. You will find more on the interpretation of the slope coefficient in multiple linear regression models in Lunneborg (1994).

Is β_2 positive? How large is it? How stable is it as a description of the alcohol-adjusted relation of cirrhosis mortality to pork consumption? In particular, how dependent is the population description on the data for one particular province? Form structured subsamples, leaving out one province each time. How does the contribution of pork consumption to cirrhosis mortality change when you do that?

Table 16.7: *Lung Cancer Frequencies Among Smokers and Nonsmokers*

		Cancer	
		Yes	No
Beijing	Smokers	126	100
	Nonsmokers	35	61
Shanghai	Smokers	908	688
	Nonsmokers	497	807
Shenyang	Smokers	913	747
	Nonsmokers	336	598
Nanjing	Smokers	235	172
	Nonsmokers	58	121
Harbin	Smokers	402	308
	Nonsmokers	121	215
Zhengzhou	Smokers	182	156
	Nonsmokers	72	98
Taiyuan	Smokers	60	99
	Nonsmokers	11	43
Nanchang	Smokers	104	89
	Nonsmokers	21	36

4. The data in Table 16.7 are cited in Agresti (1996). The data sets for the 8 Chinese cities are nonrandom. Collapse the eight subtables into one and compute an aggregate odds ratio:

$$\text{Odds Ratio(Smoking)} = \frac{f(\text{Smoker \& Yes}) \times f(\text{Nonsmoker \& No})}{f(\text{Smoker \& No}) \times f(\text{Nonsmoker \& Yes})}$$

where $f(A \& B)$ is the frequency with which A and B occurred together. An odds ratio greater than 1.0 indicates that the odds of contracting lung cancer are greater among smokers than among nonsmokers. Then recompute the odds ratio leaving out one city at a time. Is the aggregate odds ratio a stable description, or does it depend heavily upon the data from one city?

PART II

Resampling Applications

In the balance of this book, we illustrate the application of our three resampling approaches. These applications are organized by study design, beginning with one group of cases and moving through two and more independent group designs to studies with multiple treatment factors and covariates and within-cases designs. Finally, we consider two popular analysis techniques, linear and logistic regression: the use of linear models where the response attribute is, first, measured and then dichotomous.

For each design, we briefly describe the classical parametric analysis, with its assumptions and requirements, and then present our resampling alternatives. Our emphasis is on developing resampling alternatives. If you are interested in a fuller understanding of the classical approaches, you should consult a standard data analysis text. Again, we use our three statistical packages, Resampling Stats, SC, and S-Plus. Exercises, with additional data sets, accompany each of the Applications.

Applications 17, A Single Group of Cases, begins with the problem of assessing the typical size or location of a score distribution. The classic approach is to use the cdf for one of the t-random variables to develop a confidence interval (CI) for a population mean. The assumptions are that we have a random sample that is sufficiently large or was obtained from a normal population distribution. Where only the random sample requirement is met, we can use the bootstrap-t approach to CI estimation. Rerandomization plays no role in one-group studies, but we can use subsampling to assess the stability of a typical value description.

We then move to assessments of the variability of a score distribution, beginning with a parametric example, estimating a CI for the variance of a normal distribution. Bootstrap resampling allows us to estimate CIs for more robust scale parameters and for nonnormal population distributions. We illustrate with a bias-corrected and accelerated (BCA) CI for the population median absolute deviation (MAD). As a nonrandom one-group example, we establish the stability of the robust Winsorized variance as a variability description for a small data set.

We conclude the unit with bivariate association. The parametric example uses Fisher's transformation of the product moment correlation. This produces a pivotal statistic with a normal sampling distribution cumulative distribution function (cdf) if the bivariate data were randomly sampled from a bivariate normal population distribution. Without the assumption of bivariate normality, we illustrate the use of bootstrap resampling to estimate a CI for a population correlation. We finish with a stability assessment of the correlation coefficient computed from a nonrandom data set.

Applications 18, Two Independent Groups of Cases, focuses on what might be the most common study design. The classical two-independent-groups t-test with its assumption of random samples from population distributions that have normal cdfs and a common variance is contrasted with the use of bootstrap resampling to estimate a CI for a more robust and nonparametric comparison of locations, a difference in trimmed means. Rerandomization is used to assess the magnitude of treatment-effect differences for a pair of two-group available-case experimental designs, completely randomized (CR) and randomized blocks (RB), and subsampling is used to assess the stability of a difference in median scores between two halves of a population.

We conclude the unit by showing how resampling can be used to determine the size of groups needed for a randomization study to have acceptable power.

Applications 19, Multiple Independent Groups, treats of observations on more than two groups of cases. Classically, the one-way analysis of variance (ANOVA) has been used for statistical inferences. We begin with a review of parametric CI estimation for a contrast among population distribution means. We then develop the bootstrap-t equivalent, still requiring random samples but neither normality nor homogeneity of variance, and apply this equivalent to contrasts among trimmed population means, a more robust alternative.

Where the K treatment groups are formed by random assignment of available cases, we illustrate the testing of an omnibus null treatment-effect hypothesis in a CR design and focussed-treatment comparisons in a RB design. An example illustrating the assessment of the stability of an ordering of treatments in a nonrandom study is followed by the presentation of an approach to the adjustment of p-values in multiple comparison analyses.

Applications 20, Multiple Factors and Covariates, looks at three important study designs, those with two treatment factors, those with a blocking factor as well as one or more treatment factors, and those with a covariate and treatment factors. For each design we contrast the classical ANOVA or analysis of covariance approach—which depends on random samples from normal population distributions that have homogeneous variances—with (a) bootstrap resampling (for random samples of cases from arbitrary populations), (b) rerandomization (for the random allocations among treatments of available cases), and (c) subsampling (for nonrandom cases). We pay particular attention to the concept of an interaction between factors, the difference between treatment and blocking factors, and the estimation of adjusted means.

Applications 21, Within-Cases Treatment Comparisons, treats repeated measures designs, each case assessed on M occasions, perhaps receiving different treatments. There are two classical parametric models for such data. The univariate repeated measures ANOVA requires that—in the M-variate population distribution from which we have randomly sampled—each of the M measures has a normal cdf and a common variance. The bivariate correlations among all pairs of measures are restricted as well. The multivariate ANOVA model requires that the M-variate population distribution be multivariate normal in form. Nonparametric bootstrap resampling based CIs or hypothesis tests make no such parametric assumptions about the M-variate population distributions. We illustrate with the analysis of data from a fixed-order, multiple-treatment random sample study. This is followed by the rerandomization analysis of a study in which the order of presentation of tasks is randomly determined for each case. Finally, share price changes, for a population of shares, are assessed for stability using subsamples.

Applications 22, Linear Models: Measured Response, covers some inferential tasks associated with that most popular of all statistical models, the linear model. In its classic parametric form, the linear model subsumes multiple linear regression and fixed-effects ANOVA. The requirements of the parametric linear model are random samples from population response distributions that have (a) normal cdfs, (b) a common variance, and (c) means that are linear functions of one or more explanatory attributes. We illustrate inference under that model and then develop two nonparametric bootstrap resampling alternatives, depending on whether the set of design points in a study is fixed or sampled.

We present a bootstrap-t methodology for testing multiple constraints on the parameters of a linear model; this Q-test methodology serves as a nonparametric alternative to the parametric F-ratio. Because researchers use model comparisons extensively, we develop the nonparametric Q-test with some care. The S-Plus code provided for the Q-test is reusable while the algorithms are sufficiently detailed that code can be produced easily for other compter packages.

Continuing with random sample applications, we outline an approach to obtaining better estimates of the accuracy of predictions made from linear models. We conclude the unit with illustrations of linear models where cases have not been sampled randomly.

Applications 23, Categorical Response Attributes, looks at two of the more important designs involving categorical responses. The first is the cross-classification of cases by a pair of categorical attributes. The classical analysis, the chi-squared test of row-by-column independence, requires random samples large enough that the sampling distribution of the usual test statistic is sufficiently close to its asymptotic form. We present a small sample alternative, based on rerandomization ideas but appropriate for random samples as well as for random assignment designs. We also look at resampling inference for measures of association between the rows and columns of the two-way table of cross-classified frequencies.

The second categorical data design discussed in the unit is logistic regression, the use of linear models for a dichotomous response. We review and illustrate the parametric model. We present a nonparametric, bootstrap resampling alternative that permits us to estimate CIs for the parameters of the linear log-odds model and for functions of those parameters. We also examine the linear log-odds function in studies that use available rather than randomly sampled cases. In particular, we illustrate testing the significance of the odds-multiplier in a random case assignment study and assessing odds-multiplier stability in a nonrandom study.

Part II is followed by a Postscript that summarizes the principal ideas of this book.

Applications 17

A Single Group of Cases

We now have all the concepts needed to allow us to choose among and use the most appropriate resampling technique. Beginning with this unit we see how to apply these techniques to a succession of data analysis goals associated with case-based studies. We start with the analysis of data from a one-group study. Our cases, random or available, have not been grouped by treatment nor blocked by any characteristic they possess.

Random Sample or Set of Available Cases

The single group of cases is either a random sample from some known population or a set of available cases, presumably homogeneous. As the cases are not divided, they cannot have been randomly allocated among two or more treatment or treatment sequence groups. Thus, rerandomization will not be an appropriate resampling technique. Rather, we shall focus on using bootstrap inference for the random sample and subsampling for the nonrandom set of available cases.

Typical Size of Score Distribution

The first question asked of many score distributions is "How large are the scores?" This is usually refined to "What is the magnitude of the typical score?," if we intend simply to describe a data set, or to "What is the location of the population sampled?," if we want to make an inference from the scores for a random sample of cases. The latter question sometimes takes the form "Could the population be centered at μ_0?," where we want to test a hypothesis about the location of the population, rather than estimate it.

Parametric Inference of Population Mean

Parametric inference about the location of a population almost always takes the form of estimating a confidence interval (CI) for the population mean or testing some hypothesis about its value. The vehicle for doing this is the sampling distribution of the statistic

$$t(\overline{x}) = \frac{\overline{x} - \mu_0}{\widehat{\text{SE}}(\overline{x})}$$

where $(x_i, i = 1, 2, \ldots, n)$ is the score distribution for a random sample of n cases, $\overline{x} = (1/n)\sum x_i$ is the mean of that sample distribution, and

$$\widehat{\text{SE}}(\overline{x}) = \sqrt{(1/n)[1/(n-1)]\sum(x_i - \overline{x})^2}$$

is the estimated standard error (SE) of $\overline{x}$.

If the distribution for the population sampled has a normal cumulative distribution function (cdf), then the sampling distribution of $t(\overline{x})$ has the cdf of a t-random variable, one with $(n-1)$ degrees of freedom (df). When this requirement is met, we can form a $(1 - 2\alpha)100\%$ equal tails exclusion CI that will have limits

$$\left[\overline{x} - \{\widehat{\text{SE}}(\overline{x}) \times F_{t,(n-1)}^{-1}(1-\alpha)\}, \overline{x} + \{\widehat{\text{SE}}(\overline{x}) \times F_{t,(n-1)}^{-1}(1-\alpha)\}\right]$$

where $F_{t,(n-1)}^{-1}(1-\alpha)$ is the $(1-\alpha)$ quantile of the t-distribution with $(n-1)$ df, a quantity readily available from statistical tables or from statistical computing packages. For an appropriate choice of α, we can test the hypothesis that $\mu = \mu_0$; if μ_0 falls outside the CI, we customarily reject the hypothesis.

We first encountered the data of Table 5.1 in Applications 5.

Table 5.1:   *Weights (in grams) of 50 Randomly Sampled Bags of Pretzels*

464	450	450	456	452	433	446	446	450
447	442	438	452	447	460	450	453	456
446	433	448	450	439	452	459	454	456
454	452	449	463	449	447	466	446	447
450	449	457	464	468	447	433	464	469
457	454	451	453	443				

The data are taken from Daly et al. (1995) and give the weight in grams of a random sample of 50 bags of pretzels. The bags were sampled from the output of a packaging machine.

```
> pretzels<-c(464,450,450,456,452,433,446,446,450)
> pretzels<-c(pretzels,447,442,438,452,447,460,450,453,456)
> pretzels<-c(pretzels,446,433,448,450,439,452,459,454,456)
> pretzels<-c(pretzels,454,452,449,463,449,447,466,446,447)
> pretzels<-c(pretzels,450,449,457,464,468,447,433,464,469)
> pretzels<-c(pretzels,457,454,451,453,443)
> n<-length(pretzels)
> n
[1]   50
> xbar<-mean(pretzels)
> xbar
[1] 451.22
> sexbar<-sqrt((1/n)*var(pretzels))
> sexbar
[1] 1.187448
> q1malf<-qt(0.975,(n-1))
> q1malf
[1] 2.009575
> xbar-(sexbar*q1malf)
[1] 448.8337
> xbar+(sexbar*q1malf)
[1] 453.6063
```

Figure 17.1: *CI Limits for a Normal Population Mean (S-Plus)*

If we had reason to believe that, in the long-term, the pretzel bags produced by this machine have weights that follow a normal distribution, we could estimate a 95% CI for the mean of that long-term distribution via the S-Plus computational dialog shown in Figure 17.1.

Or we could use a shortcut. The S-Plus command `t.test(pretzels)` produces the same 95% confidence bounds: [448.83 gm, 453.61 gm].

Nonparametric Location Inference

It would be exceptional for us to know the population score distribution X to be normal. Very rarely do we know the parametric form of the population distribution we've sampled. We can, however, form a nonparametric estimate of X from the sample distribution x. By drawing a large number of random bootstrap samples from this estimated population and computing our estimate of the population location in each sample, we build a bootstrap sampling distribution from which we can estimate, by the techniques of Concepts 9 or 10, an appropriate CI for the population location.

```
> vector(pretwts,50)
> read(pretwts)
464 450 450 456 452 433 446 446 450
447 442 438 452 447 460 450 453 456
446 433 448 450 439 452 459 454 456
454 452 449 463 449 447 466 446 447
450 449 457 464 468 447 433 464 469
457 454 451 453 443
     50
> matrix(PRETZ,50,1)
> setcol(PRETZ,pretwts,1)
> proc tmean_mn(){}     # User-written functions for 20%
> proc tmean_se(){}     # trimmed mean and its SE
> vector(gsiz,1)
> read(gsiz)
50
     1
> iboot_t(tmean_mn,tmean_se,PRETZ,gsiz,1,2500,FBOOT,95)
_____theta:
  450.93333       # Sample 20% trimmed mean
_____means of  2500  bootstrap estimates:
  451.02249
_____SDs of  2500  bootstrap estimates:
  1.0459703
 Parameter: 1   95 % CI: 448.83469 to  453.17407
     Total iboot_t() running time: : 23.18
> bundle("pretzels",pretwts,PRETZ,gsiz)
[ done ]
```

Figure 17.2: *Bootstrap-t CI for 20% Trimmed Population Mean (SC)*

In bootstrap inference, we are free to choose the population location parameter and sample estimate that most interest us. Because the mean, both sample and population, are not robust against very small or very large scores in the distribution, the mean can give misleading information about where the center of the bulk of the distribution lies. In consequence, we will use, in general, a 20% trimmed population mean as the population location and the 20% trimmed sample mean as the point estimate.

In Figure 17.2, we use the SC procedure `iboot_t` to estimate a 95% bootstrap-*t* CI for the 20% trimmed mean of the pretzel weight population distribution. In doing so, we apply again the user-written procedures of Applications 9 (Figure 9.2) to compute a 20% trimmed mean (`tmean_mn`) and the estimated SE of a 20% trimmed mean (tmean_se). The CI is estimated by iboot_t rather than fboot_t because we assume the population sampled contained a very large number of bags, more than 5,000. (In fact, we do not know the population size. Were we to know, fboot_t might be a more appropriate choice.)

For our pretzel bag weights example, the 95% CI for the 20% trimmed population mean has almost the same limits as the parametric normal 95% CI for the untrimmed population mean. In Figure 17.3, we explore the bootstrap sampling distribution of $t(\overline{x}_{t,0.20}^*)$ stored by the iboot_t procedure as the 2,500 elements of the first column of the matrix FBOOT.

```
> vector(tout,2500)
> getcol(FBOOT,tout,1)
> desc(tout)

_____tout   (# 2500 )   (not integer)
    range=        8.09094    (-4.50908   to   3.58186 )
    mean=         7.88008e-006
    median=       0.0322821
    sd=           1.0418
    iqr=          1.37658    (-0.677445  to   0.699132 )
    MAD=          0.680454
```

Figure 17.3: *Description of Bootstrap Sampling Distribution, $t(\overline{x}_{t,0.20}^*)$ (SC)*

The bootstrap sampling distribution is slightly negatively skewed (median greater than mean) but only in the tails. The middle 50% of the bootstrap sample Studentized trimmed means fell between -0.677 and 0.699, the limits of iqr, the interquartile range.

Stability of Typical Value Description

If the single set of cases is not a random sample, then we have no basis for estimating the location of any population. We can describe the typical size of scores for the available cases and assess that description for stability.

Table 17.1 is taken from Dusoir (1997) and gives the percentage of total sleep time spent in stage-0 sleep by 16 healthy males aged between 50 and 60. The 16 men are not a random sample.

Table 17.1:   *Percentage of Sleep Time at Stage-0 for 16 Adult Males*

0.07	0.69	1.74	1.9	1.99
2.41	3.07	3.08	3.1	3.53
3.71	4.01	8.11	8.23	9.1
10.16				

The 20% trimmed mean of these stage-0 percentages is 3.491%. In Figures 17.4 and 17.5, Resampling Stats is asked to assess the stability of this description of the typical percentage by generating a series of 100 random half-samples and computing a 20% trimmed mean in each of those.

```
copy (0.07 0.69 1.74 1.9 1.99 2.41 3.07 3.08) dset
concat dset (3.1 3.53 3.71 4.01 8.11 8.23 9.1 10.16) dset
copy 0.2 gama
copy 100 resamps
size dset n                  ' data set size
multiply 0.5 n temp
round temp hn
if hn > temp
   subtract hn 1 hn          ' half-sample size
end
multiply gama n temp
round temp tn
if tn > temp
   subtract tn 1 tn       .  ' trim size, full data set
end
multiply gama hn temp
round temp thn
   if thn > temp
   subtract thn 1 thn       'trim size, half-sample
end
```

Figure 17.4:   *Part 1, Resampling Stats, Stability of a γ-Trimmed Mean*

Figure 17.4 inputs the data set, the trimming proportion, and the desired number of half-samples to be drawn. Trimming 20% of the observations from each tail of the data set is a frequent choice. And where half-samples are to be randomly chosen, 100 will be sufficient to appraise stability of description. The script then computes the size of each half-sample and the number of small and large scores to trim from the original data set and from each half-sample.

Figure 17.5 computes and reports the trimmed mean of the full data. Then, 100 random half-samples are constructed and a trimmed mean found for each. Finally, the distribution of 100 half-sample trimmed means is evaluated. The mean, the limits to the middle 50% (the interquartile range limits), the .025 and 0.975 quantiles, and the smallest and largest half-sample trimmed means are reported.

```
add tn 1 bot
subtract n tn top
sort dset dset
take dset bot,top tset        'the trimmed data set
mean tset tmean               ' and its mean
print tmean
'
add thn 1 bot
subtract hn thn top
repeat resamps
  shuffle dset dset
  take dset 1,hn hset         'a random half-sample
  sort hset hset
  take hset bot,top hset
  mean hset thmean            'half-sample trimmed mean
  score thmean stabil
end
mean stabil xbar
percentile stabil (25 75) iqr    'mid 50% of h-s tmeans
percentile stabil (2.5 97.5) mid95   'mid 95%
max stabil top
min stabil bot
concat bot top extrm              'smallest and largest
print xbar iqr mid95 extrm
```

Figure 17.5: *Part 2, Resampling Stats, Stability of a Trimmed Mean*

The output of the Resampling Stats script appears in Figure 17.6.

The 20% trimmed mean was 3.5%. How stable is that description? Nearly 95% of the random half-samples had trimmed means that were within 1.4% of this value. Let's contrast this with what we would have found for the untrimmed mean. In a second random selection of 100 half-samples, the width of the MID95 range for untrimmed half-sample means was 3.27%, nearly 20% wider than the 2.77% range for our trimmed half-sample means. The trimmed mean is that much more stable than the untrimmed one.

```
TMEAN     =       3.491

XBAR      =       3.5596

IQR       =       2.935        4.075

MID95     =       2.1533       4.9233

EXTRM     =       2.0433       5.665
```

Figure 17.6:   *Half-Sample Trimmed Mean Properties (Resampling Stats)*

Variability of Attribute Scores

After we have found or estimated the typical value or location of a score distribution, the next most often asked question is "How much do the scores vary, either from one another or about the typical value?"

Parametric Variance Estimate

The classical measure of population score variability is the variance

$$\sigma^2 = (1/N)\sum(X_i - \mu)^2$$

the average of the squared distances of the scores in a population distribution from their mean. It has a sample or data set analog,

$$S^2 = (1/n)\sum(x_i - \bar{x})^2$$

Mathematical analysis has established that, when X is a normal distribution, the plug-in estimator, S^2, is not unbiased. The unbiased estimator of the normal distribution variance,

$$\hat{\sigma}^2 = [(1/(n-1)]\sum(x_i - \bar{x})^2$$

for $(x_i, \; i = 1, 2, \; \ldots \; , n)$ a random sample from X, has a known sampling distribution related to the chi-squared distribution with $(n-1)$ df. This relationship allowed us in Applications 5 to develop lower and upper bounds to the $(1 - 2\alpha)100\%$ CI for the normal distribution variance:

$$\text{LB} = \hat{\sigma}^2 \times \left[(n-1)/F^{-1}_{\chi^2,(n-1)}(1-\alpha) \right]$$

and

$$\text{UB} = \hat{\sigma}^2 \times \left[(n-1)/F^{-1}_{\chi^2,(n-1)}(\alpha) \right]$$

where $F^{-1}_{\chi^2,(n-1)}(\alpha)$ is the α quantile of the chi-squared distribution with $(n-1)$ df.

We used these formulas in Applications 5 to develop a 90% CI for the distribution of weights in the population of pretzel bags sampled for Figure 4.1. Those limits were [52.07 gm^2, 101.81 gm^2].

Because the variance is a value in a metric that is the square of that for the sample data, x, we often are more comfortable talking about the square root of the variance, the standard deviation (SD). In the pretzel bag weight example, we would have as the 90% CI for σ, the population SD, [$\sqrt{52.07 \text{ gm}^2}$, $\sqrt{101.81 \text{ gm}^2}$] = [7.22 gm, 10.09 gm].

Nonparametric Variability Estimates

The family of chi-squared distributions provides a model for the sampling distribution of $\hat{\sigma}^2$ only when our random sample distribution, x, is obtained from a normal population distribution. For other population distributions, the sampling distribution of $\hat{\sigma}^2$ or of the plug-in estimator S^2 is not known to have a neat mathematical form. We can, however, use bootstrap resampling from an estimated population and the bias-corrected and accelerated (BCA) approach of Concepts and Applications 10 to produce useful estimates of a $(1 - 2\alpha)100\%$ CI for the population variance.

Robust Measures of Variability

The variance and standard deviation, whether sample or population, are subject to the same lack of robustness as the mean. One very small or very large value in the distribution can cause σ^2 (and σ, S^2, S, or $\hat{\sigma}^2$ as well) to take an arbitrarily large value. For this reason, we find it better to use a measure of variability that is sensitive to differences in the bulk of the attribute score values, but resists the influence of a few very large or very small values. Statisticians have proposed a number of robust scale measures, and the interested reader should consult a source such as Wilcox (1997). We illustrate two of these robust measures of variability or scale. The first is our choice when a measure like the variance is wanted, a measure sensitive to squared distances from the location of a distribution, and we'll use the second when we want a scale measure in the same metric as the attribute.

Winsorized Variance

We've already met the first robust variability measure in connection with the trimmed mean of Applications 9. The γ-Winsorized variance reduces the influence of the g [$= \text{int}(\gamma \times n)$] smallest and g largest values in a distribution by replacing the former with the next larger value and the latter with the next smaller value before computing a mean and squared deviations from that mean. Thus, the score distribution

$$[8, 12, 24, 40, 42, 44, 44, 48, 50, 55, 58, 60, 66, 70, 75, 80]$$

would be Winsorized—for $\gamma = 0.20$ and $g = \text{int}(0.2 \times 16) = 3$—to

$$[40, 40, 40, 40, 42, 44, 44, 48, 50, 55, 58, 60, 66, 66, 66, 66]$$

before the variance is computed.

Median Absolute Deviation

Our second robust scale measure is the median absolute deviation (MAD), which is the median of the absolute values of the deviations of n (or, N) scores from their median. We can illustrate the computation. For the distribution

$$[8, 12, 24, 40, 42, 44, 44, 48, 50, 55, 58, 60, 66, 70, 75, 80]$$

we first find the median, $(48 + 50)/2 = 49$. Then, we compute the absolute value of the deviation of each score from this median,

$$[41, 37, 25, 9, 7, 5, 5, 1, 1, 6, 9, 11, 17, 21, 26, 31]$$

Finally, we sort these absolute deviations,

$$[1, 1, 5, 5, 6, 7, 9, 9, 11, 17, 21, 25, 26, 31, 37, 41]$$

and find their median, $(9 + 11)/2 = 10$, the MAD.

The MAD is easily interpreted. Half of the scores in the distribution are at least this close to the median, the other half are further removed from that value.

Because it depends on the median of absolute deviations rather than on the average of squared deviations, as does the computation of the SD, the MAD cannot be thrown off by a few scores lying a large distance from the center of the distribution.

Here is an example in which we use the BCA approach to estimate a 95% upper confidence bound for the MAD of a population score distribution. That is, we want to find a value, u, such that we have 95% confidence that the population MAD is no larger than u.

Reproduced from Applications 9 is Table 9.1, taken from Wilcox (1997) and giving the time in seconds that a laboratory apparatus is kept in contact with a target by each of a random sample of 19 university students. We earlier used these data to illustrate the estimation of a CI for a trimmed mean, Figure 9.3.

We'll use the SC procedure fbootci2 to compute the BCA confidence bound because our sample of 19 students was drawn from a small population of 100 volunteers.

Table 9.1: *Time on Target for a Random Sample of University Students*

77	87	88	114	151	210	219	246	253
262	296	299	306	376	428	515	666	1310
2611								

The computational flow is presented in Figure 17.7. Our plug-in estimator of the population MAD is $t = 114$. Before calling `fbootci2`, we create a user-written procedure to find the MAD for the scores that make up the first column of a matrix (1) and return that result as the first element of a vector (3).

```
> load(wilcox31)
  [TTARG nn NN ]
> nn
  19
> NN
  100
> mad(TTARG')
  114
> proc madp(){
    $3$[1]= mad($1$')[1]
    }
> fbootci2(madp,madp,TTARG,nn,NN,1,2500,FBOOT,90)
_____tstat:
  114
_____thetahat:
  114
_____means of  2500  bootstrap estimates:
  114.8132
_____SDs of  2500  bootstrap estimates:
  44.956912
_____z0 (bias corrections):
  -0.061199912
_____a (acceleration constants):
  -0.009089560
  Theta 1  90 % BC & Accelerated CI:   44    to   182
     Total fbootci2() running time:  10.27
```

Figure 17.7: *BCA Confidence Bound for Population MAD (SC)*

This procedure is passed to `fbootci2` together with the data matrix, sample and population sizes, the number of bootstrap samples to draw, the name of a matrix to hold the bootstrap sampling distribution, and the level of confidence to be used in CI estimation. Because we want an 95% upper confidence bound, we specify 90% for the equal tails exclusion CI. In both instances, we exclude the upper 5% of the sampling distribution from θ-coverage.

Our BCA estimate of the 95% upper confidence bound, based on one random sequence of 2,500 bootstrap samples, is 182. We have 95% confidence, then, that the population MAD is no larger than 182.

Nonrandom Score Variability

The γ-Winsorized variance is robust against the presence of very small or very large scores in a data set. Let's assess its stability as a descriptor of the variability of a nonrandom data set. Here, again, are the percentages of total time spent in stage-0 sleep by a nonrandom set of 16 healthy males aged between 50 and 60.

Table 17.1:   *Percentage of Sleep Time at Stage-0 for 16 Adult Males*

0.07	0.69	1.74	1.9	1.99
2.41	3.07	3.08	3.1	3.53
3.71	4.01	8.11	8.23	9.1
10.16				

Figure 17.8 gives the text of a user-written procedure `var_wvar` that computes the 20% Winsorized variance of a set of scores. The scores are obtained from the first column of a matrix (1) and the Winsorized variance is returned as the first entry to a vector (3).

This procedure is used by `subsam_k` to compute Winsorized variances for a sequence of 100 randomly chosen random half-samples from the original data set, as described in Figure 17.9.

```
proc var_wvar(){
  local(gama,n,v)
  gama= 0.20
  n=rows($1$)
  vector(v,n)
  getcol($1$,v,1)
  winsor(v,(floor(gama*n)/n)*100)
  $3$[1]=var(v)
}
```

Figure 17.8: *User-written SC Procedure for* γ*-Winsorized Variance*

```
> vector(stage0,16)
> read(stage0)
0.07 0.69 1.74 1.9 1.99 2.41 3.07 3.08
3.1 3.53 3.71 4.01 8.11 8.23 9.1 10.16
     16
> matrix(SLP,16,1)
> setcol(SLP,stage0,1)
> proc var_winvar(){}
> nn:=16
> hn:=8
> subsam_k(var_winvar,1,SLP,nn,hn,100,SS)
     Full Sample Statistics:    5.9092359
     Description of Statistic Number:   1
____v1  (# 100 )
     range=        11.1795    ( 0.341169  to   11.5207 )
     mean=         5.79754
     median=       6.46345
     sd=           3.28207
     iqr=          6.69534    ( 1.48324  to   8.17858 )
     MAD=          1.83333
     Total subsam_k time:   0.16
>bundle("stg0slp",SLP,stage0,nn,hn)
[done]
```

Figure 17.9: *Stability of Winsorized Variance of Stage-0 Sleep (SC)*

The 20% Winsorized variance for the full data set, $5.9\%^2$, is solidly in the middle of the 100 half-sample Winsorized variances, lying between the mean and median of that distribution. Fifty percent of the half-sample descriptions was within $\pm 1.8\%^2$ of $6.5\%^2$, those two values being the MAD and median of this particular half-sample distribution.

Association Between Two Attributes

Measures of location and scale characterize the distribution of a single attribute, a univariate distribution. When two attributes are measured on each case, we are often interested in whether or by what amount the two attributes are associated with one another. Measures of association between two attributes are characteristics of a bivariate distribution, each case contributes a pair of scores.

Normal Product Moment Correlation

The best-known statistical measure of the strength of linear association between two measured attributes is the product moment correlation coefficient. It is defined for the sample or data set as

$$r_{xy} = \frac{(1/n)\sum_{i=1}^{n}(x_i - \overline{x})(y_i - \overline{y})}{S_x \times S_y}$$

where S_x and S_y in the denominator are the SDs of the distributions of scores on the two attributes, x and y, and the numerator is known as the covariance between the two sets of scores.

When computed over a random sample of n cases, r_{xy} is a plug-in estimator of the population correlation,

$$\rho_{XY} = \frac{(1/N)\sum_{i=1}^{N}(X_i - \mu_X)(Y_i - \mu_Y)}{\sigma_X \times \sigma_Y}$$

a parameter of a population bivariate distribution: (X_i, Y_i), $i = 1, \ldots, N$. The parameter ρ_{XY} takes values between -1 and $+1$. Values close to zero indicate little or no association between the two attributes. Negative values of ρ_{XY} indicate a negative relation between the two attributes and positive values of ρ_{XY} indicate a positive relation between the two attributes.

If we want to use r_{xy} as an estimate of ρ_{XY}, we must know something of the sampling distribution of our estimator. Mathematical statistics provides a way of getting at the sampling distribution of r_{xy} whenever we know that the population distribution, (X_i, Y_i), is bivariate normal. This requires, among other things, that the univariate population distributions X and Y both have normal cdfs.

When our bivariate sample is drawn from a bivariate normal distribution, a transformation of the sample correlation, known as the Fisher transformation,

$$\widehat{\zeta}_{xy} = (1/2)\log\left(\frac{1 + \mathrm{r}_{xy}}{1 - \mathrm{r}_{xy}}\right)$$

is very nearly pivotal. Closely enough, the sampling distribution of $\widehat{\zeta}_{xy}$ has

1. A mean of

$$\zeta_{XY} = (1/2)\log\left(\frac{1 + \rho_{XY}}{1 - \rho_{XY}}\right)$$

2. A SD of $\left[1/\sqrt{n-3}\right]$

3. A normal cdf

The parametric sampling distribution of $\widehat{\zeta}_{xy}$ makes it easy to write down the estimated limits of a $(1 - 2\alpha)100\%$ equal tails exclusion CI for ζ_{XY}:

$$[\zeta_{\mathrm{lo}}, \zeta_{\mathrm{up}}] =$$

$$[\widehat{\zeta}_{xy} - z_{[1-\alpha]}/\sqrt{n-3}, \ \widehat{\zeta}_{xy} + z_{[1-\alpha]}/\sqrt{n-3}]$$

where $z_{[1-\alpha]}$ is the $(1 - \alpha)$ quantile for the standard normal distribution (e.g., for $\alpha = 0.05$, $z_{[1-\alpha]} = 1.65$).

Once confidence limits have been estimated for ζ_{XY}, we can apply the inverse of the Fisher transformation to convert $[\zeta_{\mathrm{lo}}, \zeta_{\mathrm{up}}]$ to the corresponding confidence limits for ρ_{XY}, for example:

$$\rho_{\mathrm{lo}} = 1 - \frac{2}{1 + e^{2\zeta_{\mathrm{lo}}}}$$

Some data from Applications 7 (Table 7.1) are reproduced below. They are from Efron and Tibshirani (1993) and give the average Law School

Admissions Test (LSAT) scores and undergraduate grade point averages (GPA) for the 1973 entering classes at 15 randomly sampled U.S. law schools.

Table 7.1:   *LSAT and GPA Scores for Randomly Sampled Law Schools*

School	LSAT	GPA	School	LSAT	GPA	School	LSAT	GPA
1	576	3.39	6	580	3.07	11	653	3.12
2	635	3.30	7	555	3.00	12	575	2.74
3	558	2.81	8	661	3.43	13	545	2.76
4	578	3.03	9	651	3.36	14	572	2.88
5	666	3.44	10	605	3.13	15	594	2.96

How strong is the association between LSAT and GPA in the population of law schools sampled? If we assume, for the moment, that the bivariate distribution of LSAT and GPA scores for the 82 U.S. law schools sampled was bivariate normal, we can follow a six-step procedure to estimate a $(1 - 2\alpha)100\%$ equal tails exclusion CI for the population correlation:

1. Select α, based on your desired degree of confidence.

2. Find $z_{[1-\alpha]}$, the $(1 - \alpha)$ quantile of the standard normal distribution.

3. Compute the sample correlation coefficient, r_{xy}.

4. Apply the Fisher transformation, converting r_{xy} to $\widehat{\zeta}_{xy}$.

5. Compute the CI limits for ζ_{XY}: $\widehat{\zeta}_{xy} \pm (z_{[1-\alpha]}/\sqrt{n-3})$.

6. Invert the transformation, converting $[\zeta_{lo}, \zeta_{up}]$ to $[\rho_{lo}, \rho_{up}]$.

We have SC apply this algorithm to the LSAT-GPA example in Figure 17.10.

The sample correlation between the two variates was $r_{xy} = 0.776$ providing an estimated 90% CI for the population correlation, ρ_{XY}, of $[0.509, 0.907]$.

Nonparametric Population Correlation

We cannot be certain, in most instances, that our bivariate sample distribution was drawn from a normal bivariate population distribution and, hence, that we should estimate confidence bounds for ρ_{XY} by the above algorithm.

```
> vector(lsat,15,gpa,15)
> read(lsat)
576 635 558 578 666 580 555 661 651
605 653 575 545 572 594
      15
> read(gpa)
3.39 3.30 2.81 3.03 3.44 3.07 3.00 3.43
3.36 3.13 3.12 2.74 2.76 2.88 2.96
      15
> alfa=0.05
> r=pmcorr(lsat,gpa)
> r
     0.77637449
> zeta= 0.5*(log((1+r)/(1-r)))
> zeta
     1.0361785
> z1ma=normQ(alfa,0,1)
> z1ma
      1.6448536
> zetlo=zeta-(z1ma/sqrt(12))
> zethi=zeta+(z1ma/sqrt(12))
> zetlo
     0.56135013
> zethi
     1.5110068
> rholo= 1 - (2/(1+exp(2*zetlo)))
> rhohi= 1 - (2/(1+exp(2*zethi)))
> rholo
     0.50897849
> rhohi
     0.90711756
> bundle("etlawsch",lsat,gpa)
[ done ]
```

Figure 17.10: *CI for a Bivariate Normal Product Moment Correlation (SC)*

What we can do, however, is to estimate BCA confidence bounds for ρ_{XY}, making no parametric assumptions about the form of our bivariate population distribution. We'll work out two such estimates here for the U.S. law school population LSAT-GPA correlation. We'll find BCA limits both for ρ_{XY} and for ζ_{XY}, then use the inverse Fisher transformation to convert the latter to a second set of BCA limits for ρ_{XY}.

```
> load(etlawsch)
  [lsat gpa ]
> matrix(LAWSCH,15,2)
> setcol(LAWSCH,lsat,1)
> setcol(LAWSCH,gpa,2)
> nn:= 15
> NN:= 82
> proc zetacorr(){}
> fbootci2(zetacorr,zetacorr,LAWSCH,nn,NN,2,2500,FBOOT,90)
_____tstat:
  0.77637449   1.0361785
_____thetahat:
  0.78071143   1.0475943
_____means of  2500  bootstrap estimates:
  0.77711815   1.1235375
_____SDs of  2500  bootstrap estimates:
  0.11963213   0.35405085
_____z0 (bias corrections):
 -0.085328795 -0.081304243
_____a (acceleration constants):
 -0.075671565 -0.10060452

  Theta 1  90 % BC & Accelerated CI: 0.48288996   to   0.92173144
  Theta 2  90 % BC & Accelerated CI: 0.51359364   to   1.5821492

> zetalo= 0.51359364
> zetahi= 1.5821492
> rholo= 1- (2/(1+exp(2*zetalo)))
> rhohi= 1- (2/(1+exp(2*zetahi)))
> rholo
      0.47274046
> rhohi
      0.91893687
```

Figure 17.11: *Nonparametric BCA CIs for a Population Correlation (SC)*

Because the population is small ($N = 82$), we use the SC procedure `fbootci2` to generate our bootstrap samples and sampling distribution. Figure 17.11 shows the interaction with SC.

The 90% BCA confidence bounds for ρ_{XY} are reported as $[0.483, 0.922]$ and those for ζ_{XY} as $[0.514, 1.582]$. When we back-transform the ζ_{XY} limits, we have a second BCA CI estimate for ρ_{XY} of $[0.473, 0.919]$. The two nonparametric BCA CI estimates are quite close to each other, and both are slightly wider than the bivariate normal limits. This suggests the bivariate normal limits are narrower than appropriate to this problem.

Why did we produce two BCA CIs here? The theory behind the BCA estimation procedure is that it employs an implicit pivotal transformation for the general estimator t. The Fisher transformation is an explicit pivotal transformation of the particular estimator, r_{xy}, at least when the population distribution is bivariate normal. Although the Fisher transformation may not be pivotal for r_{xy} in the nonnormal setting, the near-agreement of the raw and Fisher-transformed CIs provides evidence that the BCA procedure did use a good implicit pivotal transformation of r_{xy}. The two CIs are not identical. Should we have a preference between them? We might have helped BCA find a better implicit pivot by prepivoting, by starting out with an approximate pivot, that of the Fisher transformation. Using an approximate pivot as a starting point when one is available improves CI estimation.

Although the implicit-pivotal BCA approach generally provides good CI estimates, we should take advantage of an explicit pivotal transformation if we know of one that might work for our estimator. We follow this principle when we compute bootstrap-t confidence limits for a location parameter, first Studentizing the estimate. It is a good idea to use the Fisher transformation of the correlation coefficient when you compute BCA confidence limits for ρ_{XY}.

In Figure 17.11, a user-written procedure zetacorr was declared and then used by fbootci2 to compute both r_b^* and $\widehat{\zeta}_b^*$ from each bootstrap sample it forms. The text of this procedure appears in Figure 17.12

```
proc zetacorr(){
   local(n,x,y,r)
   n= rows($1$)
   vector(x,n,y,n)
   getcol($1$,x,1)
   getcol($1$,y,2)
   r= pmcorr(x,y)
   $3$[1]=r
   $3$[2]=0.5*log((1+r)/(1-r))
}
```

Figure 17.12: *Providing SC fbootci2( ) with r and its Fisher Transformation*

Correlation in Nonrandom Data

When our bivariate data set, (x_i, y_i), $i = 1, \ldots, n$, is for an available set of cases, rather than for a random sample, no population correlation exists to be estimated. Rather, we concentrate on describing the association between attributes x and y in the data set.

Table 17.2 is taken from Risebrough (1972) and shows, for a nonrandom set of 65 Anacapa pelican eggs, the thickness in millimeters of the shell, and the concentration in parts-per-million of PCB (polychlorinated biphenyls, industrial pollutants). The theory suggests that the two attributes should be negatively correlated, with PCB concentration inhibiting shell growth.

Table 17.2:   *PCB Concentration and the Thickness of Pelican Eggs*

```
> pcb
   452       184       115       315       139
   177       214       356       166       246
   177       289       175       296       205
   324       260       188       208       109
   204        89       320       265       138
   198       191       193       316       122
   305       203       396       250       230
   214        46       256       204       150
   218       261       143       229       173
   132       175       236       220       212
   119       144       147       171       216
   232       216       164       185        87
   216       199       236       237       206

> thick
  0.14      0.19      0.2       0.2       0.21
  0.22      0.22      0.22      0.23      0.23
  0.23      0.23      0.24      0.25      0.25
  0.26      0.26      0.26      0.26      0.27
  0.28      0.28      0.28      0.29      0.29
  0.29      0.29      0.29      0.29      0.3
  0.3       0.3       0.3       0.3       0.3
  0.3       0.31      0.31      0.32      0.34
  0.34      0.34      0.35      0.35      0.36
  0.36      0.36      0.37      0.37      0.37
  0.39      0.39      0.39      0.4       0.41
  0.41      0.42      0.42      0.42      0.44
  0.46      0.46      0.47      0.49      0.49
```

In Figure 17.13, we trace a stability analysis of the correlation between these two egg attributes using the SC procedure `subsam_k`.

The correlation between PCB concentration and shell thickness is negative, $r_{xy} = -0.253$, over the full set of 65 eggs. This description is near the median and mean of the 100 random half-sample descriptions generated by subsam_k and solidly within the middle 50% of those, between -0.348 and -0.180. The r_{xy} of -0.253 would appear to be a stable description of the degree of association between PCB concentration and shell thickness in this data set. The user-written procedure proccorr used in this script is the zetacorr procedure of Figure 17.12 modified to return only the correlation and not the Fisher transformation as well.

```
> load(anacapa)
[  thick[65]  pcb[65]   ]
> matrix(ANAC,65,2)
> setcol(ANAC,pcb,1)
> setcol(ANAC,thick,2)
> nn:=65
> hn:=33
> proc proccorr(){}
> subsam_k(proccorr,1,ANAC,nn,hn,100,SS)
     Full Sample Statistics:   -0.25281265
     Description of Statistic Number:   1
_____v1   (# 100 )
     range=        0.641566    (-0.597508  to  0.0440587 )
     mean=        -0.26355
     median=      -0.284723
     sd =          0.116978
     iqr=          0.167632    (-0.347691  to -0.180059 )
     MAD=          0.0742171
```

Figure 17.13: *Subsample Analysis of PCB-Shell Thickness Correlation (SC)*

Exercises

1. This exercise is based on a random sample of $n = 32$ employed adult males. We want to know how blood pressure is related to body size in the population sampled.

Table 17.3 gives the data as reported in Kleinbaum, Kupper, and Muller (1988). SBP is the systolic blood pressure, and QUET is the Quetelet index of body size, computed as

$$QUET = 100 \times \left[\frac{\text{Weight in pounds}}{(\text{Height in inches})^2} \right]$$

The Quetlet index assesses stockiness!

Table 17.3:   *Blood Pressure and Body Size Data*

SBP	135	122	130	148	146	129	162	160
QUET	2.876	3.251	3.100	3.768	2.979	2.790	3.668	3.612
SBP	144	180	166	138	152	138	140	134
QUET	2.368	4.637	3.877	4.032	4.116	3.673	3.562	2.998
SBP	145	142	135	142	150	144	137	132
QUET	3.360	3.024	3.171	3.401	3.628	3.751	3.296	3.210
SBP	149	132	120	126	161	170	152	164
QUET	3.301	3.017	2.789	2.956	3.800	4.132	3.962	4.010

An equation of the form

$$\text{Pred}(\text{SBP}_i) = \beta_0 + \beta_1(\text{QUET}_i)$$

predicting SBP from QUET was fit to these sample data. The estimated values for the intercept and slope were $\widehat{\beta}_0 = 70.576$ and $\widehat{\beta}_1 = 21.492$. This second estimate reflects the impact of QUET on SBP; on average, an increase of 1.0 in QUET is associated with an increase of about 21.5 (mm of mercury, Hg) in SBP.

But our $\widehat{\beta}_1 = 21.492$ is the slope coefficient estimate just for this sample. How much might the estimate differ for a second or third sample of men from this same population? An important measure of this sample-to-sample variability is the SE of the estimate. Find a bootstrap estimate of this SE.

2. A May 1992 Wall Street Journal presidential poll of a random sample of 750 voters put Bush at 36% (270 voters), Perot at 30% (225), and Clinton at 28% (210). The estimate for Bush's margin over his closest rival was 6%. Establish a 95% CI for this estimate.

3.   Table 17.4 gives the examination marks (out of 75) for 134 students (Basak, Balch, and Basak, 1992). The students are a set of available cases, not a random sample. Use half-sample analysis to describe the stability of the

untrimmed and 20% trimmed means. Do the same for the variance and Winsorized variance.

Table 17.4:   *Examination Marks for 134 Candidates*

49	49	70	55	52	55	61	65	57	71
49	48	49	69	44	53	49	52	53	36
61	68	67	53	33	64	57	56	41	40
42	40	51	53	62	61	49	54	57	71
45	70	58	62	28	72	37	67	51	55
68	58	61	43	60	53	51	51	60	64
66	52	45	48	51	73	63	32	59	68
35	64	62	51	52	44	64	65	56	52
59	66	42	67	48	56	47	68	58	59
45	31	47	56	38	47	65	61	45	63
66	44	57	56	56	54	61	58	46	62
68	58	47	66	61	58	45	55	54	54
54	41	65	66	38	51	49	49	51	42
61	69	42	53						

Applications 18

Two Independent Groups of Cases

Certainly the most common statistical analyses are those in which we compare the performances of two independent groups of cases. The two groups might be the only ones in a study or, as in the studies we cover in Applications 19, the two might be singled out for comparison by a specific research hypothesis. The two groups might be random samples from two populations. The two groups might have been formed by the random division of one or more groups of available cases. Or they might be simply two groups of available cases, distinguished one from the other by some nonrandom selection or division. Depending on how our cases were selected or divided, one of our three modes of inference—population, causal, or descriptive—is appropriate. And one of our three resampling techniques—bootstrap sampling, rerandomization, or subsampling—provides the needed statistical basis. Any of Concepts 7 through 13 and Concepts 16, then, can be important in the analysis of the two independent groups study.

Constitution of Independent Groups

How the two groups of cases are constituted provides the key to the appropriate statistical comparison of their performances. Here we describe the different possibilities.

Explicit Samples from Two Populations

We can identify two natural populations and then draw a random sample of cases from each. For example, we might have a random sample of 50 Open University graduates of 1997 and a random sample of 50 Open University graduates of 1987. By sampling each population randomly, you create two random and independent samples.

Our statistical interest would center on inferences about the two population score distributions. Where parametric assumptions are unwarranted, we make these population inferences from a bootstrap resampling perspective.

Splitting a Single Random Sample

A random sample of patients—from, say, a well-defined diagnostic population—might be divided, randomly, into two treatment groups. The two groups are then given different treatments, and the responses of the two groups are those of random samples from two prospective populations with, possibly, different population score distributions on one or more response to treatment attributes.

A single random sample can be divided arbitrarily as well. For example, a random sample of patients, all receiving the same treatment, might be divided between males and females. The result again would be two random samples, one sample from the population of male patients receiving treatment and a second sample from the population of female patients receiving the same treatment. The male and female populations, here, are each contained within the combined population initially sampled.

As we have score distributions for random samples of cases from two populations, our treatment comparisons again take the form of population inferences, based on either classical parametric methods or bootstrap resampling.

Random Split of a Set of Available Cases

If our set of patients, sharing a diagnosis, is not a random sample but an available set of cases, such as all those patients with that diagnosis receiving treatment at a particular clinic, then their random division into two treatment groups cannot produce two random samples. However, the random allocation of available cases among treatments does enable local causal inference. Specifically, we can make use of our rerandomization techniques to test treatment-effect hypotheses.

We can first sort the available cases into two or more blocks on the basis of some shared attribute, and then randomly divide each block of cases into two treatment groups. Such a blocked design permits the same causal inferences, but requires that we attend to the blocking structure in our rerandomization analysis.

Nonrandom Separation of Available Cases

If the two groups are neither random samples—either by explicit sampling from two populations or by the splitting of a single random sample—nor the result of the random allocation of available cases, then we cannot use either population or causal inference. We can describe the difference in response to treatments and assess that description for stability. Stability assessment is local descriptive inference.

The nonrandom separation can occur in either of two ways. The separation could be the result of the nonrandom splitting of a single set of available cases, for example, a group of 11-year-old school children is divided between two classrooms and, hence, two methods of teaching mathematics, on the basis of the alphabetic ordering of their surnames. Or the separation might be the result of the nonrandom selection of two sets of cases, for example, a researcher compares the spatial memory abilities of members of an orienteering club with those of participants in a square-dance contest.

Location Comparisons for Samples

We typically compare the performances of two groups by comparing the magnitudes of the scores on one or more attributes for the cases in one group with those for the other group. When the two groups are independent, we must aggregate scores over cases within a group to make the comparison.

Two Independent Groups *t*-test

Taught in all introductory statistics courses, the classical technique for comparing the magnitudes of scores for two independent groups of cases is the independent groups *t*-test. The *t*-test assumes that the two sample score distributions, x and y, are for random samples from two populations with score distributions, X and Y, that:

1. Have normal cdfs

2. Have the same variance, $\sigma_X^2 = \sigma_Y^2 = \sigma^2$

3. Might have different means, μ_X and μ_Y

The independent groups *t*-test is a test of the null hypothesis that the difference in population means takes a particular value, $(\mu_X - \mu_Y) = (\mu_X - \mu_Y)_0$. More often than not, our research hypothesis leads to a null-hypothesized value of zero for the difference; we want to test the hypothesis that $\mu_X = \mu_Y$.

The test statistic has the form

$$t(\bar{x} - \bar{y}) = \frac{(\bar{x} - \bar{y}) - (\mu_X - \mu_Y)_0}{\widehat{SE}(\bar{x} - \bar{y})}$$

where $\widehat{SE}(\bar{x} - \bar{y})$ is the estimated standard error (SE) of the difference in sample means. This estimate can be computed as

$$\widehat{SE}(\bar{x} - \bar{y}) = \sqrt{\left(\frac{1}{n_x} + \frac{1}{n_y}\right)\left(\frac{\sum\limits_{i=1}^{n_x}(x_i - \bar{x})^2 + \sum\limits_{i=1}^{n_y}(y_i - \bar{y})^2}{n_x + n_y - 2}\right)}$$

where n_x and n_y are the sizes of the two samples.

If the two population distributions have normal cdfs and the same variance and if $(\mu_X - \mu_Y)_0$ correctly specifies the difference in population means, then the sampling distribution of the test statistic, $t(\bar{x} - \bar{y})$, has as its cumulative distribution function (cdf) that of the t-random variable with $(n_x + n_y - 2)$ degrees of freedom (df). The quantiles of this cdf are well-known. The known sampling distribution allows us to reject the null hypothesis—depending on the alternative—if $t(\bar{x} - \bar{y})$ is too large or too small .

Table 18.1 is taken from Daly et al. (1995) and gives the birth weights in Kilograms of a random sample of 50 infants who displayed severe idiopathic respiratory distress syndrome (SIRDS), a life-threatening condition.

Table 18.1: *Birth Weights (in kg) of 50 Randomly Sampled Infants with SIRDS*

Infants Did Not Survive			Infants Survived		
1.050	1.175	1.230	1.130	1.575	1.680
1.310	1.500	1.600	1.760	1.930	2.015
1.720	1.750	1.770	2.090	2.600	2.700
2.275	2.500	1.030	2.950	3.160	3.400
1.100	1.185	1.225	3.640	2.830	1.410
1.262	1.295	1.300	1.715	1.720	2.040
1.550	1.820	1.890	2.200	2.400	2.550
1.940	2.200	2.270	2.570	3.005	
2.440	2.560	2.730			

Twenty-three of the infants survived their first year, 27 did not. Because the initial 50 were a random sample from a SIRDS population, we now have two random samples—a random sample of 23 from the population of SIRDS who survived their first year, and a random sample of 27 from the population of SIRDS who failed to survive.

The research hypothesis was that higher birth weight SIRDS infants would be more likely to survive than would those with lower birth weights. For our *t*-test, then, the null hypothesis is that $(\mu_{surv} - \mu_{nonsurv})_0 = 0$, while the research or alternative hypothesis is that $(\mu_{surv} - \mu_{nonsurv}) > 0$.

```
> diffmean<-mean(surv)-mean(nonsurv)
> diffmean
[1] 0.6156506
> sedf<-sqrt(((1/27)+(1/23))*(26*var(nonsurv)+22*var(surv))/48)
> sedf
[1] 0.1673084
> tval<-diffmean/sedf
> tval
[1] 3.679735
> pval<- 1-pt(tval,48)
> pval
[1] 0.000295094

> t.test(surv,nonsurv,alternative="greater")

     Standard Two-Sample t-Test
data:   surv and nonsurv
t = 3.6797, df = 48, p-value = 0.0003
alternative hypothesis: true mean difference is greater than 0
95 percent confidence interval:
 0.3350369            NA
sample estimates:
 mean of x mean of y
  2.307391  1.691741
```

Figure 18.1: *Two Sample* t-*test, Survival of SIRDS Infants (S-Plus)*

Figure 18.1 traces the computation of the *t*-test in S-Plus, the birth weights in the two samples having been stored as the vectors `surv` and `nonsurv`. The mean birth weight of the surviving infants is, indeed, greater than that for the nonsurvivors, consistent with our research hypothesis. We compute $(\bar{x} - \bar{y})$, $\widehat{SE}(\bar{x} - \bar{y})$, and, from these, the *t*-statistic: $t(\bar{x} - \bar{y}) = 3.680$. The proportion of the sampling distribution of the *t*-distribution with 48 df greater than 3.680 is evaluated to be only p $= 0.0003$. We safely can reject the null hypothesis in favor of the alternative; the population mean infant weight of survivors is greater than that of nonsurvivors.

In the lower part of Figure 18.1, the S-Plus t.test command does all of the computing for us and arrives at the same results.

Nonparametric Comparison of Locations

The *t*-test requirements that both population distributions have normal cdfs and a common variance are very unlikely to be met in practice. We can replace the parametric *t*-test with a nonparametric bootstrap alternative, one that requires neither normality nor variance homogeneity.

From Concepts 9 and 11, we can use the bootstrap-*t* approach to estimate a $(1 - 2\alpha)100\%$ confidence interval (CI) for $(\mu_X - \mu_Y)$ with α chosen appropriate to our hypothesis testing strategy. Then, depending on our alternative hypothesis, we can reject our null hypothesis if $(\mu_X - \mu_Y)_0$ lies below, above, or simply outside that CI.

The bootstrap-*t* approach to CI estimation depends on building the bootstrap sampling distribution of the Studentized plug-in estimator, t^*,

$$t(t^*) = \left[\frac{t^* - t}{\widehat{SE}(t^*)}\right]$$

Adapting this general formulation to our present problem, we have

$$t^* = (\bar{x}^* - \bar{y}^*)$$
$$t = (\bar{x} - \bar{y})$$
$$\widehat{SE}(t^*) = \widehat{SE}(\bar{x}^* - \bar{y}^*)$$

This last is the estimated SE of the difference in sample means. We don't use the formula employed for the denominator of the parametric *t*-test because that assumes a common variance for the two population distributions. Rather, we use a formula that allows the two population distributions to have different variances:

$$\widehat{SE}(\bar{x}^* - \bar{y}^*) = \sqrt{\frac{\hat{\sigma}_{x^*}^2}{n_x} + \frac{\hat{\sigma}_{y^*}^2}{n_y}}$$

Figure 18.2 shows the text of two user-written procedures, p_mndiff and p_sediff, that can be employed with the SC procedures fboot_t and iboot_t to estimate small or large population nonparametric bootstrap-*t* CIs for $(\mu_X - \mu_Y)$.

We exercise these user-written procedures in the SC session described in Figure 18.3.

```
proc p_mndiff(){
  local(n,nx,x)
  n= rows($1$)
  nx= $2$[1]
  vector(x,n)
  getcol($1$,x,1)
  $3$[1]= mean(x[1:nx])-mean(x[(nx+1):n])
}

proc p_sediff(){
  local(n,nx,x)
  n= rows($1$)
  nx= $2$[1]
  vector(x,n)
  getcol($1$,x,1)
$3$[1]=sqrt((vare(x[1:nx])/$2$[1])+(vare(x[(nx+1):n])/$2$[2]))
}
```

Figure 18.2:   *SC User Procedures for a Difference in Means and its SE*

```
> ghq
32 34 49 33 42 49 17 48 37 48 37 22
13 13 11 21  8 24 28  9 24 22 11
> gsiz
12 11
> matrix(MGHQ,23,1)
> setcol(MGHQ,ghq,1)
> proc p_mndiff(){}
> proc p_sediff(){}
> iboot_t(p_mndiff,p_sediff,MGHQ,gsiz,1,2500,FBOOT,95,99)
 Parameter: 1    95 % CI: 12.489364 to 27.364165
 Parameter: 1    99 % CI:  9.297134 to 29.896189
```

Figure 18.3:   *A Lower Confidence Bound for Difference in Population Means (SC)*

The vector ghq is taken from Leach (1991) and contains scores on the General Health Questionnaire (GHQ) for two random samples of police officers. The first 12 scores belong to a random sample of officers present at the 1989 Hillsborough football stadium disaster. The remaining 11 scores are for a random sample of those officers from the same police force who were not directly involved. We would expect the Hillsborough officers to have higher scores on the GHQ, subscribing to more somatic complaints, than those not present.

We have 97.5% confidence that the difference in population distribution means, $\mu(\text{GHQ} \mid \text{Hillsborough}) - \mu(\text{GHQ} \mid \text{Not Hillsborough})$, is greater than 12.49 and 99.5% confidence that it is greater than 9.30. If we were inclined to report one, the p-value for a test of the null hypothesis of a zero mean difference versus the alternative of a positive difference would be something less than 0.005, the lower bound of the 99% CI being distinctly greater than zero.

Difference in Trimmed Means

As we noted in Applications 9, the mean, both sample and population, can be influenced very strongly by the presence in the distribution of a tiny number of very small or very large scores. There we preferred a more robust measure of location, the γ-trimmed mean of the distribution—the mean of the distribution after the smallest and largest $(\gamma \times 100)\%$ of the scores have been removed. In general, we take $\gamma = 0.20$, giving as our location measure the mean of the middle 60% of the score distribution.

The difference between means clearly will be subject to the same lack of resistance to disparate scores and can give a misleading picture of the separation of the location of two population distributions. Again, we prefer to estimate a CI for a more robust population characteristic, the difference in 20% trimmed population means. We'll base this CI on the bootstrap sampling distribution for the corresponding plug-in estimator, the difference in 20% trimmed sample means.

We can make this move to a more robust comparison because bootstrap inference provides a mechanism, previously unavailable, for estimating the relevant sampling distributions and CIs.

Because we remain interested in population locations, we'll again use our bootstrap-t approach, based on developing a bootstrap sampling distribution for the Studentized form of our plug-in estimator, t^*,

$$t(t^*) = \left[\frac{t^* - t}{\widehat{\text{SE}}(t^*)} \right]$$

where we now have

$$t^* \quad = (\bar{x}_{t,\gamma}^* - \bar{y}_{t,\gamma}^*)$$

$$t \quad = (\bar{x}_{t,\gamma} - \bar{y}_{t,\gamma})$$

$$\widetilde{SE}(t^*) = \widetilde{SE}(\bar{x}_{t,\gamma}^* - \bar{y}_{t,\gamma}^*)$$

This last is the estimated SE of the difference in sample γ-trimmed means.

It can be computed as

$$\widetilde{SE}(\bar{x}_{t,\gamma}^* - \bar{y}_{t,\gamma}^*) = \sqrt{\frac{S_{w,\gamma}^2(x^*)}{(1 - 2\gamma)n_x} + \frac{S_{w,\gamma}^2(y^*)}{(1 - 2\gamma)n_y}}$$

where $S_{w,\gamma}^2(x^*)$ is the γ-Winsorized variance of the bootstrap sample distribution x^*. This SE estimate does not require symmetry of the sampled populations (Wilcox, 1997).

Figure 18.4 shows the text of the two user-written SC procedures that are required for the bootstrap-t procedure. The first, p_tmndif, computes the difference in trimmed means, $t^* = (\bar{x}_{t,\gamma}^* - \bar{y}_{t,\gamma}^*)$, and the second, p_tsedif, estimates the SE of this difference, $\widetilde{SE}(t^*) = \widetilde{SE}(\bar{x}_{t,\gamma}^* - \bar{y}_{t,\gamma}^*)$.

In Figure 18.5, we develop bootstrap-t CIs for both the difference in population means and the difference in 20% trimmed population means for a study reported by Wilcox (1997).

The study involves two random samples, each consisting of 20 men. The first sample was drawn from a Control population; the second sample was drawn from a population of sons of Alcoholic fathers. The subjects consumed a fixed amount of alcohol, and on the following morning rated the severity of their hangover symptoms. Researchers hypothesized that the sons of alcoholics would rate their reactions less severely.

The hangover symptom ratings appear in the SC dialog as the vector hngovr, the first 20 for the control subjects, the second 20 for the subjects with alcoholic fathers. Our plug-in estimate of the difference in population means is

$$\bar{x}(\text{hngovr} \mid \text{Control}) - \bar{x}(\text{hngovr} \mid \text{Alcoholic}) = 4.55$$

The lower bound of the 90% CI for the difference in population means is estimated at -0.11, running beyond, but barely, the null hypothesized value of zero.

```
proc p_tmndif() {
  local(gama,nx,ny,n,xy,x,y,gamx,gamy)
  gama= 0.20
  nx= $2$[1]
  ny= $2$[2]
  n= nx+ny
  vector(xy,n,x,nx,y,ny)
  getcol($1$,xy,1)
  copy(xy[1:nx],x)
  copy(xy[(nx+1):n],y)
  gamx=floor(gama*nx)/nx
  gamy=floor(gama*ny)/ny
  $3$[1]= tmean(x,gamx)-tmean(y,gamy)
}


proc p_tsedif() {
  local(gama,nx,ny,n,xy,x,y,gpcx,gpcy,dx,dy)
  gama= 0.20
  nx= $2$[1]
  ny= $2$[2]
  n= nx+ny
  vector(xy,n,x,nx,y,ny)
  getcol($1$,xy,1)
  copy(xy[1:nx],x)
  copy(xy[(nx+1):n],y)
  gpcx=100*(floor(gama*nx)/nx)
  gpcy=100*(floor(gama*ny)/ny)
  winsor(x,gpcx)
  winsor(y,gpcy)
  dx= nx*(1-(2*gama))^2
  dy= ny*(1-(2*gama))^2
  $3$[1]= sqrt((vare(x)/dx)+(vare(y)/dy))
}
```

Figure 18.4: *SC Procedures for a Difference in Trimmed Means and its SE*

In contrast, the plug-in estimate of the difference in 20% trimmed population means is

$$\bar{x}_{t,0.20}(\text{hngovr} \mid \text{Control}) - \bar{x}_{t,0.20}(\text{hngovr} \mid \text{Alcoholic}) = 3.67$$

while the estimated 90% CI for the difference in 20% trimmed population means has a lower bound of 0.48, stopping just before reaching zero. The distance from point estimate to lower confidence bound estimate was reduced from $4.55 - (-0.11) = 4.663$, for the difference in population means, to $3.67 - 0.48 = 3.184$, for the difference in population trimmed means. We have a more precise estimate of the difference in trimmed means than of the difference in untrimmed means.

```
> hngovr[1:20]
      0         32         9         0         2
      0         41         0         0         0
      6         18         3         3         0
     11         11         2         0        11
> hngovr[21:40]
      0          0         0         0         0
      0          0         0         1         8
      0          3         0         0        32
     12          2         0         0         0
> matrix(MHO,40,1)
> setcol(MHO,hngovr,1)
> gsiz:= 20,20
> proc p_mndiff(){}
> proc p_sediff(){}
> proc p_tmndif(){}
> proc p_tsedif(){}
> iboot_t(p_mndiff,p_sediff,MHO,gsiz,1,2500,FBOOT,90)
_____theta:
  4.55
_____means of  2500  bootstrap estimates:
  4.49416
_____SDs of  2500  bootstrap estimates:
  3.0040772
 Parameter: 1    90 % CI: -0.11308432  to 10.536137
     Total iboot_t() running time: : 17.58
> iboot_t(p_tmndif,p_tsedif,MHO,gsiz,1,2500,FBOOT,90)
_____theta:
  3.6666667
_____means of  2500  bootstrap estimates:
  3.7543
_____SDs of  2500  bootstrap estimates:
  2.1551844
 Parameter: 1    90 % CI:  0.48254203  to 9.5107487
     Total iboot_t() running time: : 25.04
```

Figure 18.5: *CIs for Differences in Means and Trimmed Means (SC)*

The two CIs would lead to different decisions about the null hypothesis of a common population location if we were to test that hypotheis at the 5% significance level. We would reject that null hypothesis if the population location were the trimmed mean, and not reject it if the population location were the 20% trimmed mean.

Magnitude Differences, CR and RB Designs

We turn now to a second class of two-group designs, those in which available cases are randomly divided between two treatment groups. We'll look at two examples, one in which a single block of cases is randomized, a completely randomized (CR) design, and a second in which randomization takes place separately for each of several blocks of cases, a randomized blocks (RB) design.

Completely Randomized Two-Group Studies

In Applications 12, we examined a prototypic two-group CR study. A collection of Alloxan-diabetic mice was randomly divided between a Control group and a group receiving Insulin. Serum albumen levels in the mice were measured. The research hypothesis was that the insulin mice should have lower amounts of albumen.

Analyzing the two-group CR study involves rerandomizing cases between the two treatment groups. Under the null treatment-effect hypothesis, a case would respond identically to the two treatments. Thus, cases' responses accompany them on rerandomization. Recomputing the treatment comparison statistic over all such rerandomizations leads to a null reference distribution against which the actual treatment-comparison statistic can be evaluated. If the proportion of the reference distribution with values at least as extreme as the actual treatment statistic is small enough, we can dismiss the possibility that the observed treatment comparison was a chance result of the randomization of cases. Rather, again because of the random assignment of cases to treatments, we can attribute the observed difference in treatment response to differences in treatment. We infer the treatments caused the differences in treatment response.

Randomized Blocks Two-Group Studies

The analysis of the RB two-groups design is very much the same as for the CR two-groups design. The only difference is that in rerandomizing cases for the purpose of building a reference distribution of treatment-comparison statistics, we must honor the block structure of the original randomizations. That is, we rerandomize cases by blocks, rather than as a whole.

Figure 18.6 presents a Resampling Stats script for a four-block, two-treatment study. Within each block of eight cases, four were randomized to the High and four to the Low treatment groups.

The response scores for the 64 cases are arranged on input by blocks. Within each block of scores, those of the low-treatment cases precede those of the high-treatment cases. The difference in mean response is calculated and reported as MDIFF.

```
copy (13 16 14 15 17 18 17 18) BL1
copy (10 8 11 9 14 12 16 17) BL2
copy (8 8 9 10 12 11 16 15) BL3
copy (8 7 8 8 11 11 13 16) BL4
take BL1 1,4 LO1
take BL1 5,8 HI1
take BL2 1,4 LO2
take BL2 5,8 HI2
take BL3 1,4 LO3
take BL3 5,8 HI3
take BL4 1,4 LO4
take BL4 5,8 HI4
concat LO1 LO2 LO3 LO4 LO
concat HI1 HI2 HI3 HI4 HI
mean LO MLO
mean HI MHI
subtract MHI MLO MDIFF
print MLO MHI MDIFF
copy 1 TAIL
repeat 999
  shuffle BL1 BL1
  shuffle BL2 BL2
  shuffle BL3 BL3
  shuffle BL4 BL4
  take BL1 1,4 LO1
  take BL1 5,8 HI1
  take BL2 1,4 LO2
  take BL2 5,8 HI2
  take BL3 1,4 LO3
  take BL3 5,8 HI3
  take BL4 1,4 LO4
  take BL4 5,8 HI4
  concat LO1 LO2 LO3 LO4 LO
  concat HI1 HI2 HI3 HI4 HI
  mean LO MLO
  mean HI MHI
  subtract MHI MLO HMDIFF
  if HMDIFF >= MDIFF
    add TAIL 1 TAIL
  end
end
divide TAIL 1000 PROP
print PROP
```

Figure 18.6:   *Reference Distribution for Randomized Blocks (Resampling Stats)*

Under the null treatment hypothesis, cases and their response scores can be rerandomized between the low- and high-treatment groups. In accordance with the RB design of the study, this rerandomization (`shuffle` and `take`) is carried out separately on each block of cases (`BL1`, `BL2`, and so on).

The script tallies the number of times the treatment comparison statistic computed for a rerandomization of the cases, s^H, equals or exceeds the original treatment comparison statistic, s. Such a tally is appropriate where, as here, the alternative to the null hypothesis is that the response scores for high-treatment cases should exceed those for the low-treatment cases.

Magnitude Differences, Nonrandom Designs

We can describe the difference in the magnitudes of two distributions when obtained from two nonrandom groups of cases, even though we cannot make population or causal inferences. Here is an example. The data in Table 18.2 are taken from Hand et al. (1994, Data Set 44).

Table 18.2:   *Oral Socialization Anxiety Scores for 39 Nonliterate Societies*

Oral Explanation of Illness Absent Societies

   6   7   7   7   7   7   8   8   9   10 10 10 10 12 12 13

Oral Explanation of Illness Present Societies

   6   8   8   10 10 10 11 11 12 12 12 12 13 13 13 14 14
   14 15 15 16 17

In a study of child-rearing practices in nonliterate cultures, each of 39 societies was characterized in two ways. First, an Oral Socialization Anxiety score was assigned to each society, reflecting the severity and rapidity of oral socialization in child-rearing. Second, the cultures were dichotomized between those in which Oral Explanations of Illness are present and those in which they are absent.

Do the data provide support for the hypothesis that oral-socialization and illness-explanations are positively related, that higher levels of oral socialization are to be found among societies with illness explanations than among societies that do not have such explanations?

The Applications 16 Resampling Stats script (Figure 16.2) provides a prototype for the half-sample stability analysis of a two-group comparison where the two groups are design-divided and we want to half-sample each group separately.

```
copy (6 7 7 7 7 8 8 9 10 10 10 10 12 12 13) attrib
concat attrib (6 8 8 10 10 10 11 11 12 12 12 12) attrib
concat attrib (13 13 13 14 14 14 15 15 15 16 17) attrib
copy 16 n1
size attrib nn
copy 1,nn idvec
add n1 1 n1p
take idvec 1,n1 id1              'case numbers, group 1
take idvec n1p,nn id2           'case numbers, group 2
take attrib id1 grp1            'scores, group 1
take attrib id2 grp2            'scores, group 2
median grp1 mdn1               'instructions for
median grp2 mdn2               'making the
subtract mdn2 mdn1 diffmdn     'group comparison
print mdn1 mdn2 diffmdn
divide nn 2 x
round x nh
if nh> x
  subtract nh 1 nh              'half-sample size
end
repeat 100
  shuffle idvec idvec
  take idvec 1,nh ihalf         'case numbers,
  sort ihalf ihalf             'random half-sample
  count ihalf <= n1 t1         'no. in group 1
  add t1 1 b2
  take ihalf 1,t1 id1          'case numbers, group 1
  take ihalf b2,nh id2         'case numbers, group 2
  take attrib id1 grp1         'scores, group 1
  take attrib id2 grp2         'scores, group 2
  median grp1 mdn1             'develop the between
  median grp2 mdn2             'groups comparison for
  subtract mdn2 mdn1 hadiff    'the half-sample
  score hadiff dist
end
percentile dist (25 75) iqr_dist
mean dist m_dist
median dist md_dist
stdev divn dist sd_dist
print iqr_dist m_dist md_dist sd_dist
histogram dist
```

Figure 18.7:   *Half-Samples for Design-Divided Data (Resampling Stats)*

In the present example, the grouping of societies relative to the presence or absence of illness-explanations played no role in their selection for the study; the societies were aggregated into those groups after they had been selected for study. Therefore we'll want to draw our random half-samples from the full set of societies, and then group the societies included in any half-sample as to the presence or absence of illness-explanations.

The Resampling Stats script of Figure 18.7 shows how such half-samples might be drawn, and then subdivided into two groups to be compared.

The oral-socialization scores are input to a vector, the scores for explanation-present societies following those for explanation-absent societies. The size of the first group of societies is also supplied. In the half-sample loops, we sort the randomly selected case identifications. This allows us to use the information that case identification from one through n1 belong to the first group while case identifications larger than that belong to the second group.

Notice that in making the group comparisons, we look here at the difference in medians rather than the difference in means. The reason for this is that the oral-socialization ratings might only order the societies rather than measure the strength of that attribute in the cultures.

The results of running the script of Figure 18.7 appear in Figure 18.8.

```
MDN1      =          8.5
MDN2      =           12
DIFFMDN   =          3.5

IQR_DIST  =         2.75        4.5
M_DIST    =        3.675
MD_DIST   =          3.5
SD_DIST   =       1.2296
```

Figure 18.8: *Stability of Median Comparison of Oral-Socialization*

The median oral-socialization score for the explanation-present societies was 3.5 points higher than was the median for the explanation-absent societies. This description of the difference lies solidly in the middle of the distribution of half-sample descriptions—very near the median and mean of that distribution—suggesting that the description is a stable one for this data set.

The histogram of the 100 half-sample median differences produced by Resampling Stats is reproduced as Figure 18.9. The median in each half-sample is the average of two integers and, hence, can only take one of two forms, $I.0$ or $I.5$ where I is an integer. The difference between the two medians is similarly limited, as is shown by Figure 18.9. Note that the median difference was not negative for any of the half-samples.

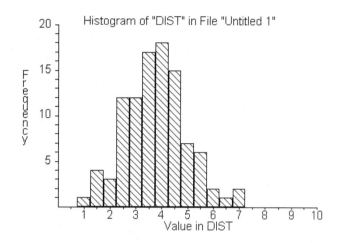

Figure 18.9: *Histogram of Half-Sample Median Differences*

Study Size

Our ability to detect statistically a difference in population distribution locations is influenced by the size of the random samples we use for detection. Similarly, in studies that depend on the random assignment of available cases to different treatments, our ability to detect a treatment effect depends on the number of available cases employed. In designing a study, we should try to ensure that we sample or recruit enough cases to have a good chance of detecting any important treatment effects.

Some approaches to this problem are based on parametric assumptions, but we can use a nonparametric resampling approach to help us answer the question "How many cases do I need?"

The technique to be described is one in which we simulate what we believe might occur in the study we are designing. To carry out the simulation, we need to specify certain things about the study:

   1. The design of the study. How many treatment levels? How are cases to be allocated to treatments; is it a CR, a RB design, and so on?

2. Some reasonable sizes for the study. What is the possible range of sizes for the study? Is there some upper limit, fixed either by availability of potential cases, cost of experimentation, availability of facilities, and so forth?

3. The effects of treatment. What experimental effects do you want to be able to detect? What response data would you expect to see at the different levels of treatment, if the experiment works?

The first of these requirements tells us how we would use rerandomization to assess the chance nature of any observed treatment effect. The second helps us choose some realistic alternative study sizes for the simulation. And, most important, the third requirement gives us the grist for our resampling mill, information that we can use in simulation. The first two factors structure the simulation, and we might want to investigate both the influence of different study sizes and of choice of design. We are free to vary these and study the results. The third factor, though, uniquely determines how successful our simulations will be. The more accurately we can predict the treatment effects to be seen in the actual study, the more we can rely on the results of our simulations.

This last point means that, typically, we should have gathered some data already. We might have done a pilot study, involving only a few cases, or we might have accessed data from similar, earlier studies. The more information we have the better.

We illustrate this technique of study simulation through resampling with a simple example. We want to study the effectiveness of a medication for hypertension. This will be a CR two-group study: Some number of volunteer hypertensive patients will be randomly allocated, in equal numbers, to a Control group (to receive a placebo as the medication) or to a Treatment group (to receive the new medication). We have observations of systolic blood pressure (SBP), the response to be measured in our study, for a number of untreated hypertensive patients, individuals whom we believe are similar to those who would be recruited into our study. These data give us an idea of what the control-group SBPs might be like. We don't have the same kind of data for treatment cases, patients receiving the new medication. But we do have an idea of what the effect of the treatment should be, or of what results the treatment would have to produce if we were to judge it effective. Let's say that we want the medication to lower SBP for the typical patient by about 10 points (mm Hg). The statistic we will use to assess the difference in SBPs is the difference in means for the two groups:

$$s = \text{Mean(SBP | Control)} - \text{Mean(SBP | Treatment)}$$

Our goal is to select a size for this CR two-group design that is large enough that we have a good chance of deciding the medication has a real effect (i.e., that an observed difference is not a chance result) when the medication has the desired effect, lowering SBP among hypertensive patients by 10 mm Hg. Before going any further, we will need to quantify what constitutes a good chance of detecting an effect, and what criteria we will use to decide a difference is not a chance result.

Let's start with the latter. We link our decision about whether an effect is chance or not to the p-value of our test statistic, s. Let's choose a p-value of 0.05. A p-value of this magnitude or smaller gives strong enough evidence that the difference in treatments is not a chance result. We could choose a more stringent criterion, 0.01 for instance, but the smaller we take this critical p-value to be, the greater proof we require and the larger we will have to make our study.

Now, for the last item. We can specify what hypertensive SBPs look like in general, from our earlier observations, but we cannot say exactly what we will see in this next study. We might, for example, recruit patients with overall lower SBPs, or with less variable SBPs.

Although we are going to do only one study, we can take the uncertainty about results into account by thinking of what would happen if we were to do that same study, over and over, each time with a different group of patients. Some of these studies would give results favorable to the medication (i.e., yield a p-value of 0.05 or smaller), and others would not.

The idea of giving ourselves a good chance of deciding in favor of a real effect for the medication is linked to the proportion of studies that would lead to that decision. We'd like the proportion of studies detecting a real effect to be high. But the larger we require this proportion to be, the larger we must make our study.

What is a reasonable value for this good chance? Most researchers want that to be toward the higher end of a range from about 60% to 90%. A common choice is 75% or 80%. We'll use 80% in our illustration.

Everything is in place now for simulating the outcomes of a series of studies. We know enough that we can ask a statistics package to do the work for us. Figure 18.10 is a Resampling Stats script tailored to our specific study, but it can serve as a template for other designs.

```
copy (210 169 187 160 167 176 185 206 173) SYSCON   'Control
concat SYSCON (146 174 201 198 148 154) SYSCON      'group SBP
subtract SYSCON 10 SYSTRT  'H(A): Reduction in SBP of 10 points
copy 0.05 CRIT             'Critical p-value
copy (20 40 60) NN         'Potential study sizes
copy 100 RPT               'Trials at each study size
copy 500 MONTE             'Resamples for p-value
size NN NL                 'No of study sizes
subtract MONTE 1 MM1
copy 0 J                   'Initialize study size choice
repeat NL
  add J 1 J
  take NN J N
  copy 1,N CPIK            'First N to Control, after shuffle
  add N 1 NP1
  multiply N 2 N2
  copy NP1,N2 TPIK         '(N+1) to 2N to Treatment
  copy 0 CRITCT            'Initialize count of p-vals
  repeat RPT
    sample N SYSCON CONGRP     'A Control group of size N
    sample N SYSTRT TRTGRP     'A Treatment group of size N
    mean CONGRP MC
    mean TRTGRP MT
    subtract MC MT S          'S, the statistic from study
    copy 1 PCOUNT             'S is >= S
    concat CONGRP TRTGRP HYPGRP
    repeat MM1
      shuffle HYPGRP HYPGRP
      take HYPGRP CPIK CONH    'Rerandomized groups under the
      take HYPGRP TPIK TRTH    'treatment effect hypothesis
      mean CONH MC
      mean TRTH MT
      subtract MC MT SH        'SH, statistic from rerandomized
      if SH >=S
        add PCOUNT 1 PCOUNT    'Counts towards upper tail p
      end
    end
    divide PCOUNT MONTE PVAL   'The rerandomized p-value
    if PVAL <=CRIT             'Is it 0.05 or smaller?
      add CRITCT 1 CRITCT
    end
  end
  divide CRITCT RPT PROPDET    'Proportion of effect detections
  print N PROPDET
end
```

Figure 18.10: *Evaluating Alternative Study Sizes (Resampling Stats)*

The computations are organized into three loops. The outermost loop allows us to try out NL different values of N, the number of hypertensive patients to assign to each of the two levels of treatment. For each value of N, the middle loop repeats our two-group study some large number of times, RPT. The innermost loop finds a p-value for the statistic computed in each of those studies by resampling MONTE times from the data in accordance with the null treatment-effect hypothesis.

We have SBPs from 15 nonmedicated hypertensive patients in the vector SYSCON. These blood pressures provide our best guess about what kinds of SBPs we might see in our control group, following the administration of a placebo.

We want to detect a treatment effect when the new medication decreases SBP by 10 mm Hg. To simulate that outcome, we create a second set of SBPs by subtracting 10 from each of those in our nonmedicated group, and store these as SYSTRT. These SBPs are our guess about what we would see in our treatment group if the medication has the desired effect. A pilot study with our medication is an alternative source for SYSTRT.

The variable CRIT is set equal to our critical p-value of 0.05, three different study sizes are to be evaluated, either 20, 40, or 60 patients randomized to each of the control and treatment groups and, for each of these study sizes, we are asking the computer to generate 100 (RPT) simulated repetitions of the two-group study, drawing patients and their SBPs for each simulation from SYSCON and SYSTRT.

In each of these simulated studies, the chance nature of a treatment effect is to be evaluated by rerandomizing the patients, in accordance with the null treatment-effect hypothesis, 500 (MONTE) times and estimating the upper tail of the reference distribution for s.

How should Resampling Stats be directed to create the simulated patient groups? At the top of the loop beginning with repeat(RPT), these two commands appear:

```
sample N SYSCON CONGRP
sample N SYSTRT TRTGRP
```

The first directs that the vector SYSCON be randomly sampled N times and the results stored as CONGRP. This sampling is done with replacement. That is, each sample is chosen at random from among the 15 SBPs making up SYSCON. The resulting simulated control group contains N (20, 40, or 60) patients with SBPs sampled from those we think are characteristic of nonmedicated hypertensive patients. Similarly, our simulated treatment group, TRTGRP, contains that same number of patients with SBPs sampled from those we think are characteristic of hypertensive patients treated with a medication that lowers SBP, on average, by 10 mm Hg.

For each of the simulated studies, a p-value for the difference in mean-SBPs is found in the `repeat(MM1)` loop by resampling under the null treatment-effect hypothesis and counting the number of times that the resulting s^H equals or exceeds s. Finally, for each of the three values of N, the proportion of p-values less than or equal to 0.05 is computed and reported.

```
N         =   20
PROPDET   =   0.52

N         =   40
PROPDET   =   0.65

N         =   60
PROPDET   =   0.79
```

Figure 18.11: *Proportions of Effects Detected (Resampling Stats)*

Figure 18.11 shows the result of executing the study size script. We need a total of 120 patients, 60 in each treatment arm, to have an 80% chance of detecting, as real, a medication effect when the medication decreases SBP by 10 mm Hg. With only 20 patients in each group, we would have only a 50/50 chance of detecting a treatment effect of this magnitude.

This illustration of study size exploration involved a simple design with but one statistic of interest. Multitreatment designs, as we see in Applications 19, usually involve more than one comparison. For those designs, you can examine study size from one of these three perspectives:

1. Carry out the analysis of the proportions-of-effect-detection for each comparison to be made.

2. If some comparisons are more important, substantively, than others, you will want to study the consequences of study-size only for the most important comparisons.

3. Where both main and simple effects are to be evaluated, the simple ones involve fewer observations and, hence, are less likely to detect effects. Thus these should be the target of study-size analysis, rather than the main effects.

Exercises

1. Find a 90% CI for the difference in 20% trimmed means for the Cornell Medical Index scores for the two neighborhood samples shown in the Applications 9 exercises as Table 9.2.

Table 18.3: *Response of Lymphocytic Lymphoma Patients to Two Treatments*

Response	Treatment: BP	CP
Complete Response	26	31
Partial Response	51	59
No Change	21	11
Progression	40	34
Totals	138	135

2. The data in Table 18.3 are taken from Hand et al. (1994, Data Set 30) and report the responses of available patients with lymphocytic lymphoma randomly assigned to one of two treatments: CP (cytoxin and prednisone), or BP (BCNU and prednisone). Patient responses are recorded on a graded, qualitative scale. Complete Response is the most favorable outcome, and Progression is the least favorable. For this analysis, replace the graded scale with a numeric one of your choice. Assume a CR design for the study. Is there evidence for a difference in the efficacy of the two treatments?

Table 18.4: *Two-Weeks Growth of Plant-Pairs*

Pair:	1	2	3	4	5	6	7	8	9	10
Treated	7	10	9	8	7	6	8	9	12	13
Control	4	6	10	8	5	3	10	8	8	10

3. In the RB design study reported in Table 18.4, each block consists of exactly two experimental units. This specialized form is sometimes known as a matched-pair or paired-comparison design, the idea being that each block is composed of two carefully matched experimental units. Here, in a study reported in Mead, Curnow, and Hasted (1993), the blocks consist of closely matched plants. Within each pair, one plant was randomly chosen for experimental treatment, and the other was raised as a control. The data recorded are the increases in height (in cm) over a two-week period. Does the experimental treatment lead to faster growth?

4. The data in Table 18.5 are from Hand et al. (1994, Data Set 243) and report the results of a study of rocking as a way of reducing infant crying. They summarize the study in this way:

> "An investigation was pursued to explore the possible beneficial effects of rocking on babies' crying. On each of 18 days, babies not crying at a specified time in a hospital ward were the subjects. These groups varied in size from six to ten. One of the subjects was chosen at random and rocked; the remainder (the controls) were not rocked. At the end of a specified time it was noted whether babies were crying or not."

On Day 1 of the study there were 9 non-crying infants on the ward at the beginning of the observation period. One was randomly chosen to be rocked. At the conclusion of the observation period, the rocked infant was not crying but 5 of the 8 control infants were. At issue is whether the rocked infants are less likely to be crying at the end of the observation periods. We might assess this with a relative rate of crying statistic, defined as

$$\text{Rel Rate of Crying} = \frac{\text{Prop of Rocked Infants Crying}}{\text{Prop of Non-rocked Infants Crying}}$$

This relative rate will be smaller than 1.0 if rocked babies cry less often than do non-rocked ones.

Because a random selection of the infant to be rocked is made each day, these data can be treated as obtained from a RB design. Each day defines a different block of infants to be randomly assigned, a single one to the rocked treatment, all others to the control treatment. Does the result of these rerandomizations suggest that the observed Rel Rate of Crying is smaller than one might expect as a function of chance assignment alone?

Table 18.5:   *Crying Frequencies among Rocked and Control Infants*

Day	Controls	Controls Crying	Rocked Crying
1	8	5	0
2	6	4	0
3	5	4	0
4	6	5	1
5	5	1	0
6	9	5	0
7	8	3	0
8	8	4	0
9	5	2	0
10	9	1	1
11	6	1	0
12	9	1	0
13	8	3	0
14	5	1	0
15	6	2	0
16	8	1	0
17	6	2	1
18	8	3	0

5. In a study reported by Regan, Williams, and Sparling (1977), shoppers in a mall were recruited randomly into either a Guilt or Control treatment. In both instances the shoppers were approached and asked to use the experimenter's camera to photograph the experimenter. In the control treatment, the shopper was thanked but in the guilt treatment, the shopper was led to believe (only for the duration of the experiment) that the shopper had damaged the camera. Shopper-subjects then encountered a second experimenter who was clearly having trouble with a disintegrating shopping bag. The shoppers either Helped or Did not help the experimenter. The data for this complete randomization of available cases are given in Table 18.6. Is the increased rate of helping among the guilt cases greater than would be expected by chance? The odds ratio is be an appropriate statistic.

Table 18.6:   *Number of Shoppers Who Helped the Second Experimenter*

	Experimental Condition	
	Guilt	Control
Helped	11	3
Did not help	9	17

6. Table 18.7 shows General Health Questionnaire (GHQ) and Impact of Events (IES) scores from a study (Leach, 1991) of two random samples of police officers, one sample taken from among those present at the Hillsborough Football Stadium disaster in 1989 and the other sample from among officers in the same force who were not directly involved. You should expect the Hillsborough officers to show signs of their traumatization. The GHQ data were analyzed earlier in the unit (Figure 18.3). Repeat the analysis using the IES data to further test this hypothesis.

Table 18.7:   *Health and Event Impact Scores for Two Police Officer Samples*

Stadium Sample

GHQ: 32 34 49 33 42 49 17 48 37 48 37 22

IES: 38 17 49 45 41 49 29 43 51 48 32 51

Comparison Sample

GHQ: 13 13 11 21 8 24 28 9 24 22 11

IES: 6 14 13 36 0 23 19 0 34 7 5

Applications 19

Multiple Independent Groups

We move now to the analysis of studies involving $K > 2$ groups of cases. The case-groups can be (a) random samples from one or more case populations; (b) randomly assigned available cases; or (c) arbitrary groups of available cases. Our resampling approaches differ for the three possibilities. We begin with a review of the classical, parametric approach to the K-groups design.

Multiple Group Parametric Comparisons

The one-way analysis of variance (ANOVA) model for case-based studies assumes random samples of sizes $n_1, n_2, \ldots, n_K$ from K populations. The treatment-response distributions, $Y_1, Y_2, \ldots, Y_K$, for these populations are required to have normal cdfs and a common variance, σ^2, though they can have distinct means, $\mu_1, \mu_2, \ldots, \mu_K$.

Population inference for the ANOVA model takes the form of testing hypotheses about the population means. It is convenient to phrase these hypotheses in terms of linear contrasts on the population means.

Let $c_1, c_2, \ldots, c_K$ be a set of numeric weights chosen so that $\sum c_k = 0$, then $\sum c_k \mu_k$ is a linear contrast on the K population means. The linear contrast becomes a hypothesis when it is set equal to a numeric constant: $C_0 = \sum c_k \mu_k$.

Here is a simple example. For a three-group study ($K = 3$), the null hypothesis $\mu_1 = \mu_3$ can be expressed as $C_0 = \sum c_k \mu_k$, for $k = 1, 2, 3$, where $c_1 = 1$, $c_2 = 0$, $c_3 = -1$, and $C_0 = 0$.

The test statistic for the null hypothesis, $C_0 = \sum c_k \mu_k$, is the plug-in estimator of the linear contrast, $\widehat{C} = \sum c_k \bar{y}_k$, where $\bar{y}_k$ is the mean of the treatment-responses in the k-th treatment group.

If $\widehat{C}$ differs enough from C_0, and in the direction stipulated by the alternative hypothesis, we reject the null hypothesis. To make this judgment on statistical grounds, of course, we need to know the null sampling distribution for $\widehat{C}$.

Under the ANOVA sampling assumptions, the null sampling distribution for $\widehat{C}$ has:

1. A mean of C_0

2. A standard deviation of

$$\text{SE}(\widehat{C}) = \sqrt{\sigma^2 \sum_{k=1}^{K} \frac{c_k^2}{n_k}}$$

3. A normal cumulative distribution function (cdf)

Typically, we do not know the common population distribution variance, σ^2, and our sampling distribution knowledge is incomplete. We can, however, estimate σ^2 from the K sample distributions,

$$\widehat{\sigma}^2 = \frac{\displaystyle\sum_{k=1}^{K}\left[\sum_{i=1}^{n_k}(y_{ki} - \bar{y}_k)^2\right]}{\displaystyle\sum_{k=1}^{K}(n_k - 1)}$$

where y_{ki} is the treatment-response of the i-th case in the k-th treatment group.

If we use the estimated σ^2 to estimate, in turn, the standard error (SE) of the linear contrast

$$\widetilde{\text{SE}}(\widehat{C}) = \sqrt{\widehat{\sigma}^2 \sum_{k=1}^{K} \frac{c_k^2}{n_k}}$$

we can Studentize the contrast

$$t(\widehat{C}) = \frac{\widehat{C} - C_0}{\sqrt{\widehat{\sigma}^2 \sum [c_k^2/n_k]}}$$

Under the ANOVA sampling assumptions and the null hypothesis, the sampling distribution for $t(\widehat{C})$ is that for the t-random variable with $\sum(n_k - 1) = (n - K)$ degrees of freedom, df, (where $n = \sum n_k$).

Large positive or large negative values of $t(\widehat{C})$ can be used as support for rejecting the null hypothesis, depending on the alternative hypothesis.

The cdf for $t(\widehat{C})$ can also be used to estimate a confidence interval (CI) for the linear contrast, $\sum c_k \mu_k$. The bounds to a $(1 - 2\alpha)100\%$ CI are given by

$$\widehat{C} \pm \left[\widehat{SE}(\widehat{C}) \times F_{t,(n-K)}^{-1}(1-\alpha) \right]$$

where $F_{t,(n-K)}^{-1}(1-\alpha)$ is the $(1-\alpha)$ quantile of the t-distribution with $(n-K)$ df.

Table 19.1 gives data for a study described in Box, Hunter, and Hunter (1978). A random sample of 48 devices was selected from those in a warehouse. This sample, in turn, was randomly divided into three smaller groups. Nothing was done to the devices in the Control group. Each device in the Water group was immersed in water. Each of those in the Drop group was dropped from a certain height. All 48 devices were then tested, producing the data of Table 19.1. Larger scores indicate poorer performance. Did the water or drop treatments worsen performance?

Table 19.1:   *Performance Data for Control and Stressed Devices*

Control	0.38 0.26 0.41 0.33 0.33 0.37 0.54 0.76
	0.51 0.55 0.53 0.41 0.47 0.49 0.42 0.34
Water	0.53 0.35 0.38 0.45 1.09 0.46 0.57 0.47
	0.39 0.74 0.32 0.74 0.48 0.37 0.52 0.44
Drop	0.51 0.63 0.46 0.47 0.42 0.45 0.41 0.39
	0.35 0.41 0.49 0.40 0.58 0.46 0.38 0.48

In Figure 19.1, we have the record of an interactive session with S-Plus in which the Studentized-contrast test just described is applied to the two hypotheses $C_w = (\mu_{water} - \mu_{control}) = 0$ and $C_d = (\mu_{drop} - \mu_{control}) = 0$.

Both contrasts are in the predicted direction, $\widehat{C}_w = (\bar{y}_{water} - \bar{y}_{control}) = 0.075$ and $\widehat{C}_d = (\bar{y}_{drop} - \bar{y}_{control}) = 0.012$, so we can continue with the test procedure.

We ask what proportions of the t-distribution with $(n - K) = (48 - 3) = 45$ df are as larger or larger than $t(\widehat{C}_w)$ and $t(\widehat{C}_d)$. The two p-values are found to be 0.067 and 0.405. The experiment provides no evidence that dropping the devices leads to decreased performance and not very strong evidence that immersing them in water has a deleterious effect.

```
> nc<-length(congrp)
> nw<-length(watgrp)
> nd<-length(drpgrp)
> mc<-mean(congrp)
> mw<-mean(watgrp)
> md<-mean(drpgrp)
> mc
[1] 0.44375
> mw
[1] 0.51875
> md
[1] 0.455625
> c1<-mw-mc
> c2<-md-mc
> c1
[1] 0.075
> c2
[1] 0.011875
> df<-nc+nd+nw-3
> varest<-var(congrp,SumSquares=T)
> varest<-varest+ var(watgrp,SumSquares=T)
> varest<-varest+ var(drpgrp,SumSquares=T)
> varest<-varest/df
> varest
[1] 0.01932764
> se1<-sqrt(varest*((1/nc)+(1/nw)))
> se2<-sqrt(varest*((1/nc)+(1/nd)))
> c1/se1
[1] 1.525868
> c2/se2
[1] 0.2415957
> 1-pt(c1/se1,df)
[1] 0.06702062
> 1-pt(c2/se2,df)
[1] 0.4050955
```

Figure 19.1:   *S-Plus p-Values for Two Contrasts*

Nonparametric K-Group Comparisons

The assumptions that our randomly sampled populations have attribute score distributions that have normal cdfs and a common variance quite often are unsustainable. We simply do not know enough about typical empirical population distributions to warrant such leaps of faith.

The bootstrap-t approach to the comparison of two population distributions outlined in Applications 18 can be extended to provide CI estimates for contrasts among K population means.

Bootstrap-t CIs for Mean Contrasts

If the mean contrast is between just two of the K populations that were sampled randomly—as were the control-water and control-drop comparisons we've already analyzed parametrically—then a nonparametric bootstrap-t CI for the contrast can be estimated in the way developed in Applications 18 and illustrated in Figures 18.2 and 18.3. Nonparametrically, such contrasts are two-group comparisons, even though they are embedded in a K-groups design.

For contrasts involving more than two populations, it will be helpful to adapt the linear contrast orientation to the nonparametric bootstrap-t.

Now, we want to estimate one or both bounds to a $(1 - 2\alpha)100\%$ CI for a linear contrast among population distribution means, $C = \sum c_k \mu_k$, using a plug-in estimator of the contrast, $\widehat{C} = \sum c_k \bar{y}_k$.

Our bootstrap-t strategy requires that we build up a bootstrap sampling distribution for the Studentization of the plug-in estimator,

$$t(\widehat{C}^*) = \frac{\widehat{C}^* - \widehat{C}}{\widetilde{SE}\left(\widehat{C}^*\right)}$$

In our nonparametric bootstrap world, we need not assume that the K population distributions have a common variance. Instead, we can estimate the SE for our contrast allowing the K distributions to have different variances,

$$\widetilde{SE}\left(\widehat{C}^*\right) = \sqrt{\sum_{k=1}^{K}\left(\frac{c_k^2 \, \widehat{\sigma}_k^{2*}}{n_k}\right)}$$

where $\widehat{\sigma}_k^{2*}$ is a variance estimate computed in a bootstrap sample from the k-th estimated population distribution:

$$\widehat{\sigma}_k^{2*} = [1/(n_k - 1)] \sum_{i=1}^{n_k} (y_{ki}^* - \overline{y}_k^*)^2$$

```
proc p_mncont(){
  local(cwts,unit,beg,end,y,c,j)
  cwts:= -1,0.5,0.5; unit:=1
  beg= vcat(unit,$2$); end= $2$
  cum(beg); cum(end)
  vector(y,rows($1$)); getcol($1$,y,1)
  c=0; j=0
  repeat(sizeof($2$)){
    j++
    c= c + (cwts[j]*mean(y[beg[j]:end[j]]))
  }
  $3$[1]=c
}
proc p_secont(){
  local(cwts,n,unit,beg,end,y,se,j)
  cwts:= -1,0.5,0.5
  n= $2$; unit:= 1
  beg= vcat(unit,n); end= n
  cum(beg); cum(end)
  vector(y,rows($1$)); getcol($1$,y,1)
  se=0; j= 0
  repeat(sizeof($2$)){
    j++
    se= se + (cwts[j]^2*vare(y[beg[j]:end[j]])/n[j])
  }
  $3$[1]=sqrt(se)
}
```

Figure 19.2:   *SC Prototype Procedures for a Mean Contrast and SE*

To illustrate the approach we'll return to the random samples of Table 19.1 and estimate a CI for a third linear contrast, involving all three population distribution means: $C_s = (1/2)(\mu_{\text{water}} + \mu_{\text{drop}}) - \mu_{\text{control}}$. The comparison is between the control or nonstressed-device population mean on the one hand and the average of the two stressed-device population means on the other. The contrast weights are $c = (-1, 0.5, 0.5)$, given in the order in which the three sample distributions are presented in Table 19.1.

Figure 19.2 shows the text of two user-supplied SC procedures, p_mncont and p_secont that can be used to supply iboot_t with values of $\hat{c}_s^*$ and $\widehat{SE}\left(\hat{c}_s^*\right)$ respectively. In Figure 19.3, these procedures are deployed.

```
> load(bhhperf)
 [cvec wvec dvec pvec MPERF gsiz ]
> proc p_mncont(){}
> proc p_secont(){}
> iboot_t(p_mncont,p_secont,MPERF,gsiz,1,2500,FBOOT,90,98)
_____theta:
 0.0434375
_____means of  2500  bootstrap estimates:
 0.043623625
_____SDs of  2500  bootstrap estimates:
 0.038786585

 Parameter: 1    90 % CI: -0.023789785    to  0.10938554
 Parameter: 1    98 % CI: -0.050822239    to  0.13533001
    Total iboot_t() running time: : 28.5
```

Figure 19.3: *Bootstrap-t Confidence Bounds on a Mean Contrast (SC)*

The matrix MPERF contains in its first (and only) column, the performance scores of Table 19.1, those for control devices followed by those for the water-immersed devices and for the dropped devices. The vector gsiz contains the three samples sizes, all sixteen. The sample-mean contrast takes the value,

$$\hat{C}_s = (1/2)\left(\bar{y}_{water} + \bar{y}_{drop}\right) - \bar{y}_{control} = 0.043$$

indicating, at least at the sample level, decreased performance in the stressed devices. However, the bootstrap estimate of the SE for this contrast is quite large by comparison, 0.039, and we should not be surprised to find the lower bound of our 90% CI becomes negative, -0.024. At the 5% significance level, then, we could not reject the null hypothesis of equal performance score means for unstressed and stressed devices.

Bootstrap-*t* CIs for Trimmed Mean Contrasts

We have noted that the mean, either population or sample, is not robust against the presence of disparate-sized scores in the distribution and have argued that a γ-trimmed mean might may be a better measure, and estimator, of the location or typical size of a score distribution. This argument applies, equally, to the estimation of a contrast among population locations. Fortunately, our bootstrap-t approach to inference for mean contrasts easily adapts to inference for trimmed mean contrasts.

Again, we want to estimate one or both bounds to a $(1 - 2\alpha)100\%$ CI for a linear contrast, among population γ-trimmed means, $C_{t,\gamma} = \sum c_k \mu_{t,\gamma}(Y_k)$, using a plug-in estimator, $\widehat{C}_{t,\gamma} = \sum c_k \bar{y}_{t,\gamma,k}$. For this, we need a bootstrap sampling distribution for the Studentization of the plug-in estimator,

$$t(\widehat{C}^*_{t,\gamma}) = \frac{\widehat{C}^*_{t,\gamma} - \widehat{C}_{t,\gamma}}{\widetilde{SE}\left(\widehat{C}^*_{t,\gamma}\right)}$$

where the SE estimate, $\widetilde{SE}\left(\widehat{C}^*_{t,\gamma}\right)$, can be expressed as

$$\widetilde{SE}\left(\widehat{C}^*_{t,\gamma}\right) = \sqrt{\sum_{k=1}^{K}\left[\frac{c_k^2\, S^2_{w,\gamma}(y_k^*)}{(1 - 2\gamma)\, n_k}\right]}$$

and $S^2_{w,\gamma}(y_k^*)$ is the γ-Winsorized variance of the bootstrap sample distribution, y_k^*.

In Figure 19.4, we have adapted the SC contrast and SE procedures of Figure 19.2 to be appropriate for trimmed mean contrasts.

When these procedures are applied to the sample performance data for stressed and nonstressed devices, we find that our contrast among trimmed means is only

$$\widehat{C}_{t,\gamma} = (1/2)\left(\bar{y}_{t,0.20,\text{water}} + \bar{y}_{t,0.20,\text{drop}}\right) - \bar{y}_{t,0.20,\text{control}} = 0.024$$

with an estimated SE of 0.037. With the distance from estimate to zero only about 70% of the SE, we have little reason to estimate a confidence bound. The lower bound of a 90% CI is certain to be negative.

The untrimmed nonparametric mean contrast and the earlier parametric control-water comparisons were inflated, possibly unrealistically, by the presence in the water sample distribution of at least one atypically large score, 1.09.

```
proc p_tmncon(){
   local(gama,cwts,unit,nvec,beg,end,nugam,y,c,j)
   gama= 0.20; cwts:= -1,0.5,0.5
   unit:=1; nvec= $2$
   beg= vcat(unit,nvec); end= nvec
   cum(beg); cum(end)
   nugam= floor(gama*nvec)/nvec
   vector(y,rows($1$)); getcol($1$,y,1)
   c=0; j=0
   repeat(sizeof(nvec)){
      j++
      c= c + (cwts[j]*tmean(y[beg[j]:end[j]],nugam[j]))
   }
   $3$[1]=c
}

proc p_tsecon(){
   local(gama,cwts,n,unit,beg,end,gpc,y,se,j,wy,dy)
   gama= 0.20; cwts:= -1,0.5,0.5
   n= $2$; unit:=1
   beg= vcat(unit,n); end= n
   cum(beg); cum(end)
   gpc= 100*(floor(gama*n)/n)
   vector(y,rows($1$)); getcol($1$,y,1)
   se=0; j=0
   repeat(sizeof($2$)){
      j++
      wy= y[beg[j]:end[j]]
      winsor(wy,gpc[j])
      dy= n[j]*(1-(2*gama))^2
      se= se + (cwts[j]^2*vare(wy)/dy)
   }
   $3$[1]=sqrt(se)
}
```

Figure 19.4: *SC Procedures for Trimmed Mean Contrast and SE*

All three analyses were shown here for illustrative and not for comparative purposes. The decision to use a parametric or nonparametric analysis, or to base the comparisons on trimmed or untrimmed means, always should be made before you inspect the data. For all the reasons advanced earlier, the general recommendation of the present text is to carry out a nonparametric comparison of trimmed means.

Comparisons among Randomized Groups

If the K treatment groups are formed by the random allocation of available cases, then the focus of inference shifts from population inference to local but causal inference. We seek statistical evidence that one or more observed differences in response to treatment are the result of differences in treatment rather than of the random presence of certain cases in certain treatment groups.

Essentially, all randomized treatment group hypotheses call for a comparison of the responses to one treatment, or to one set of treatments, with that to a second treatment or set of treatments. These hypotheses become, in effect, two-group comparisons. The one exception is testing the so-called omnibus null hypothesis, that all K treatments have equal impact. A rerandomization approach to the omnibus test was described in Applications 13, illustrated by the analysis detailed in Figures 13.5 and 13.6.

The one trick to carrying out two-group comparisons in the context of the K-groups study is to ensure that the rerandomization of cases is consistent with both the original randomization strategy, including any pre-randomization blocking of cases, and the particular null treatment-effect hypothesis being tested.

Completely Randomized Example

We return to the three treatment CR study of Applications 13 to illustrate these ideas. The data were taken from Loftus and Loftus (1988), described in Figure 13.5, reproduced here.

```
'Omnibus, three group comparison via permutation test
'
'The Loftus & Loftus (1988) Misguided Memory example.
'12 students randomly allocated, 4 apiece, among 3 conditions.
'After watching a filmed simulated auto accident Ss read an
'account of the accident:
'    g0: non-attributed article with no misinformation
'    g1: NY Daily News attributed article with misinformation
'    g2: NY Times attributed article with misinformation
'Score: number of errors in test of memory for film contents.
COPY (4 7 1 1) g0
COPY (4 9 10 5) g1
COPY (9 13 15 7) g2
'
```

Figure 13.5: *Part 1, Resampling Stats Omnibus Test of Treatments*

Rather than the omnibus test of Applications 13, there are, arguably, more interesting hypotheses to investigate in this study. Here are two examples:

1. Does misinformation attributed to a newspaper distort memory?

2. Does misinformation attributed to a respected newspaper distort memory more than misinformation attributed to a less-respected newspaper?

The null form of the first hypothesis is that the presumed newspaper accounts should not mislead and, as a result, a case's memory-test score should be the same under newspaper-misinformation as under no misinformation. A test statistic sensitive to a departure from this null hypothesis would contrast, in summary form, the scores of newspaper-misinformation cases with those of no-misinformation cases.

The null form of the second hypothesis is that a case would respond the same to an article attributed to the NY *Daily News* as to one attributed to the NY *Times*. This null hypothesis is silent about the responses of cases reading a no-misinformation account. A test statistic sensitive to departure from this second null hypothesis would contrast the responses of NY *Daily News* cases and NY *Times* cases.

Figure 19.5 traces tests of these two hypothesis using the `perm2` function in SC.

For the first test, we pool the NY *Daily News* and NY *Times* cases to form a newspaper-misinformation group (np). The mean number of memory test errors for the np group is greater by 5.75 than the mean number for the no-misinformation group (nnp) is. This is the direction required by the alternative hypothesis and we proceed with the significance test for this mean difference.

Under the null hypothesis, cases can be rerandomized between the np and nnp groups, carrying their memory test scores with them. The only restriction is that in this rerandomization, 8 cases are to be assigned to the np group and 4 cases to the nnp one. For the test, we ask what proportion of the possible rerandomizations yields a mean difference of 5.75 or greater, in favor of the np group.

The reference distribution for the mean difference in number of memory errors is generated and evaluated by `perm2`, which reports that the proportion of mean differences that are 5.75 or larger is $p = 0.014$ (and, as well, that the proportion of mean differences that are 5.75 or smaller is 0.994).

```
> nnp
    4 7 1 1
> nydn
    4 9 10 5
> nyt
    9 13 15 7
> np= vcat(nydn,nyt)
> mean(np)-mean(nnp)
    5.75
> perm2(np,nnp)
    / 0.014141414 / 0.99393939 /
> mean(nyt)-mean(nydn)
    4
> perm2(nyt,nydn)
    / 0.085714288 / 0.94285714 /

> 495*0.014141414
    6.9999999
> 70*0.085714288
    6.0000002
```

Figure 19.5: *Exact Tests for Two Treatment Comparisons (SC)*

The second null hypothesis is that the error score for a case would be the same following exposure to misinformation attributed either to the NY *Daily News* or to the NY *Times*. My alternative hypothesis is that errors should be greater for the NY *Times* group. The observed mean difference of 4 is in that direction. `perm2` rerandomizes the scores for the 8 cases originally assigned to one of the misinformation treatments between the two, four to each, and computes the mean difference between treatments, $s^H = \bar{x}^H_{\text{NYT}} - \bar{x}^H_{\text{NYDN}}$, for all such rerandomizations. The proportion of randomizations yielding an s^H greater than or equal to $s = 4$ was p $= 0.086$.

In neither invocation of `perm2` did we specify the number of randomly chosen rerandomizations to be used in forming the reference distribution. As a result, `perm2` evaluates all possible randomizations of the cases. For this study, the numbers of permutations are small enough that `perm2` completes its work in a very short time. For the first hypothesis, the 12 cases are to be rerandomized 4 to the no-misinformation group and 8 to the newspaper-misinformation group. The number of different assignments that can be made is given by

$$\frac{12!}{8! \times 4!} = \frac{12 \times 11 \times 10 \times 9}{4 \times 3 \times 2 \times 1} = 495$$

For the second hypothesis, the 8 cases in the newspaper-misinformation group are to be rerandomized 4 to each of the NY *Daily News* and NY *Times* groups. The number of different assignments is only

$$\frac{8!}{4! \times 4!} = \frac{8 \times 7 \times 6 \times 5}{4 \times 3 \times 2 \times 1} = 70$$

Knowing these results, we can restate the p-values produced by `perm2`. Exactly 7 of the 495 possible random assignments of cases between newspaper-misinformation and no-misinformation groups would produce, under the first null hypothesis, mean differences favoring the newspaper group by 5.75 or more errors. And exactly 6 of the 70 possible random assignments of misinformation-cases between the two newspapers would produce, under the second null hypothesis, mean differences favoring the NY *Times* by 4 or more errors. The two p-values are exact, not Monte Carlo approximations.

Randomized Blocks Example

Animal research is an area in which randomized blocks designs are widely used to provide control over the influence of one or more nuisance factors or variables. This example is taken from Mead, Curnow, and Hasted (1993).

The effect of each of three experimental drugs (labeled A, B, and C) on the lymphocyte count in the blood of mice, as compared with that of a placebo, is to be evaluated. To minimize the possibility that any drug-level effect is confounded with genetic differences among the mice used as subjects, four mice are randomly chosen from among each of five litters. The five litter-sources provide blocks of experimental subjects. The mice in each block, all from the same litter, are then randomly allocated among the four levels of drug (a placebo and experimental drugs A, B, and C).

The data are reproduced in Table 19.2.

The average lymphocyte counts for the different drug levels are

    A: 6.42   B: 5.72   C: 6.06   Placebo: 5.66

and, for purposes of illustration, we take as our treatment-comparison measure for a particular experimental drug the ratio of the mean lymphocyte count for the mice receiving that drug to the mean lymphocyte count for the placebo mice. The three ratios are

$$s_A = \frac{6.42}{5.66} = 1.1343, \; s_B = 1.0106, \text{ and } s_C = 1.0707$$

All are greater than 1.0, suggesting an increased lymphocyte count for each of the experimental drugs. Are ratios this size or larger what we might expect just on the basis of the random assignment of mice to drug levels? Or are one or more of the ratios so large as to provide evidence of a drug effect?

Table 19.2: *Mouse Lymphocyte Counts (thousands/cubic mm)*

Litter	1	2	3	4	5
Drug Level					
A	7.1	6.1	6.9	5.6	6.4
B	6.7	5.1	5.9	5.1	5.8
C	7.1	5.8	6.2	5.0	6.2
Placebo	6.7	5.4	5.7	5.2	5.3

In Figure 19.6, we have the log of an S-Plus session in which we develop p-values for the three ratios using the permutation resampling option of the `bootstrap` function.

The vectors `testa`, `testb`, and `testc` contain scores for the cases contributing to the three hypotheses. In each instance, we are comparing one drug against the placebo.

The vector `permblok` identifies the block origins of the cases involved in each test, and the user-supplied function `f_mnrat` computes a ratio of means for data organized as `testa`, `testb`, or `testc`.

By including the qualifier `group=permblok` in the `bootstrap` command, we ensure that the rerandomization of cases takes place block by block, that is, within litters, rather than en masse. Based on 1,001 randomizations of the cases under each null hypothesis, including the initial randomization in each instance, the proportions of the (s_A^H, s_B^H, s_C^H) distributions that are as large or larger than $(s_A = 1.13, s_B = 1.01, s_C = 1.07)$ work out to be $(p_A = 0.032, \; p_B = 0.376, \; p_C = 0.068)$.

```
> druga<-c(7.1,6.1,6.9,5.6,6.4)
> drugb<-c(6.7,5.1,5.9,5.1,5.8)
> drugc<-c(7.1,5.8,6.2,5.0,6.2)
> placebo<-c(6.7,5.4,5.7,5.2,5.3)
> blok<-c(1,2,3,4,5)
> testa<-c(druga,placebo)
> testb<-c(drugb,placebo)
> testc<-c(drugc,placebo)
> permblok<-c(blok,blok)
> f_mnrat<- function(x){
+ mean(x[1:5])/mean(x[6:10])
+ }
> x<-f_mnrat(testa)
> x
[1] 1.134276
> x<-f_mnrat(testb)
> x
[1] 1.010601
> x<-f_mnrat(testc)
> x
[1] 1.070671
> arefdist<-
bootstrap(testa,f_mnrat,group=permblok,sampler=samp.permute)
> x<-f_mnrat(testa)
> length(arefdist$replicates[arefdist$replicates >= x])
[1] 31
> (31+1)/1001
[1] 0.03196803
> brefdist<-
bootstrap(testb,f_mnrat,group=permblok,sampler=samp.permute)
> x<-f_mnrat(testb)
> length(brefdist$replicates[brefdist$replicates >= x])
[1] 375
> (375+1)/1001
[1] 0.3756244
> crefdist<-
bootstrap(testc,f_mnrat,group=permblok,sampler=samp.permute)
> x<-f_mnrat(testc)
> length(crefdist$replicates[crefdist$replicates >= x])
[1] 67
> (67+1)/1001
[1] 0.06793207
```

Figure 19.6: *Comparisons of Drugs A, B, and C with Placebo (S-Plus)*

In thinking about these proportions, note the very restricted nature of the reference distributions. Under any of the three null treatment-effect hypotheses, how many different randomizations of the cases are possible? From each block of cases, only two mice are involved in a particular hypothesis. The two cases can be randomized between the two treatments in only two ways. The randomization of cases takes place independently from one block to another. As a result, we can compute the number of possible randomizations as the product:

$$2 \times 2 \times 2 \times 2 \times 2 = 2^5 = 32$$

With only 32 randomizations possible under a null hypothesis, the p-value can never be smaller than $1/32 = 0.03125$. This p-value is attained only if the original randomization yields, uniquely, the largest of the 32 values of s^H (appropriate to our alternative hypothesis in this example) or, uniquely, the smallest of the values (if our alternative hypothesis called for us to count values of s^H as small as or smaller than s). If our alternative hypothesis were nondirectional, then the smallest value of p would be even larger, $2 \times 0.03125 = 0.0625$ for this design.

The restricted number of rerandomizations is illustrated in Figure 19.7. This histogram describes a Monte Carlo approximation to the null reference distribution equating the effects of drug A and the placebo, an approximation based on 5,000 rerandomizations. The histogram is the result of the following S-Plus commands:

```
> arefdist<- bootstrap(testa,f_mnrat,group=permblok,
+ sampler=samp.permute,B=5000,trace=F)
> ap<-arefdist$rep
> histogram(ap,breaks=c((845:1155)/1000))
```

The `breaks` qualifier sets bin widths of 0.01 across the range of the reference distribution. Our histogram describes the occurrence of 20 distinct values of the test statistic. Twelve of the values are each associated with a single rerandomization; each of these shortest bars has height approximating 3.125%. Four additional values, with intermediate height bars approximating 6.25%, are each associated with two rerandomizations. The final four values, with the highest bars approximating 9.375%, are each associated with three randomizations. Thus, the 20 possible values of the test statistic are linked with the 32 possible rerandomizations.

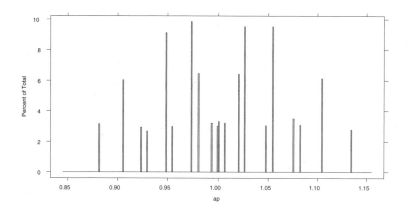

Figure 19.7: *Histogram of Null Reference Distribution, Drug A versus Placebo*

The p-value obtained for the test of drug A versus the placebo in the rerandomizations of Figure 19.6, $p_A = 0.032$, is an approximation to $1/32 = 0.03125$. Our test statistic took the largest possible value under rerandomization.

The small size—limited number of cases—of this design makes it difficult to detect departures from the null hypothesis. If, instead, experimental animals had been recruited from eight rather than five litters, then the number of randomizations under any one of the three null hypotheses would be increased from $2^5 = 32$ to $2^8 = 256$, and p-values of the order of 0.01 would be possible.

Comparisons among Nonrandom Groups

Where available cases have been distributed among K-groups in a nonrandom way, we still can make group comparisons. We have no statistical basis, however, for drawing causal or population inferences from these comparisons. The group comparisons are local comparisons, which can and should be evaluated for stability.

The data for the following example are taken from a larger study reported by Albert and Harris (1987) and consist of a measure of the amount of the enzyme ornithine carbonyl-transferase (oc-t) present in each of 218 patients distributed over four diagnostic categories: acute viral hepatitis (57 patients), persistent chronic hepatitis (44 patients), aggressive chronic hepatitis (40 patients), and postnecrotic cirrhosis (77 patients). The diagnostic groups are not random samples.

```
proc p_tmnord(){
   local(gam,unit,gsiz,beg,end,k,nugam,j,tm,ot,k,ctm)
   gam= 0.20; unit:= 1; gsiz= $2$
   beg= vcat(unit,gsiz); end= gsiz; cum(beg); cum(end)
   nugam= floor(gam*gsiz)/gsiz; vector(x,rows($1$))
   getcol($1$,x,1); k= sizeof(gsiz); vector(tm,k,ot,k)
   ords(ot); j=0
   repeat(k){
     j++
     tm[j]= tmean(x[beg[j]:end[j]],nugam[j])
   }
   ctm= tm; pairsort(ctm,ot)
   copy(vcat(tm,ot),$3$)
}
```

Figure 19.8:   *An SC Procedure for Ordering a Set of Trimmed Means*

At issue is whether the levels of oc-t differ across the diagnostic groups. The 20% trimmed means for the four groups are as follows:

Group 1: Acute viral hepatitis (avh):	2.585
Group 2: Persistent chronic hepatitis (pch):	1.959
Group 3: Aggressive chronic hepatitis (ach):	2.624
Group 4: Postnecrotic cirrhosis (pnh):	2.260

One comparative summary of these results is that the oc-t level is lowest in group 2, intermediate in magnitude in group 4, and largest in groups 1 and 3. How stable is that description?

To answer that question, the trimmed means were recomputed for each of 100 random half-samples of the data and the ordering of the four trimmed means was recorded. The user-written SC procedure p_tmnord of Figure 19.8 performs these calculations for each data set (complete or half-sample) supplied by the procedure subsam_k.

For our example, p_tmnord will return the trimmed means for half-samples from groups 1, 2, 3, and 4 as the first four elements of the output vector, 3. The ordering of these trimmed means will be returned as well, in elements 5 through 8 of that output vector. The entry in element 5 identifies by number the group with the smallest trimmed mean, element 8 identifies the group with the largest trimmed mean, and elements 6 and 7 identify the two intermediate groups.

Figure 19.9 is an extract of an SC interactive session in which the returns of `p_tmnord` for a randomly chosen sequence of 100 random half-samples are evaluated.

```
> load(blood)
[   avh[57]   pch[44]   ach[40]   pnc[77]   ]
> allcas= vcat(avh,pch,ach,pnc)
> matrix(MALL,sizeof(allcas),1)
> setcol(MALL,allcas,1)
> gsiz:= 57,44,40,77
> hsiz= floor(0.5*gsiz)
> proc p_tmnord(){}
> subsam_k(p_pairs,8,MALL,gsiz,hsiz,100,HH8)
> mlimit(HH8,1,100,5,8)
> pcount= mcount(HH8,UP)
> UP
# [1][]
  2              4              3              1
# [2][]
  2              4              1              3
> pcount
  31             69
```

Figure 19.9: *Patterns of Half-Sample Trimmed Mean Orders (SC)*

We don't show in this display the output of `subsam_k`. Rather we focus on the analysis of the last four columns of the matrix HH8 saved by `subsam_k` and containing the group order of the trimmed means for each of 100 half-samples.

The command `pcount= mcount(HH8,UP)` does two things. First, the unique rows of HH8 (now limited to columns 5 through 8) are extracted and stored as the rows of UP. Second, the frequency with which each of these unique rows was encountered in HH8 is saved in the vector `pcount`. When we inspect these two results we see that:

1. HH8 has only two distinct rows. These correspond to the two orderings of the trimmed means, (2, 4, 3, 1) and (2, 4, 1, 3).

2. The first of these two orders occurred for 31 of the half-samples; the second order occurred for the remaining 69 half-samples.

In each of the 100 half-samples, then, the smallest of the oc-t trimmed means was for group 2 (persistent chronic hepatitis), and the next largest was for group 4 (postnecrotic cirrhosis). The largest oc-t trimmed mean was either for group 1 (acute viral hepatitis), in 31% of the half-samples, or for group 3 (aggressive chronic hepatitis), in the other 69%.

The half-sample results offer support for the stability of our proposed summary of the results on the full data set.

In composing half-samples, `subsam_k` randomly samples from each group separately. This is appropriate here because we assume that the four diagnostic groups were recruited into the study. That is, they did not result from a subsequent division of a single set of 218 patients.

Adjustment for Multiple Comparisons

The greater the number of treatments that are in the random sample or random assignment study, the greater the number of treatment comparisons we are likely to want to make. We often want to attach a p-value to these comparisons, to express how unlikely it is that treatment differences could have occurred solely by the chance selection or assignment of cases to treatment. However, the logic of p-values dictates that the more comparisons we make, the more likely it is that we will encounter one or more small p-values, even if all our null hypotheses are correct and there are no treatment effects.

If we make just one comparison and the null hypothesis for that comparison is correct, there is a 5% chance that our p-value will be 0.05 or smaller. If we make two independent comparisons and both null hypotheses are correct, the chances that at least one of the two p-values will be 0.05 or smaller is greater than 5%. In fact, it is 9.75%. Where does that come from? There is a 95% chance that the first comparison will have a p-value greater than 0.05, and the same is true for the second comparison. Then the chances that both comparisons will have p-values greater than 0.05 will be 95% of 95% or 90.25%. And if both p-values are not greater than 0.05, then one of them must be 0.05 or smaller. The chances of at least one of the two p-values being as small as 0.05 must be $100\% - 90.25\% = 9.75\%$.

We can write this result a bit more economically as $(1 - 0.95^2)100\%$. In that form it makes it easy for us to compute the chances of at least one p-value of

0.05 or smaller when three independent comparisons are made and all three null hypotheses are correct: $(1 - 0.95^3)100\% = 14.26\%$. Or if we make five independent treatment comparisons and there truly are no treatment effects, the chances of obtaining at least one p-value no greater than 0.05 increases to $(1 - 0.95^5)100\% = 22.62\%$. We can state this logic in a more general form and in a slightly different way. If the null hypotheses for each of m independent treatment comparisons are all correct, the chances that the smallest of the m p-values will be 0.05 or smaller is $(1 - 0.95^m)100\%$.

The importance of all this is that we should be less impressed by a p-value below 0.05 when it is the smallest of several, than when it is the only one we have computed. A variety of techniques have been proposed to help us de-emphasize the importance of a small p-value among many. You can read more about them in a source like Westfall and Young (1993), but a technique worth serious consideration is the Šidák stepdown p-value adjustment, which takes this form:

1. Compute the p-values for the m comparisons in the usual way.

2. Arrange these observed p-values in descending value from largest to smallest $(p_{obs[1]}, p_{obs[2]}, \ldots, p_{obs[m]})$; $p_{obs[1]}$ is the largest, $p_{obs[m]}$ is the smallest, and $p_{obs[j]}$ is the j-th largest.

3. Adjust each of the observed p-values using this equation:

$$p_{adj[j]} = 1 - (1 - p_{obs[j]})^j, j = 1, \ldots, m$$

The amount of adjustment depends on the position of the particular observed p-value among the m. For example, if 0.05 is the smallest of 10 p-values then it is adjusted to $1 - (1 - 0.05)^{10} = 0.401$, but if 0.05 is the largest of 10 p-values, then it is not adjusted at all: $1 - (1 - 0.05)^1 = 0.05$.

How should these adjustments be understood? $p_{obs[j]}$ is the proportion of resamples, under the j-th null treatment-effect hypothesis, that would yield a test statistic with a value at least as favorable to the alternative hypothesis as the one you obtained. The interpretation of $p_{obs[j]}$ takes only the j-th hypothesis into account, as if it were the only one tested.

$p_{adj[j]}$ is the proportion of resamples, under null treatment-effect hypotheses 1 through j, that would yield at least one test statistic with a value at least as favorable to the alternative hypothesis as the one you obtained for the j-th comparison. The interpretation of $p_{adj[j]}$ takes into account all the j null hypotheses for which there is at least as much support as for the j-th null hypothesis itself.

Let's apply these ideas to some real multiple comparisons.

For the lymphocyte count example, we had as the observed p-values for the comparison of drugs A, B, and C with the placebo,

$$p_A = 0.03197, p_B = 0.37562, p_C = 0.06793$$

Adjusting these for the multiplicity of comparisons made, we have

$$p_{A,adj} = 1 - (1 - 0.03197)^3 = 0.093$$

$$p_{B,adj} = 1 - (1 - 0.37562)^1 = 0.376$$

$$p_{C,adj} = 1 - (1 - 0.06793)^2 = 0.131$$

The $p_{A,adj}$ tells us that there was nearly one chance in ten (0.093) that at least one of the three comparisons would yield an observed p-value of 0.032 or smaller if all three null hypotheses had been correct. This gives a fairer interpretation of having found the smallest of three p-values to be 0.032. Similarly, the $p_{C,adj}$ reminds us that there was a 13% chance of having found a p-value of 0.068 or smaller in at least one of the two comparisons, drug B versus the placebo, and drug C versus the placebo, if both null hypotheses had been correct. Again, this is a fairer interpretation of having found the second smallest p-value to be 0.068.

No adjustment is required for $p_{B,adj}$ because the observed p-value conveys the chances of the largest p-value being no larger than the one observed, when only that single null hypothesis is correct.

The Šidák stepdown p-value adjustment is a slight simplification because it does not account for any dependence among the hypotheses. The adjustment is slightly conservative (i.e., producing larger p-values) where treatment comparisons are dependent. That is true for the example just considered. Each of the three comparisons involves the same placebo treatment. However, correcting for patterns of dependence is a nontrivial task and conservatism might be the price to be paid. The Šidák adjustments are easily undertaken and well worth making if we are to avoid overemphasizing the importance of the occasional small p-value in the company of many larger ones.

The interpretation of the observed p-values refers to the "proportion of resamples." These resamples may be either rerandomizations or bootstrap random samples. That is, the observed multiple comparison p-values might have been obtained from a rerandomizaton reference distribution or from bootstrap CIs, depending on whether the treatment groups were randomly allocated available cases or random samples.

Exercises

1. The data in Table 19.3 are organized as a series of input statements for a Resampling Stats script. This data set from Westfall and Young (1993, p. 95) gives the weights of mice randomly allocated among seven treatments. One of the treatments is a control one and the weights of mice randomized to each of the six actual treatments should be compared to those of the control animals. Compare the observed and Šidák stepdown rerandomization p-values.

Table 19.3: *Weights of Mice Randomized among Seven Treatments*

```
copy ( 89.8 93.8 88.4 112.6) CONTROL
copy ( 84.4 116.0 84.0 68.6) TRT1
copy ( 64.4 79.8 88.0 69.4) TRT2
copy ( 75.2 62.4 62.4 73.8) TRT3
copy ( 88.4 90.2 73.2 87.8) TRT4
copy ( 56.4 83.2 90.4 85.6) TRT5
copy ( 65.6 79.4 65.6 70.2) TRT6
```

2. The data in Table 19.4 are from Mead, Curnow, and Hasted (1993) and report the egg-laying performance of pullets under different lighting conditions. The case or experimental unit is a pen of six pullets and the response is the number of eggs laid by a pen in an 80-day winter period.

The study was carried out as a RB design. The pens formed four blocks of three pens each. Pens within each block were randomized among the three lighting conditions: Natural Winter Daylight, Extended (14 hour) Daylight, and Flash Lighting (natural daylight augmented by two 20-second flashes of light during the night). The authors do not report the basis on which the pens were blocked. Make three comparisons of the effect of lighting, natural versus extended, natural versus flash, and extended versus flash. Find adjusted p-values by the Šidák stepdown method.

Table 19.4: *Pullet Eggs Laid as Function of Light Treatment*

Blocks	1	2	3	4
Treatments				
Natural	330	288	295	313
Extended	372	340	343	341
Flash	359	337	373	302

3. Table 19.5 gives blood coagulation times for laboratory animals receiving one of four diets. The laboratory animals were not a random sample and they were not allocated among diets at random. Carry out a subsample stability analysis to support your description of diet-related differences in coagulation time.

Table 19.5: *Blood Coagulation Times for Four Laboratory Diets*

Diet	Coagulation Times
A	62 60 63 59
B	63 67 71 64 65 66
C	68 66 71 67 68 68
D	56 62 60 61 63 64 63 59

4. The London *Independent* of 12 December 1996 reported the results of a poll of randomly sampled lecturers, 50 from each of six disciplines— mathematics (Maths), English, science, sociology (Sociol), the humanities, and engineering (Engineer). They were asked several questions about the changing nature of universities and undergraduates. One question had to do with their feeling pressure to give good grades. The results are summarized as Table 19.6.

Group the Strongly Agree and Slightly Agree responses into a single Pressured response and the other three into a Not Pressured response. Estimate bias-corrected and accelerated (BCA) confidence intervals (CIs) for several between-discipline comparisons. You could use an odds ratio or a simple difference in the proportion pressured. What do you conclude?

Table 19.6:   *Survey of UK Academics, Felt Pressure to Award Grades*

"I feel under pressure to give students good grades because of the need to compete in the marketplace."						
	A	B	C	D	E	F
Strongly Agree	6	0	2	4	1	4
Slightly Agree	13	6	4	5	7	7
Slightly Disagree	10	5	7	6	9	3
Strongly Disagree	21	39	36	35	32	34
Don't Know	0	0	1	0	1	2
A: Maths, B: English, C: Science, D: Sociol, E: Humanities, F: Engineer						

Applications 20

Multiple Factors and Covariates

The analysis of multiple group studies, begun in Applications 19, is continued here. In particular, we look at three standard parametric extensions of the multiple treatment groups design:

1. Designs with two treatment factors

2. Designs with treatment and blocking factors

3. Designs with one or more covariates

Our interest is in how to carry out the resampling analyses that would be appropriate when one or more assumptions of these parametric approaches are unwarranted.

Two Treatment Factors

The hallmark of these designs is that the treatment delivered to a particular group of cases has two or more facets; it is the combination of levels of two or more treatment factors. We saw an example of this design in Applications 15.

Table 15.1: *Estimated IQ Scores for Targets*

Attractiveness of Target			
High		Low	
Sex of Target		Sex of Target	
Male	Female	Male	Female
112	124	120	122
114	124	121	123
116	126	123	125
118	126	124	126

The data in Table 15.1 are taken from Woodward, Bonet, and Brecht (1990) and describe a study in which the cases, 16 college students, were randomly allocated, 4 apiece, to 4 treatments. In each treatment, the case is shown a photograph of a college-age person and asked to estimate the IQ of the target. The 4 treatments result from the pairing of the target's sex (male or female) with the target's attractiveness (high or low).

There were two treatment factors in this study, Sex and Attractiveness of target, each at two levels.

The design of the IQ Estimation study is said to be a fully-crossed two-factor one. That is each of the I levels of the first factor is paired with each of the J levels of the second. The number of treatments is the product of the number of levels of the two factors, $(I \times J)$.

In the typical analysis of the fully-crossed two-factor design, we want to study the effects of the two factors, and we also might be interested in whether the effects of one factor change as we move from level to level of a second factor. We refer to this latter possibility as an interaction between two factors.

Parametric Analysis of the Two-Factor Study

The analysis of variance (ANOVA) approach to the fully-crossed two-factor case-based study takes as its starting point that we have random samples of cases from each of $(I \times J)$ populations. The population score distributions for the attribute of interest are assumed to have the following:

1. Normal cumulative distribution functions (cdfs)

2. The same variance, σ^2

3. Potentially different means, μ_{ij}, $i = 1, \ldots, I$, $j = 1, \ldots, J$

We want to test hypotheses about the pattern of the $(I \times J)$ population means. Referring to the two treatment factors arbitrarily as A and B, these hypotheses can be phrased as contrasts:

1. Among the I levels of factor A.

2. Among the J levels of factor B.

3. To assess whether a particular contrast among the levels of factor B is to be evaluated across all levels of factor A or separately at each level of factor A.

These last contrasts determine if there is an interaction between the two factors and, consequently, how to address the factor B contrasts in the right way. Thus, we usually test these interaction contrasts first.

The data in Table 20.1 show the average butterfat percentage in the content of milk from ten random samples, each of ten cows. The samples are drawn from populations of mature (5 years old or older) and two-year-old cows of five different breeds. The data are taken from Hand et al. (1994, Data Set 23).

Table 20.1: *Butterfat Percentages by Breed and Age of Cow*

Ayreshire		Canadian		Guernsey		Holstein		Jersey	
Mat	2-yr	Mat	2-yr	Mat	2-yr	Mat	2-yr	Mat	2-yr
3.74	4.44	3.92	4.29	4.54	5.30	3.40	3.79	4.80	5.75
4.01	4.37	4.95	5.24	5.18	4.50	3.55	3.66	6.45	5.14
3.77	4.25	4.47	4.43	5.75	4.59	3.83	3.58	5.18	5.25
3.78	3.71	4.28	4.00	5.04	5.04	3.95	3.38	3.49	4.76
4.10	4.08	4.07	4.62	4.64	4.83	4.43	3.71	5.24	5.18
4.06	3.90	4.10	4.29	4.79	4.55	3.70	3.94	5.70	4.22
4.27	4.41	4.38	4.85	4.72	4.97	3.30	3.59	5.41	5.98
3.94	4.11	3.98	4.66	3.88	5.38	3.93	3.55	4.77	4.85
4.11	4.37	4.46	4.40	5.28	5.39	3.58	3.55	5.18	6.55
4.25	3.53	5.05	4.33	4.66	5.97	3.54	3.43	5.23	5.72

Let the mean butterfat percentage in the several populations be represented by

$$\mu_{AM}, \mu_{CM}, \mu_{GM}, \mu_{HM}, \mu_{JM}$$

for the mature populations, and

$$\mu_{A2}, \mu_{C2}, \mu_{G2}, \mu_{H2}, \mu_{J2}$$

for the two-year-old populations. Of interest is whether butterfat content changes from breed to breed and between young and mature cows. Specifically, we look at four breed contrasts, comparing the butterfat production of each of the other four breeds with that of the Jersey.

First, though, we test the null interaction hypothesis, that the difference in butterfat content of milk between mature and young cows is the same for all five breeds. This null hypothesis implies four linear constraints on the ten population means. The four constraints can be expressed as follows:

$$(\mu_{A2} - \mu_{AM}) - (\mu_{C2} - \mu_{CM}) = 0$$

$$(\mu_{A2} - \mu_{AM}) - (\mu_{G2} - \mu_{GM}) = 0$$

$$(\mu_{A2} - \mu_{AM}) - (\mu_{H2} - \mu_{HM}) = 0$$

$$(\mu_{A2} - \mu_{AM}) - (\mu_{J2} - \mu_{JM}) = 0$$

Taken together, these constraints establish that the young-mature butterfat difference is to be the same for all breeds.

In Applications 19, we learned that, in the parametric model, a null hypothesis stated as a single linear constraint on the $(I \times J)$ population means, $C_0 = \sum\sum c_{ij}\mu_{ij}$, could be tested via the Studentization of $\widehat{C} = \sum\sum c_{ij}\bar{y}_{ij}$,

$$t(\widehat{C}) = \frac{\widehat{C} - C_0}{\sqrt{\widehat{\sigma}^2 \times \sum\sum[c_{ij}^2/n_{ij}]}}$$

Under the null hypothesis, the sampling distribution for $t(\widehat{C})$ is that of the t-random variable with $\sum_i\sum_j(n_{ij} - 1)$ degrees of freedom (df).

This provides us with a statistical basis for deciding whether $t(\widehat{C})$ is too large (or, too small) to be consistent with the null hypothesis or constrained model.

When our null hypothesis implies two or more linear constraints on the population means, the parametric ANOVA test takes an apparently different form. The hypothesis test has the following form and basis:

1. Compute

$$\text{SSR}_U = \sum_{i=1}^{I}\sum_{j=1}^{J}\sum_{k=1}^{n_{ij}}(y_{ijk} - \bar{y}_{ij})^2$$

where $\bar{y}_{ij}$ is the mean attribute score for the ij-th sample. Because $\bar{y}_{ij}$ is a plug-in and, hence, unconstrained estimate of μ_{ij}, this sum-of-squares is called the unconstrained sum of squared-residuals (SSR$_U$).

2. Let C_Q represent a system of Q (nonredundant) linear constraints on the $I \times J$ population means. The q-th constraint has the form $\sum\sum c_{q,ij}\mu_{ij}$.

3. Compute a constrained SSR:

$$\text{SSR}_{C_Q} = \sum_{i=1}^{I}\sum_{j=1}^{J}\sum_{k=1}^{n_{ij}}(y_{ijk} - \widehat{\mu}_{C_Q,ij})^2$$

where $\widehat{\mu}_{C_Q,ij}$ is the estimate of μ_{ij}, now constrained by the Q linear contrasts of C_Q. We refer to SSR$_{C_Q}$ as a constrained sum of squared-residuals.

4. Form the ratio

$$F = \frac{(1/Q)\left[\text{SSR}_{C_Q} - \text{SSR}_U\right]}{[1/(n-IJ)]\text{SSR}_U}$$

where $IJ = (I \times J)$ and $n = \sum_{i=1}^{I}\sum_{j=1}^{J} n_{ij}$.

5. Under the null hypothesis—when the Q linear constraints correctly describe relations among the $(I \times J)$ population distribution means—this F-ratio has a sampling distribution whose cdf is that of the F-random variable with Q and $(n - IJ)$ df. A statistically large value of the F-ratio, caused by a large increase in SSR when the Q linear constraints are imposed on the mean estimates, provides evidence against the null hypothesis. The alternative to the multiple-contrast null hypothesis is nondirectional.

How the constrained population estimates, the $\widehat{\mu}_{C_Q,ij}$, are computed depends on the particular Q linear constraints making up the hypothesis, and the algebra involved need not concern us here.

For our null hypothesis of no interaction between age and breed, we can express the constrained population mean estimates for the i-th breed as

$$\widehat{\mu}_{C_Q,i,M} = \left[\frac{\bar{y}_{iM} + \bar{y}_{i2}}{2} \right] + \left(\frac{1}{2} \right) \left[\frac{\sum_{\ell=1}^{5} (\bar{y}_{\ell M} - \bar{y}_{\ell 2})}{5} \right]$$

for the mature population, and

$$\widehat{\mu}_{C_Q,i,2} = \left[\frac{\bar{y}_{iM} + \bar{y}_{i2}}{2} \right] - \left(\frac{1}{2} \right) \left[\frac{\sum_{\ell=1}^{5} (\bar{y}_{\ell M} - \bar{y}_{\ell 2})}{5} \right]$$

for the two-year old population.

When the population means are constrained to show no breed-by-age interaction, the estimated mean butterfat productions of mature and young cows are constrained to differ by the same amount for all breeds:

$$(1/5)\sum_{\ell=1}^{5} (\bar{y}_{\ell M} - \bar{y}_{\ell 2})$$

We use the F-ratio approach to test our no-interaction hypothesis, with the help of S-Plus, in Figure 20.1.

The unconstrained SSR is obtained by taking sums of squares about the means of the ten different sample distributions and summing those. This gives us $SSR_U = 18.30$.

We use the sample distribution means (`xbarm` and `xbar2`) in the formulas given to find the constrained population mean estimates (`xbarmc` and `xbar2c`). These latter satisfy the no-interaction constraints, a constant mature-young difference in means of -0.111, and are quite close to the unconstrained mean estimates.

The constrained SSR is found by explicitly subtracting the appropriate constrained mean estimate from each observation, squaring these deviations, and finding their sum.

The constrained sum of squares of residuals, $SSR_{C_4} = 18.6145$, does not appear to be much larger than SSR_U. Indeed, the resulting F-ratio value of 0.385 is very small. It has a p-value of 0.82, the proportion of the F-distribution with 4 and 90 df with values of 0.385 or larger. This very large p-value offers no grounds for rejecting the null hypothesis of no interaction between breed and age.

```
> ayrm<- c(3.74,4.01,3.77,3.78,4.10,4.06,4.27,3.94,4.11,4.25)
> ayr2<- c(4.44,4.37,4.25,3.71,4.08,3.90,4.41,4.11,4.37,3.53)
> canm<- c(3.92,4.95,4.47,4.28,4.07,4.10,4.38,3.98,4.46,5.05)
> can2<- c(4.29,5.24,4.43,4.00,4.62,4.29,4.85,4.66,4.40,4.33)
> guem<- c(4.54,5.18,5.75,5.04,4.64,4.79,4.72,3.88,5.28,4.66)
> gue2<- c(5.30,4.50,4.59,5.04,4.83,4.55,4.97,5.38,5.39,5.97)
> holm<- c(3.40,3.55,3.83,3.95,4.43,3.70,3.30,3.93,3.58,3.54)
> hol2<- c(3.79,3.66,3.58,3.38,3.71,3.94,3.59,3.55,3.55,3.43)
> jerm<- c(4.80,6.45,5.18,3.49,5.24,5.70,5.41,4.77,5.18,5.23)
> jer2<- c(5.75,5.14,5.25,4.76,5.18,4.22,5.98,4.85,6.55,5.72)
> ssru<- var(ayrm,SumSquares=T)+var(ayr2,SumSquares=T)
> ssru<- ssru+ var(canm,SumSquares=T)+var(can2,SumSquares=T)
> ssru<- ssru+ var(guem,SumSquares=T)+var(gue2,SumSquares=T)
> ssru<- ssru+ var(holm,SumSquares=T)+var(hol2,SumSquares=T)
> ssru<- ssru+ var(jerm,SumSquares=T)+var(jer2,SumSquares=T)
> ssru
[1] 18.30117
> cowsm<- matrix(c(ayrm,canm,guem,holm,jerm),nrow=5,byrow=T)
> cows2<- matrix(c(ayr2,can2,gue2,hol2,jer2),nrow=5,byrow=T)
> xbarm<- apply(cowsm,1,mean)
> xbar2<- apply(cows2,1,mean)
> mnadj<- 0.5*mean((xbarm-xbar2))
> mnbas<- 0.5*(xbarm+xbar2)
> xbarmc<- mnbas + mnadj
> xbar2c<- mnbas - mnadj
> xbarm
[1] 4.003 4.366 4.848 3.721 5.145
> xbarmc
[1] 4.0045 4.3830 4.8945 3.6140 5.1870
> xbar2
[1] 4.117 4.511 5.052 3.618 5.340
> xbar2c
[1] 4.1155 4.4940 5.0055 3.7250 5.2980
> xbarmc-xbar2c
[1] -0.111 -0.111 -0.111 -0.111 -0.111
> ssrc<- sum((cowsm-xbarmc)*(cowsm-xbarmc))
> ssrc<- ssrc+ sum((cows2-xbar2c)*(cows2-xbar2c))
> ssrc
[1] 18.6145
> fratio<- ((0.25)*(ssrc-ssru))/((1/90)*ssru)
> fratio
[1] 0.3852172
> 1-pf(0.385,4,90)
[1] 0.8188516
```

Figure 20.1:   *Testing for Interaction of Breed and Age of Dairy Cows (S-Plus)*

Having determined that there is no interaction between breed and age, it is appropriate to test hypotheses about the main effects of each of the two factors. That is, our breed comparisons, Jersey versus each alternate breed, can be made across both levels of age, and the age comparison can be made across all five breed levels. In the absence of an interaction, we do not need to make either comparison separately, for each level of the other factor.

Our null hypotheses—for the four breed comparisons and the one age comparison—each involve a single linear constraint on the ten population means, and the tests would follow the pattern of two-group comparisons established in Applications 19.

Nonparametric Approach, Two-Factor Study

The assumption that the $(I \times J)$ population distributions we have randomly sampled all have normal cdfs and a common variance often cannot be justified. More often we do not know the parametric form of our population distributions. Indeed, where they are finite, these distributions might not possess a parametric form.

```
> allcows<- c(ayrm,ayr2,canm,can2,guem,gue2,
+ holm,hol2,jerm,jer2)
> vv<- c(1,1,1,1,1,1,1,1,1,1)
> grpx<- c(vv,2*vv,3*vv,4*vv,5*vv,6*vv,7*vv,8*vv,9*vv,10*vv)
> bfat<- data.frame(allcows,grpx)
> bxainte<- function(x)
+ {
+ d1<- mean(x[x[,2]==1,1])-mean(x[x[,2]==2,1])
+ d2<- mean(x[x[,2]==3,1])-mean(x[x[,2]==4,1])
+ d3<- mean(x[x[,2]==5,1])-mean(x[x[,2]==6,1])
+ d4<- mean(x[x[,2]==7,1])-mean(x[x[,2]==8,1])
+ d5<- mean(x[x[,2]==9,1])-mean(x[x[,2]==10,1])
+ c(d1-d2,d1-d3,d1-d4,d1-d5)
+ }
> interci<- bootstrap(bfat,bxainte,group=grpx)
> limits.bca(interci)
            2.5%       5%       95%      97.5%
bxainte1 -0.3791   -0.2930   0.35007   0.4001
bxainte2 -0.3372   -0.2613   0.52507   0.6008
bxainte3 -0.5026   -0.4584   0.05024   0.0891
bxainte4 -0.5803   -0.5000   0.58200   0.6703
```

Figure 20.2: *Nonparametric Assessment of No Interaction Hypothesis (S-Plus)*

When we have random samples from $(I \times J)$ populations, however, we can use our techniques of bootstrap inference to develop nonparametric CIs for population means or contrasts among population means.

In Figure 20.2, we repeat the test for interaction by estimating CIs for the four linear contrasts that comprise the no-interaction hypothesis.

We create a vector `grpx` to identify the population distributions from which the scores making up the vector `allcows` were randomly sampled. The `grpx` vector will be used by the `bootstrap` function to form, and then resample from, the appropriate nonparametric population estimates.

We construct a user function `bxainte` to compute the four mean contrasts from a data frame containing ten samples. The `bootstrap` command requests a 1,000 element Monte Carlo approximation to each of four bootstrap sampling distributions, one for each of the no interaction contrasts.

Finally, we notice that the null hypothesized values of zero fall within each of the four 90% bias-corrected and accelerated (BCA) CIs estimated by the last command.

Zero cannot be ruled out as a possible value of any one of the four contrasts, and, hence, we certainly have no basis for rejecting the no-interaction hypothesis.

This use of separate but simultaneous CIs to assess multiple contrasts can be misleading because the contrasts are not independent of one another. A more reliable method employs a nonparametric extension of the F-ratio test, the Q-statistic that we develop in Applications 22.

Random Two-Factor Case Assignment

The Estimation of IQ example from Applications 15 provides an illustration of the analysis of a fully-crossed two-factor design in which an available set of cases, rather than random samples, are employed. Although there is no scope for testing for an interaction between the two factors, we might have reason to suspect that there is one. If so, then we can and should assess the simple effects of one factor, evaluating that factor separately at each level of the other factor, rather than evaluating its main effects.

Nonrandom Two-Factor Studies

Table 20.2 is taken from Dusoir (1997) and reports the scores on a measure of Conformity for a set of 45 university student volunteers. The volunteers, available cases, also completed a measure of Authoritarianism and were classified by sex. The authoritarianism scores were used to divide the cases into three equal-sized groups, low, middle, and high. The conformity scores are grouped by sex- and authoritarianism-level in Table 20.2.

As the table shows, there is a clear imbalance of the sexes across levels of authoritarianism. The 45 available cases were not randomly allocated among levels of either authoritarianism or sex. As a result, any comparison of conformity scores among the (3×2) groups will be descriptive rather than causal in inference. The researcher wanted to see how conformity scores changed from one level of authoritarianism to another and whether those changes followed the same pattern for the two sex levels.

Table 20.2: *Conformity Scores Grouped by Authoritarianism and Sex*

Authoritarianism	Sex	Conformity Scores
1	1	8 7 10 6 13 12 7 6 8 12
1	2	21 23 15 12 16
2	1	12 4 4 9
2	2	12 17 14 16 20 8 14 17 7 17 15
3	1	4 8 13 9 24 7 23 13
3	2	19 9 7 11 14 13 10

```
> tmean(c11,floor(0.2*sizeof(c11))/sizeof(c11))
    8.6666667
> tmean(c21,floor(0.2*sizeof(c21))/sizeof(c21))
    7.25
> tmean(c31,floor(0.2*sizeof(c31))/sizeof(c31))
    12.166667
> tmean(c12,floor(0.2*sizeof(c12))/sizeof(c12))
    17.333333
> tmean(c22,floor(0.2*sizeof(c22))/sizeof(c22))
    15
> tmean(c32,floor(0.2*sizeof(c32))/sizeof(c32))
    11.4
```

Figure 20.3: *Trimmed Mean Conformity Scores for Two-Factor Groups (SC)*

If we examine the 20% trimmed means for the six groups of cases, as given by SC in Figure 20.3, we see evidence of an interaction between sex and authoritarianism in the response of these cases to the conformity measure.

```
> load(fox2)
 [c11 c21 c31 c12 c22 c32 MCV gsiz hsiz ]
> proc p_confor(){}
> hsiz= floor(0.5*(gsiz+1))
> hsiz
   5     2     4     3     6      4
> gsiz
   10    4     8     5     11     7
> subsam_k(p_confor,4,MCV,gsiz,hsiz,100,TT)
Full Sample Statistics:   -1.4166667    4.9166667
                          -2.3333333    -3.6
    Description of Statistic Number:   1
____v1   (# 100 )
    range=        10.5    (-6.66667  to  3.83333 )
    mean=         -1.44167
    median=       -1.66667
    sd=           2.34726
    iqr=          3     (-3.16667  to -0.166667 )
    MAD=          1.5
    Description of Statistic Number:   2
____v1   (# 100 )
    range=        16.75    (-3.5  to  13.25 )
    mean=         5.2175
    median=       5.5
    sd=           3.16596
    iqr=          4.29167    ( 3  to  7.29167 )
    MAD=          2.125
    Description of Statistic Number:   3
____v1   (# 100 )
    range=        9.91667    (-7.75  to  2.16667 )
    mean=         -2.69917
    median=       -2.91667
    sd=           1.99968
    iqr=          2.97917    (-4.09722  to -1.11806 )
    MAD=          1.5
    Description of Statistic Number:   4
____v1   (# 100 )
    range=        7.25    (-6.25  to  1 )
    mean=         -2.735
    median=       -3
    sd=           1.55717
    iqr=          2.14583    (-3.75  to -1.60417 )
    MAD=          1
```

Figure 20.4: *Half-Sample Analyses of Trimmed Mean Contrasts (SC)*

Sex level 2 shows a steady decrease in trimmed mean conformity scores from authoritarianism level 1 to level 2 to level 3. In contrast, for sex level 1, the decrease in trimmed mean conformity scores from level 1 to level 2 of authoritarianism is followed by an increase for those cases at level 3. In fact, the trimmed mean conformity scores for cases at authoritarianism level 3 are essentially the same for the two sex groups. In the presence of this apparent interaction, we should look at conformity score changes from level to level of authoritarianism separately for the two sexes.

In Figure 20.4, we use the SC command `subsam_k` to generate the four treatment comparisons over a collection of randomly selected random half-samples.

The random half-samples are drawn separately from the six authoritarianism-by-sex groups. This subsampling is not as faithful to the original selection of cases as we would like it to be. Ideally, we should form the half-samples from the entire set of 45 cases.

We have had to compromise for two reasons:

1. The raw authoritarianism scores are not available to us, meaning that we cannot divide a half-sample into thirds on the basis of those scores. We must continue to use the divisions established by the original split into thirds.

2. The imbalance of the two factors could lead to empty cells in the 3×2 design if half-sampling were carried out over the full set of cases.

The user-provided procedure `p_confor` used by `subsam_k` for computing the treatment contrasts is not described here. That procedure, however, returns four differences in trimmed means,

$$
\begin{aligned}
(\mathtt{c21} - \mathtt{c11}) &= \quad -1.4167 \\
(\mathtt{c31} - \mathtt{c21}) &= \quad\ \ 4.9167 \\
(\mathtt{c22} - \mathtt{c12}) &= \quad -2.3333
\end{aligned}
$$

and

$$
(\mathtt{c32} - \mathtt{c22}) = \quad -3.6
$$

from the original data set and corresponding quantities from each of a series of 100 random half-samples. These are the `Statistics` numbered 1 through 4 in the `subsam_k` procedure output.

The four trimmed mean contrasts are quite near the centers of their half-sample distributions, with the possible exception of $(\mathtt{c32} - \mathtt{c22}) = -3.6$, which lies 0.6 of a median absolute deviation (MAD) below the median of its half-sample distribution. Even so, more than one-quarter of the half-sample values of this statistic are smaller than -3.6, so we cannot regard the full data set description as an extreme one or as greatly unstable.

Treatment and Blocking Factors

The two-factor designs just described can be extended to include studies in which one of the two factors is a blocking factor. The distinction between treatment and blocking factors is largely in the intent of the researcher designing the study. Typically, we have research hypotheses about the values of contrasts among the levels of treatment factors. Those hypotheses motivated the study. On the other hand, cases have been grouped by levels on the blocking factor, not to test hypotheses but to ensure balance of the levels of the blocking factor across the treatment levels and to increase sensitivity.

In the random samples context, the treatment-by-blocks design can be analyzed in very much the same way as the treatment-by-treatment design. Although researchers might not be interested in contrasts among the block levels, they often do want to discover whether there is an interaction between the blocking and treatment factors, that is, whether the treatment effects are the same at all levels of blocking.

In designs where available cases have been randomly assigned among treatment levels or combinations of levels, we have seen already that we cannot test for an interaction. Furthermore, cases cannot, in general, be randomly assigned to levels of a blocking factor. This precludes testing for the significance of any contrast among blocking levels. For example, in a typical randomized blocks (RB) design you might separately randomize male and female cases among two or more levels of treatment. The random assignment to treatment levels provides the statistical basis for testing contrasts among treatments. But that randomization does not provide a statistical basis for testing for either a sex-by-treatment interaction or a sex effect. We can, if we have prior knowledge of an interaction, choose to carry out our treatment comparisons separately for the sexes.

In the nonrandom study there are no significance tests. If the blocking of available cases is by source, however, the subsampling analysis of the stability of our descriptions of treatment differences can take that source structure into account.

Covariate Adjustment of Treatment Scores

The purpose of blocking is to control for the possible confounding of two influences on the response to treatment. These influences are the differential treatments on the one hand, and some pretreatment attribute of the cases on the other. By blocking the cases on the level of that pretreatment attribute before distributing the cases among treatments, we guard against confounding.

Blocking leads to pretreatment control over possible confounding influences. In some instances, however, we cannot block on the potentially confounding attribute before making treatment assignments. Some degree of post-treatment control still might be possible, if we can adjust the treatment responses of cases for their varying levels on the pretreatment attribute.

Parametric Analysis of Covariance

In the parametric ANOVA context, such an adjustment takes the form of an analysis of covariance (ANCOVA). Similar to the ANOVA of a case-based study, we must have K random samples, each sample assigned to a different level or combination of treatments. Again, we assume that in each of the K population distributions, our treatment response attribute has the same variance and a normal cdf.

In the ANCOVA, these two assumptions are extended. We assume that for each of the K populations, the joint distribution of the response attribute and our covariate, the attribute we want to adjust for, is bivariate normal with a common variance-covariance structure and a common covariate mean. In particular, this extension requires that:

1. The covariate has a normal cdf, a common variance, and a common mean in all K population distributions.

2. The covariate and treatment response have the same correlation in all K population distributions.

The ANCOVA analysis allows us to test these assumptions before we adjust the treatment response scores.

The ANCOVA has many assumptions and a less restrictive form of adjustment is sometimes carried out in the parametric ANOVA setting. The concomitant variable approach relaxes the population distribution assumptions for the covariate, though not for the treatment response.

Nonparametric Bootstrap Adjusted Mean CIs

Bootstrap inference allows us to develop CIs for contrasts among covariate-adjusted treatment means in a wholly nonparametric way. We illustrate this with the data of Table 20.3 (Lunneborg, 1994). Four random samples of major league baseball players are represented, drawn separately from the four populations created by crossing League (American or National) with preferred batting Handedness (left or right).

The response attribute in this study is the number of Hits over the full season. Does this vary from league to league, and does it depend on the handedness of the batter? One thing the number of hits does depend on is the number of opportunities the batter has, the number of At-Bats (AB). This attribute is our covariate. Before comparing the mean number of hits for left- and right-handed hitters, or for players from the American and National leagues, we need to adjust hits for the number of times at bat.

Table 20.3:   *At Bats and Hits for Random Samples of Major League Batters*

American League				National League			
Left Handed		Right Handed		Left Handed		Right Handed	
AB	Hits	AB	Hits	AB	Hits	AB	Hits
266	55	223	51	594	168	542	145
505	156	300	71	588	196	455	110
385	103	360	86	368	74	462	130
621	205	241	62	511	121	258	64
270	74	514	127	375	86	458	106

A nonparametric version of the concomitant variable approach to adjustment can be developed from a linear model perspective in a form that keeps it as faithful as possible to the parametric version. We relax only the assumptions of normality and constant variance of the population distributions of the response attribute.

In later applications, you will have more practice resampling from linear models. Our linear model estimates the number of hits obtained by the i-th player from knowledge of the player's league, handedness, and at-bats. Actually, what the model estimates is the average number of hits for all batters with the same league, handedness, and at-bats scores as the i-th player. We write the linear model in this form:

$$\mu(y_i) = \beta_1 x_{i1} + \beta_2 x_{i2} + \beta_3 x_{i3} + \beta_4 x_{i4} + \beta_5 z_i$$

where

$$x_{i1} = \begin{cases} 1, \text{ if the } i\text{-th player bats left handed in the AL} \\ 0, \text{ otherwise} \end{cases}$$

$$x_{i2} = \begin{cases} 1, \text{ if the } i\text{-th player bats right handed in the AL} \\ 0, \text{ otherwise} \end{cases}$$

$$x_{i3} = \begin{cases} 1, \text{ if the } i\text{-th player bats left handed in the NL} \\ 0, \text{ otherwise} \end{cases}$$

$$x_{i4} = \begin{cases} 1, \text{ if the } i\text{-th player bats right handed in the NL} \\ 0, \text{ otherwise} \end{cases}$$

and z_i is the deviation of the i-th player's number of at-bats from the average number of at-bats for players in the sample.

We chose these five attributes for the model so that the parameters β_1 through β_4 would have these interpretations:

1. β_1: Mean number of hits for an AL left-handed batter with the sample average at-bats

2. β_2: Mean number of hits for an AL right-handed batter with the sample average at-bats.

3. β_3: Mean number of hits for a NL left-handed batter with the sample average at-bats.

4. β_4: Mean number of hits for a NL right-handed batter with the sample average at-bats.

Unless you are familiar with linear models, it might not be obvious why our β-parameters have these interpretations. But, as an example, our equation

$$\mu(y_i) = \beta_1 x_{i1} + \beta_2 x_{i2} + \beta_3 x_{i3} + \beta_4 x_{i4} + \beta_5 z_i$$

will take the value β_1 if $x_{i1} = 1$ and x_{i2}, x_{i3}, x_{i4}, and z_i are all zero. But that pattern of scores belongs to those AL left-handed players with average at-bats for the season.

Remember, β_1 is the estimated number of season hits not only for the i-th player but for all players having the same pattern of explanatory variable scores. For that reason, we label β_1 as the mean number of season hits for all players with this score pattern. More specifically, it is our modeled estimate of the mean hits for all players with this score pattern.

Our interpretations of β_2, β_3, and β_4 are based on three other score patterns, each having a zero for z_i, and for three of the four x_{ij}s and a one for the fourth.

The β-parameters provide the adjusted population means we require. They have been adjusted so that all refer to a common value of the covariate, the sample average at-bats.

Our substantive questions about hitting performance can be expressed as these three contrasts among β_1, β_2, β_3, and β_4:

1. League comparison: $\frac{\beta_1 + \beta_2}{2} - \frac{\beta_3 + \beta_4}{2}$

2. Handedness comparison, American League: $\beta_1 - \beta_2$

3. Handedness comparison, National League: $\beta_3 - \beta_4$

In each case, the null hypothesis is that the contrast should take a value of zero. Large values, positive or negative, provide evidence against those null hypotheses.

We have a linear model and hypothesis-based contrasts on the parameters of that model. Once we know how to estimate those parameters, we can use our bootstrap inference ideas to find CIs for the parameter contrasts or, if we want, to test the corresponding null hypotheses.

We are guided in our choice of estimator of the βs by the role those parameters play in their population distributions. If we had the complete joint distributions of hits and at bats for each of the four populations—left- and right-handed batters in the American and National Leagues—how would those joint distributions determine the values of the β population parameters? What are the properties of those parameters?

We define the linear model parameters so as to make the linear model a good model. Specifically, we want the model to be good in the following sense. If we used our model to estimate the hits for all players, left- and right-handed in the two major leagues, we'd want those estimates to be good estimates. In particular, we make our linear model as good as it can be by letting β_1 through β_5 take those values that will minimize the sum of squared differences between the estimated and actual number of hits over all players in the four populations.

```
proc p_bezbol(){
   local(n,p,yvec,temp,MM,XMAT,beta,ssr,res)
   n= rows($1$)
   p= cols($1$)
   vector(yvec,n,temp,n)
   getcol($1$,yvec,1)
   MM= $1$
   mlimit(MM,1,n,2,p)
   XMAT= MM
   mdelimit(MM)
   p= p-1
   getcol(XMAT,temp,5)
   temp= temp-mean(temp)
   setcol(XMAT,temp,5)
   vector(beta,p,ssr,p,res,n)
   lsqsregr(XMAT,yvec,res,beta,ssr)
   $3$[1]= 0.5*(beta[1]+beta[2])-0.5*(beta[3]+beta[4])
   $3$[2]= beta[1]-beta[2]
   $3$[3]= beta[3]-beta[4]
}
```

Figure 20.5: *Contrasts among Least Squares Linear Model Parameters (SC)*

We could have chosen other properties for the population βs, but minimizing the sum of squared errors of estimation is a common goal for linear models. Now that we know how the β parameters are defined, we know how to estimate them. We'll use plug-in estimators. That is, we'll find estimates, $\widehat{\beta}_1$ through $\widehat{\beta}_5$, that will minimize $\sum [y_i - \widehat{\mu}(y_i)]^2$, the sum of squares of errors, when we use our linear model to estimate the hits for the sampled ball players. This choice of estimator also keeps our nonparametric analysis close to the parametric one; for a slightly different reason, the normal, parametric analysis also provides least-squares estimates of the linear model β parameters.

Figure 20.5 shows the text of a special-purpose SC procedure p_bezbol that computes the value of the three contrasts among the estimates $\widehat{\beta}_1$ through $\widehat{\beta}_4$ obtained from a sample. The procedure requires an input matrix (1) with the sampled players' hits in the first column, their scores on x_1 through x_4 in columns 2 through 5, and their at-bats in the sixth column.

Our special purpose procedure creates the z scores by subtracting the mean at-bats from each player's at-bats, and then calls the SC procedure lsqsregr to compute the least-squares estimates of the β-parameters.

In Figure 20.6, we show the use by the SC procedure ibootci2 of p_bezbol to estimate 90% BCA CIs for the three β-parameter contrasts.

```
load(bezbol)
 [hi aleft arite nleft nrite ab BEZBOL]
> proc p_bezbol(){}
> gsiz:= 5,5,5,5
> ibootci2(p_bezbol,p_bezbol,BEZBOL,gsiz,3,2000,FBOOT,90)
_____tstat:
  11.719359          10.296722          -0.44439031
_____thetahat:
  11.719359          10.296722          -0.44439031
_____means of  2000  bootstrap estimates:
  12.045141          9.2175224          -0.056251284
_____SDs of  2000  bootstrap estimates:
  6.5145509      7.9490307      8.7942167
_____z0 (bias corrections):
 -0.050153583   0.1092553    -0.007519955
_____a (acceleration constants):
 -0.025364914   0.029257873   0.037528524
 Theta 1   90 % BC & Accelerated CI:   0.18960627   to 21.449383
 Theta 2   90 % BC & Accelerated CI:  -1.5455687    to 24.329008
 Theta 3   90 % BC & Accelerated CI: -13.136699     to 15.897435
```

Figure 20.6: *BCA Confidence Intervals for Adjusted Mean Hits Contrasts (SC)*

The sample data had been saved in the matrix BEZBOL in the form noted earlier, as they are needed by both p_bezbol and ibootci2.

The sample estimates of the three contrasts among the adjusted mean hits (the thetahat values) are reported as:

1. League Comparison: 11.72 additional hits for the AL player with the average number of at-bats.

2. Handedness Comparisons: The left-handed batter with the average number of at-bats hits an additional 10.30 times in the AL, but 0.44 fewer times in the NL.

Because of the very small sample sizes, the three CIs are 20 to 30 hits wide, giving us little guidance about how large the actual population contrasts are. In a later exercise, data are given for larger samples of baseball players.

Adjusted Contrasts: Random Assignment

In the analysis just completed, we adjusted the randomly sampled case's response, hits, for the value of a covariate, that case's at-bats. This allowed us to estimate adjusted population means—the average number of hits for players with the average number of at-bats—and contrasts among those adjusted means. In this way, we can counteract the influence on the response means of any difference in covariate scores among the populations.

When available cases are randomly allocated among treatments, it is equally desirable to adjust response attribute scores for covariate scores before comparing treatment responses.

Table 20.4:   *EFT Scores and Times to Assemble Block Design for 24 Children*

	Row Instructions Group											
Time	317	464	525	298	491	196	268	372	370	739	430	410
EFT	59	33	49	69	65	26	29	62	31	139	74	31
	Corner Instructions Group											
Time	342	222	219	513	295	285	408	543	298	494	317	407
EFT	48	23	9	128	44	49	87	43	55	58	113	7

Table 20.4 shows data collected from 24 primary school children in Sydney, Australia, and reproduced from Hand et al. (1994, Data Set 37). The children were randomly allocated to one of two groups given different instructions in how to form a pattern of colored blocks. The row group was told to start with a row of three blocks, and the corner group was told to start with one corner. The response is the time taken by each child to complete the task. The researcher was concerned that the children's performances might also be influenced by their field independence, the extent to which they can abstract the structure of a problem from its immediate context. This trait was assessed by a covariate, the Embedded Figures Test (EFT).

The analysis, outlined in Figure 20.7 for S-Plus, is of the same form as the random sample baseball example was. Again we employ a linear model

$$\widehat{y}_i = \beta_1 x_{i1} + \beta_2 x_{i2} + \beta_3 z_i$$

where

$$x_{i1} = \begin{cases} 1, & \text{if the } i\text{-th child was assigned to the row instructions group} \\ 0, & \text{otherwise} \end{cases}$$

$$x_{i2} = \begin{cases} 1, & \text{if the } i\text{-th child was assigned to the corner instructions group} \\ 0, & \text{otherwise} \end{cases}$$

and z_i is the deviation of the i-th child's EFT score from the median EFT score for study children.

Note a change in notation from the baseball example. On the left side of our linear model equation, we write the predicted time score for the i-th child in the study as $\widehat{y}_i$ and not as $\mu(y_i)$. We no longer have cases sampled from populations, so our prediction does not take the form of a mean for a population of cases with the same pattern of x-scores. Rather, the prediction is for the particular i-th case.

These explanatory attributes were chosen for the model so that, when the model is fit to the study data, the coefficients β_1 and β_2 would have these interpretations:

1. β_1: Model estimate of the time score for a child with a median EFT score and given row instructions for the block design task

2. β_2: Model estimate of the time score for a child with a median EFT score and given corner instructions for the block design task

$s = (\beta_1 - \beta_2)$ is the difference in treatment effects, adjusted for the EFT covariate.

We choose values for the linear model β-coefficients to make as small as possible the sum of squares of differences between the actual and modeled time scores,

$$\sum_{i=1}^{n}(y_i - \widehat{y}_i)^2$$

For the available case design, however, these least-squares β-coefficients are not plug-in estimates of population distribution parameters but, rather, simply data-based constants that make the linear model a good one for these local data by minimizing the sum of squared differences between actual and modeled time scores.

```
> time<-
c(317,464,525,298,491,196,268,372,370,739,430,410,342,222,219,513,
295,285,408,543,298,494,317,407)
> eft<-
c(59,33,49,69,65,26,29,62,31,139,74,31,48,23,9,128,44,49,87,43,55,
58,113,7)
> median(eft)
[1] 49
> ctreft<-eft-median(eft)
> rowgrp<-c(1,1,1,1,1,1,1,1,1,1,1,1,0,0,0,0,0,0,0,0,0,0,0,0)
> corgrp<-1-rowgrp
> mean(time[1:12])-mean(time[13:24])
[1] 44.75
> desmat<-cbind(time,ctreft)
> f20<-function(X){
+ a<-lsfit(cbind(rowgrp,corgrp,X[,2]),X[,1],intercept=F)
+ a[[1]][1]-a[[1]][2]
+ }
> f20(desmat)
  rowgrp
 44.2405
> rd20<-bootstrap(desmat,f20,sampler=samp.permute)
> length(rd20$replicates[rd20$replicates >= f20(desmat)]
+ )
[1] 162
> length(rd20$replicates[rd20$replicates <= -1*f20(desmat)])
[1] 167
> (167+162+1)/1001
[1] 0.3296703
>
```

Figure 20.7: *Finding a p-Value for the Difference in Adjusted Means (S-Plus)*

In the S-Plus dialogue, the adjusted treatment difference is computed by the user-supplied function f20. When the function f20 is applied to the observed data, we obtain $s = (\beta_1 - \beta_2) = 44.24$ seconds, indicating an advantage— faster task completion—for the corner instructions. Our statistical question, then, is whether the 44.24 seconds adjusted treatment difference is so large as to be a very unlikely outcome of the random assignment of children whose individual time scores would have been the same under either set of block design instructions.

In the dialogue, the S-Plus bootstrap function (directed to use the permutation sampler) recomputes the test statistic $s^H = (\beta_1^H - \beta_2^H)$ for each of 1,000 rerandomizations of the 24 children among the two instruction groups. In these rerandomizations, the children carry their time scores with them, in accordance with the null treatment-effect hypothesis. The children also carry with them in the rerandomizations their EFT covariate scores. These scores were obtained before treatment, and are, thus, unaffected by treatment.

The `bootstrap` function shuffles the rows of the `desmat` matrix and passes the resulting X matrix to the user-supplied function `f20` which extracts time (`X[,1]` or y) and median-centered EFT (`X[,2]` or z) scores for the linear model. The x_1 (`rowgrp`) and x_2 (`corgrp`) scores are not shuffled. The result is the rerandomization of the children (with their y and z scores) to the two instruction groups.

Of the 1,000 values of s^H produced here, 162 were equal to or larger than $s = 44.24$ seconds. Because we had not hypothesized in advance that the corner instructions would be superior, we must also ask how likely it is, under the null hypothesis, to have obtained a value of s less than or equal to -44.24 seconds. There were 167 such small values leading to an estimated p $= 0.33$ for our adjusted treatment comparison.

The number of null reference distribution elements greater than or equal to s was very nearly the same as the number less than or equal to $-s$. How symmetric is the reference distribution? Figure 20.8 describes the cdf for the 1,000 element distribution. If we were to overlay the cdf for a normal distribution with the same mean and variance as that in our reference distribution, the two would be nearly indistinguishable.

Adjusted Contrasts: Nonrandom Design

Our last example is of a multiple treatment study in which the available cases are not randomly allocated among treatments.

The illustration is based on data (Hand et al., 1994, Data Set 285) collected for "young girls receiving three different treatments for anorexia over a fixed period of time." Table 20.5 shows pretreatment and posttreatment weights in pounds for the girls.

The data will be modeled on the assumption that they are from three disconnected and nonrandom samples of anorexics. That is, each girl was enrolled into a specific treatment, rather than assigned to one of the three. At issue is whether either of two new therapies (Ther), cognitive-behavioral (Cog Beh) or family, is superior to the standard therapy.

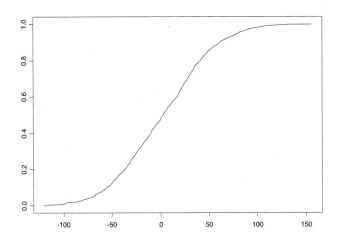

Figure 20.8: *Reference Distribution cdf, Difference in Adjusted Means*

Our response measure is posttreatment weight, adjusted for pretreatment weight. Our descriptive statistics, comparing each of the new therapies with the standard one, will adjust for the varying pretreatment weights of the study participants.

We'll achieve this again by the use of a linear model:

$$\widehat{y}_i = \beta_1 x_{1i} + \beta_2 x_{2i} + \beta_3 x_{3i} + \beta_4 x_{4i} + \beta_5 x_{5i} + \beta_6 x_{6i}$$

where

$$x_{i1} = \begin{cases} 1, \text{ if the } i\text{-th girl was in standard therapy} \\ 0, \text{ otherwise} \end{cases}$$

$$x_{i2} = \begin{cases} 1, \text{ if the } i\text{-th girl was in cognitive-behavioral therapy} \\ 0, \text{ otherwise} \end{cases}$$

$$x_{i3} = \begin{cases} 1, \text{ if the } i\text{-th girl was in family therapy} \\ 0, \text{ otherwise} \end{cases}$$

$$x_{i4} = \begin{cases} z_i, \text{ if the } i\text{-th girl was in standard therapy} \\ 0, \text{ otherwise} \end{cases}$$

$$x_{i5} = \begin{cases} z_i, \text{ if the } i\text{-th girl was in cognitive-behavioral therapy} \\ 0, \text{ otherwise} \end{cases}$$

$$x_{i6} = \begin{cases} z_i, \text{ if the } i\text{-th girl was in family therapy} \\ 0, \text{ otherwise} \end{cases}$$

z_i is the deviation of the i-th girl's pretreatment weight from the average pretreatment weight for all girls in the study, and $\hat{y}_i$ is the modeled posttreatment weight for the i-th girl.

Table 20.5:   *Pretreatment and Posttreatment Weights for Anorexic Girls*

Cog Beh Ther.		Standard Ther.		Family Ther.	
Before	After	Before	After	Before	After
84.9	85.6	89.4	80.1	83.3	94.3
80.5	82.2	80.7	80.2	83.8	95.2
81.5	81.4	91.8	86.4	86.0	91.5
82.6	81.9	74.0	86.3	82.5	91.9
79.9	76.4	78.1	76.1	86.7	100.3
88.7	103.6	88.3	78.1	79.6	76.7
94.9	98.4	87.3	75.1	76.9	76.8
76.3	93.4	75.1	86.7	94.2	101.6
81.0	73.4	80.6	73.5	73.4	94.9
80.5	82.1	78.4	84.6	80.5	75.2
85.0	96.7	77.6	77.4	81.6	77.8
89.2	95.3	88.7	79.5	82.1	95.5
81.3	82.4	81.3	89.6	77.6	90.7
76.5	72.5	78.1	81.4	83.5	92.5
70.0	90.9	70.5	81.8	89.9	93.8
80.4	71.3	77.3	77.3	86.0	91.7
83.3	85.4	85.2	84.2	87.3	98.0
83.0	81.6	86.0	75.4		
87.7	89.1	84.1	79.5		
84.2	83.9	79.7	73.0		
86.4	82.7	85.5	88.3		
76.5	75.7	84.4	84.7		
80.2	82.6	79.6	81.4		
87.8	100.4	77.5	81.2		
83.3	85.2	72.3	88.2		
79.7	83.6	89.0	78.8		
84.5	84.6				
80.8	96.2				
87.4	86.7				

This linear model, a separate-slopes and separate-intercepts model, encompasses three separate posttreatment weight prediction equations:

$$\widehat{y}_i = \beta_1 + \beta_4 z_i$$

for standard therapy girls,

$$\widehat{y}_i = \beta_2 + \beta_5 z_i$$

for cognitive-behavioral therapy girls, and

$$\widehat{y}_i = \beta_3 + \beta_6 z_i$$

for family therapy girls.

The separate intercepts, β_1, β_2, and β_3, give the modeled posttreatment weights for girls entering each of the therapies at the average pretreatment weight.

The slope coefficients, β_4, β_5, and β_6, describe the impact of an additional pound of pretreatment weight on the modeled posttreatment weight. The linear models used for the baseball and block-design analyses assumed a single slope coefficient. We could have made those models potentially more accurate by providing for, as we do here, separate slopes for each of the treatments.

Once again, we complete our linear model by choosing values for the six intercept and slope coefficients that will minimize the sum of squared discrepancies between actual and modeled posttreatment weights.

The descriptive statistics we'll use to compare therapies are these:

1. $(\beta_2 - \beta_1)$: The difference in posttreatment weights between cognitive-behavioral and standard therapies, for an average pretreatment weight girl. If positive, cognitive-behavioral therapy was more effective.

2. $(\beta_3 - \beta_1)$: The difference in posttreatment weights between family and standard therapies, for an average pretreatment weight girl.

Even though the slope coefficients do not enter explicitly into the definition of our treatment comparison statistics—for any of our three examples—their inclusion in the linear models is what allows us to adjust for the varying values of the covariate.

Figures 20.9 through 20.11 show how Resampling Stats computed the two anorexia-treatment comparison statistics and then developed half-sample distributions for each.

In Figure 20.9, the study data first are recovered from a file, `anorexia.dat`. The data are organized in this file as six columns, following the order given in Table 20.5. The average pretreatment and posttreatment weights then are computed for each therapy. Finally, the pretreatment weights for the 72 girls are centered about their average pretreatment weight.

```
read file "c:\resamp\anorexia.dat" bb ba sb sa fb fa
take sb 1,26 sb    ' Before weights, Standard therapy
take sa 1,26 sa    ' After weights, Standard therapy
take fb 1,17 fb    ' Before weights, Family therapy
take fa 1,17 fa    ' After weights, Family therapy
mean sb mnsb
mean bb mnbb
mean fb mnfb
mean sa mnsa
mean ba mnba
mean fa mnfa
print mnsb mnbb mnfb   'Mean weights, Before
print mnsa mnba mnfa   'Mean weights, After
concat bb sb fb pre
mean pre mnp           'Mean Before Weight
subtract bb mnp bb
subtract sb mnp sb
subtract fb mnp fb
```

Figure 20.9: *Part 1, Half-Sample Adjusted Contrasts (Resampling Stats)*

In Figure 20.10, the linear model attributes x_1 through x_6 are formed and the `regress` command is used to calculate the model's β-coefficients. Finally, the β-coefficients (`beta`) and the two treatment comparison statistics, $s_b = (\beta_2 - \beta_1)$ and $s_f = (\beta_3 - \beta_1)$ (`beff` and `feff`) are printed, as is the overall average pretreatment weight (`mnp`).

Finally, Figure 20.11 describes the formation of 100 random half-samples from the three therapeutic groups and the collection of values for the two test statistics from those half-samples.

The vectors `bsel`, `ssel`, and `fsel` identify the cases in each of the three therapeutic groups. Shuffling these case-lists and taking the first half of each shuffled list provides a random half-sample.

```
set 29 1 b1
set 29 0 b0
set 26 1 s1
set 26 0 s0
set 17 1 f1
set 17 0 f0
concat s1 b0 f0 si          ' x1
concat s0 b1 f0 bi          ' x2
concat s0 b0 f1 fi          ' x3
concat sb b0 f0 pr_s        ' x4
concat s0 bb f0 pr_b        ' x5
concat s0 b0 fb pr_f        ' x6
concat sa ba fa post        ' y
regress noconst noprint post si bi fi pr_s pr_b pr_f beta
print beta
take beta 1 adj_s           'Adjusted After Treatment Weights
take beta 2 adj_b
take beta 3 adj_f
subtract adj_b adj_s beff    'sb
subtract adj_f adj_s feff    'sf
print mnp beff feff
```

Figure 20.10: *Part 2, Half-Sample Adjusted Contrasts (Resampling Stats)*

Figure 20.12 displays the output of one run of the Resampling Stats script just described.

The raw posttreatment weight averages for girls in the three therapies were: standard, 81.1 pounds; cognitive-behavioral, 85.7 pounds; and family, 90.5 pounds.

The girls recruited into the three treatments, however, started at different weights. Indeed, the pretreatment weight data suggest that the standard group was slightly more anorexic than either of the experimental groups was.

When we use our linear model to obtain comparable posttreatment weights for the three therapies (i.e., that for a patient with a pretreatment weight of 82.4 lb), the picture of the relative effectiveness of the therapies is much the same as for the raw results: standard, 81.0 pounds; cognitive-behavioral, 85.5 pounds; and family, 89.7 pounds.

```
copy 1,29 bsel
copy 1,26 ssel
copy 1,17 fsel
set 13 1 ss1
set 13 0 ss0
set 15 1 bs1
set 15 0 bs0
set 9 1 fs1
set 9 0 fs0
concat ss1 bs0 fs0 ssi
concat ss0 bs1 fs0 bsi
concat ss0 bs0 fs1 fsi
repeat 100
  shuffle bsel bsel
  take bsel 1,15 btak
  shuffle ssel ssel
  take ssel 1,13 stak
  shuffle fsel fsel
  take fsel 1,9 ftak
  take sb stak ssb
  take sa stak ssa
  take bb btak bsb
  take ba btak bsa
  take fb ftak fsb
  take fa ftak fsa
  concat ssa bsa fsa post
  concat ssb bs0 fs0 sb_s
  concat ss0 bsb fs0 sb_b
  concat ss0 bs0 fsb sb_f
  regress noconst noprint post ssi bsi fsi sb_s sb_b sb_f beta
  take beta 1 ssa
  take beta 2 bsa
  take beta 3 fsa
  subtract bsa ssa xx
  score xx bgood
  subtract fsa ssa xx
  score xx fgood
end
median bgood mdnb
median fgood mdnf
percentile bgood (25 75) b_iqr
percentile bgood (10 90) b_m80
percentile fgood (25 75) f_iqr
percentile fgood (10 90) f_m80
print mdnb b_iqr b_m80
print mdnf f_iqr f_m80
```

Figure 20.11: *Part 3, Half-Sample Adjusted Contrasts (Resampling Stats)*

Based on these adjusted results, our two treatment comparison statistics take values of $s_b = 4.5$ pounds and $s_f = 8.8$ pounds. Each reflects the superiority, among these patients, of the experimental over the standard therapy. The s values are near enough to the medians of the corresponding distributions of half-sample statistics that they can be regarded as stable descriptors of this study.

```
MNSB   =   81.558
MNBB   =   82.69
MNFB   =   83.229
MNSA   =   81.108
MNBA   =   85.697
MNFA   =   90.494
BETA   =   80.994  85.458  89.748  -0.13418  0.84798  0.90923
MNP    =   82.408
BEFF   =   4.4644
FEFF   =   8.754
MDNB   =   3.8363
B_IQR  =   2.4544  5.5228
B_M80  =   1.4987  6.1458
MDNF   =   8.3155
F_IQR  =   6.7575  9.9264
F_M80  =   4.8492  11.68
```

Figure 20.12: *Half-Sample Stability of Comparisons (Resampling Stats)*

Perhaps surprisingly, our model predicts that the typical girl, weighing 82.4 pounds on entering therapy, would lose weight, to 81.0 pounds, in the standard therapy. Indeed, the average posttreatment weight was slightly lower than the average pretreatment weight for this group. A glance at the slope coefficients for our linear model—standard, -0.13 pound/pound; cognitive-behavioral, 0.85 pound/pound; and family, 0.91 pound/pound—provides further evidence of the poor performance of the standard therapy. The slope coefficients for the two new therapies are both close to $+1$ pound/pound, suggesting that girls in cognitive-behavioral therapy all gain about 3 pounds (i.e., 85.5 pounds $-$ 82.4 pounds), and those in family therapy all gain about 7 pounds (i.e., 89.7 pounds $-$ 82.4 pounds). The near-zero slope coefficient for standard therapy suggests that all posttreatment weights in this group will be close to 81 pounds, whatever their pretreatment weights were.

Exercises

1. The Business Section of London's *Independent on Sunday* for 25 February 1996 reported key economic data for 12 highly industrialized countries. Portions of those data are abstracted as Table 20.6.

Divide the countries into two or more regional groups and compare Present Unemployment in the regions, adjusting for one or more of the past economic indices. The comparison should be appropriate to nonrandom data—you have the entire population of industrialized states. Because of the small number of cases, use a single slope for each covariate in your linear model. How do your comparisons change as one country in turn is withdrawn from the study?

Table 20.6:   *The Industrialized Economies: Unemployment*

Country	Inflation		Interest Rate		Unemployment	
	Past	Present	Past	Present	Past	Present
UK	2.9	3.3	6.19	6.75	7.9	8.5
Australia	5.1	2.6	7.49	8.07	8.6	9.0
Belgium	2.0	1.9	3.38	5.50	14.7	14.2
Canada	1.8	0.2	5.23	8.33	9.6	9.7
France	2.1	1.6	4.50	5.94	11.7	12.0
Germany	1.5	2.2	3.44	5.10	10.0	9.3
Italy	5.5	3.9	10.19	9.56	12.1	11.9
Japan	-0.4	0.5	0.53	2.18	3.3	2.8
Netherlands	1.9	2.5	3.19	5.12	7.0	7.6
Spain	3.9	4.4	8.73	8.65	22.8	23.9
Sweden	2.6	2.6	7.98	8.16	8.1	8.2
USA	2.5	2.7	5.15	6.14	5.8	5.7

2.   Forty laboratory rats were randomly assigned, ten apiece, to one of four diets which differed in the amount of protein and the source of protein. The gain in weight for each rat is tabulated as Table 20.7.

How did the four diets differ? Did the higher level of protein produce greater weight gain? Was there a difference associated with cereal as compared with beef protein? These data are taken from Hand et al. (1994, Data Set 9).

Table 20.7:   *Weight Gains among Rats Randomized to Different Protein Diets*

Protein Source:	Beef		Cereal	
Protein Amount:	Low	High	Low	High
	90	73	107	98
	76	102	95	74
	90	118	97	56
	64	104	80	111
	86	81	98	95
	51	107	74	88
	72	100	74	82
	90	87	67	77
	95	117	89	86
	78	111	58	92

3. The data in Table 20.8 are from a study described by Kleinbaum, Kupper, and Muller (1988). Sixteen laboratory rats were randomly divided into four treatment groups. Three of the groups received a drug designed to retard muscular atrophy. The fourth group received a saline control. For each animal, a particular muscle was denervated. After a time, the extent of atrophy was determined by weighing the muscle.

Because muscle weight also will be influenced by the size of the animal, total animal weight was determined. Use Animal Weight as a covariate in a rerandomization analysis, comparing the adjusted treatment effect for each drug in turn against the saline treatment.

4. Table 20.9 shows data taken from an experiment described in Mead (1988). The table entries are the survival rates in colonies of *Salmonella typhimerium* seven days following the initiation of treatment. The actual survival measure is the logarithm of the density of the colony. Each treatment is a combination of the levels of three factors affecting the medium in which the salmonella colonies were maintained, the amount of sorbic acid, the relative acidity of the medium, pH, and the level of water activity. The influence of each of the three factors on survival was investigated at three

levels. (In the actual study, water activity was studied at six levels, but the design is simplified here to just three levels.)

Table 20.8:   *Drug Treatment for Experimental Muscle Atrophy*

Saline	Animal Weight	198	175	199	224
	Muscle Weight	0.34	0.43	0.41	0.48
Quinidine Sulfate	Animal Weight	233	250	289	255
	Muscle Weight	0.41	0.87	0.91	0.87
Lo Atropine Sulf	Animal Weight	204	234	211	214
	Muscle Weight	0.57	0.80	0.69	0.84
Hi Atropine Sulf	Animal Weight	186	286	245	215
	Muscle Weight	0.81	1.01	0.97	0.87

Twenty-seven salmonella colonies were randomly assigned to the resulting $3 \times 3 \times 3 = 27$ treatments, one colony to each distinct treatment. Because only one experimental unit is assigned to each treatment combination, the design is referred to as a single replicate one. Single replicate designs are feasible only with factorial studies. Here, for example, there are nine observations at each water activity level thanks to the combination of levels for the other two factors. These multiple observations allow us to use rerandomization to assess the chance nature of at least some observed differences.

Consider the three factors to be coequal in interest. Investigate the main effect of each in turn via omnibus treatment-effect hypotheses as was done in the original analysis, rather than by specific comparisons between levels. Remember to rerandomize only between levels of the factor being assessed, not between levels of either of the other two factors.

Table 20.9:   *Salmonella Survival Following Three Factor Treatment*

		Water Activity Level		
Sorbic Acid	pH	0.82	0.90	0.98
0	5.0	4.52	6.14	8.33
	5.5	4.31	5.98	8.37
	6.0	4.85	5.87	8.19
100	5.0	4.18	5.78	7.59
	5.5	4.43	5.28	7.79
	6.0	4.29	5.01	7.64
200	5.0	4.37	5.43	7.19
	5.5	4.27	5.10	6.92
	6.0	4.26	5.20	7.14

5. Table 20.10 extends the baseball study of Table 20.3 by extending the four random samples of players to 20 each. Repeat the earlier analysis. With larger random samples it is now practical to fit a model with separate slopes for the covariate, At-Bats, for each of the four populations.

Table 20.10: *At-Bats and Hits for Random Samples of Major League Batters*

American League				National League			
Left Handed		Right Handed		Left Handed		Right Handed	
AB	Hits	AB	Hits	AB	Hits	AB	Hits
266	55	223	51	594	168	542	145
505	156	300	71	588	196	455	110
385	103	360	86	368	74	462	130
621	205	241	62	511	121	258	64
270	74	514	127	375	86	458	106
373	99	399	121	253	54	572	147
253	60	535	135	276	67	510	160
498	152	391	104	604	203	570	177
475	139	564	156	335	79	304	78
217	54	484	115	215	50	482	117
233	54	375	110	355	85	325	111
409	112	461	136	374	107	377	98
455	120	545	158	361	99	259	74
361	94	396	104	428	118	230	50
320	96	541	148	266	67	537	132
375	102	519	141	287	64	549	155
245	75	441	133	408	102	606	176
239	65	239	53	343	111	482	152
353	85	414	98	476	113	573	159
631	191	378	111	251	63	432	115

Applications 21

Within-Cases Treatment Comparisons

Some research designs call for the same case to receive two or more treatments or to have an attribute assessed repeatedly, perhaps under different circumstances. In these designs, we are interested in the change in attribute score for or within a case from one occasion to another. Here, we outline how such comparisons and the inferences we draw from them can be tailored to the presence or absence of random sampling of cases or their random assignment to treatment sequences. We begin with a review of the two most common parametric models for such data.

Normal Models, Univariate and Multivariate

The study design that classically is analyzed as a within-cases analysis of variance (ANOVA) produces data consisting of M scores on some attribute for each of n cases. The n cases are a random sample from some population, and the scores are the result of assessing the same attribute on M different occasions, perhaps recording the response to each of M different treatments.

For this design, we have M population distributions of scores on some attribute. The classical data-analytic task has been to draw inferences about the means of these populations. This has taken the form of estimating confidence intervals (CIs) for one or more contrasts among the population means or testing null hypotheses about such contrasts.

Here is an example of a within-cases multiple-treatment study:

Table 21.1 from Ramsey and Schaefer (1997) reports the amount of Interleukin-1 (IL-1) in blood samples from a random sample (we shall assume) of 12 patients aged 20 to 50 years. IL-1 plays an important role in regulating the immune system and is quickly dissipated. The IL-1 samples were drawn while cases were involved in three successive and distinct activities:

1. Phase A: Neutral activity

2. Phase B: Audiotape exercises in body-imaging, recalling happy events, remembering recovery from illness

3. Phase C: Directed relaxation, positive aspects of audiotape discussed

Table 21.1: *IL-1 Levels During Three Successive Activities*

Case	Phase A	Phase B	Phase C
1	6,850	29,100	34,300
2	27,400	41,100	47,000
3	20,700	30,700	33,600
4	3,000	4,900	5,900
5	22,100	25,700	27,100
6	14,300	17,200	25,700
7	24,300	33,600	31,400
8	2,400	4,700	9,400
9	6,500	7,900	6,400
10	4,800	4,700	5,700
11	2,000	3,000	3,700
12	2,200	5,100	6,800

At issue is whether, or by what amount, the level of IL-1 changes from phase A to phase B to phase C.

The classical approach to testing hypotheses or estimating CIs for within-cases treatment comparisons is a parametric one and has a mathematical foundation that requires the population distributions of attribute scores to have certain properties:

1. The M population distributions have normal distribution functions (cdfs).

2. The M population distributions have the same variance.

3. The correlations between pairs of the M population distributions are such that the sampling distribution for any contrast among the sample means—the plug-in estimate of a contrast among the population means—has the same variance.

The first two requirements, normal cdfs with a common variance for each population distribution, are familiar to us from the between-cases ANOVA model for treatment comparisons. The third requirement imposes further restrictions on those distributions. This new requirement is strictly true only for mean contrasts whose weights sum to zero. But these are the contrasts we want to study, comparing some treatment means with others.

A second parametric approach provides an alternative to this univariate within-cases ANOVA and makes slightly different, perhaps less demanding, requirements of the M population distributions. This second mathematical model takes its name from the requirement that the joint distribution of the M attribute scores in the population be that of a multivariate normal random variable.

This overarching requirement maintains the first requirement of the ANOVA approach, each of the M population distributions must have a univariate normal cdf, but replaces the second and third requirements. Although the pattern of variances and correlations is not specified, multivariate normality requires the joint population distribution for each pair of the M attributes to have a specific mathematical form, that of a bivariate normal random variable. A similar mathematical requirement is imposed on each triplet of population distributions and, indeed, on each possible subset of the M population distributions. In addition to these marginal implications, multivariate normality has conditional implications as well. What does that mean? Say we limited our case population to just those who have a score of 50 on the first of M occasions. We now have a population conditional upon the score on the first occasion being 50. In this conditional population, the remaining $(M - 1)$-variate population distribution also would have to answer to the requirements of univariate, bivariate, trivariate, and so forth normality.

Either parametric model imposes considerable mathematical conditions on the M population distributions. For most empirical research, it is highly unlikely that either set of assumptions can be justified fully.

Bootstrap Treatment Comparisons

Nonparametric bootstrap CI estimates provide a nonparametric approach to within-cases treatment comparisons. Superficially, we develop CI estimates for contrasts among our M within-cases treatments very much as we did in Applications 19 for contrasts among treatments delivered to K independent groups. There are important differences, though, in how we form our bootstrap estimates of population distributions, and in how we sample those distributions.

Where we have sample data from K independent samples, we use those to estimate, in turn, K population distributions. Typically, these are univariate score distributions. We then sample independently from each of these distributions to produce a typical bootstrap sample. Where we have, as we now do, multivariate data for a single sample—each sampled case contributing M scores—we use those data to estimate a single multivariate population distribution. To form a bootstrap sample, we take a random sample from this single estimated population distribution, each sampled case again contributing M scores.

Although the classic approach evaluates contrasts among population distribution means, we have greater latitude in choosing our comparison statistic when we use the bootstrap as a basis for inference.

```
> il1
      phasea phaseb phasec
 [1,]   6850  29100  34300
 [2,]  27400  41100  47000
 [3,]  20700  30700  33600
 [4,]   3000   4900   5900
 [5,]  22100  25700  27100
 [6,]  14300  17200  25700
 [7,]  24300  33600  31400
 [8,]   2400   4700   9400
 [9,]   6500   7900   6400
[10,]   4800   4700   5700
[11,]   2000   3000   3700
[12,]   2200   5100   6800

> fillcmp<- function(X)
+ {
+ c(mean((X[,2]-X[,1]),trim=0.20),mean((X[,3]-X[,1]),trim=0.20))
+ }

> fillcmp(il1)
[1] 4287.5 6575.0

> phase.boot<-bootstrap(il1,fillcmp,B=2500)

> limits.bca(phase.boot)
           2.5%       5%       95%     97.5%
fillcmp1 1912.5 2087.500  8472.25  9607.978
fillcmp2 2975.0 3474.109 11407.90 12752.992
```

Figure 21.1:   *BCA CIs for Trimmed Means of Change Scores (S-Plus)*

Figure 21.1 shows an analysis using the S-Plus `bootstrap` function to obtain bias-corrected and accelerated (BCA) CIs for two treatment comparisons. IL-1 data for the 12 patients for the three phases of the study make up a three column matrix. The user-written function `fillcmp` computes our two comparison statistics.

We choose to estimate the trimmed means for two derived scores, each assessing a change in IL-1 concentrations. The first derived score is the change from phase A (neutral) to phase B (body imaging), and the second is the change from phase A to phase C (relaxation). Notice that because of the way the experiment was conducted this second change, from phase A to phase C, involves the intermediate exposure of cases to phase B (body imaging).

These change scores have population distributions, the distributions that would result if all cases in the patient population were to have their IL-1 concentrations assessed as they passed through this same series of three activities. We estimate 90% and 95% BCA CIs for the 20% trimmed means of these two population distributions. Neither of the 95% CIs overlap zero, offering evidence of real changes in IL-1 scores, changes that should be regarded as statistically significant at the 5% level. In fact, none of the bootstrap sample trimmed means is negative.

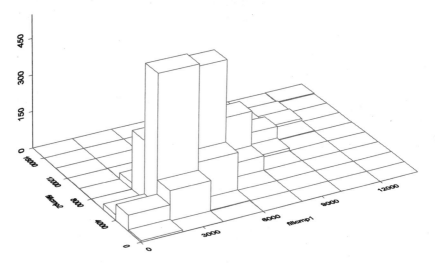

Figure 21.2: *Frequencies of Joint Occurrences of Trimmed Means*

The joint distribution of the two trimmed means over the 2,500 bootstrap samples is depicted in Figure 21.2. This three-dimensional histogram features blocks whose heights are proportional to the distribution of trimmed mean pairs over the grid. The overall picture is of a positive relation between the two trimmed means.

Randomized Sequence of Treatments

When our cases are available ones, rather than a random sample, we cannot estimate CIs for population parameters. Where allocation of these cases among treatments or among treatment sequences is on a random basis, we can use this random element of the design to test hypotheses about the impact of different treatments. Here is an example.

The data making up Table 21.2 and the description of the study which produced it are adapted from Brown and Melamed (1993). Volunteer subjects rated each of four potential changes in public policy from two perspectives. The subjects first rated a potential change relative to how that policy would benefit or cost them personally, assigning an integer between 4 (maximum benefit) and −4 (maximum cost). Next, subjects rated that same change for the extent to which they thought the (then current) Republican administration supported or opposed that policy, again assigning an integer between 4 (maximum support) and −4 (maximum opposition). The rating tasks were repeated for the four policies, with the order of presentation of the four policies randomly determined for each subject.

Table 21.2: *Subject-Administration Rating Differences on Four Public Policies*

| Subject | Policy | | | |
	Wealth	Well-Being	Respect	Rectitude
1	1	4	0	−1
2	2	−8	2	1
3	2	−7	1	2
4	0	−5	−7	1
5	−2	−8	0	−4

The scores reported in Table 21.2 are differences between the two ratings. A difference score of 8 arises if the subject judges a policy to be of great personal benefit but strongly opposed by the administration while a difference score of -8 arises if the subject judges a policy to be of great personal cost and to enjoy great support by the administration. A difference in ratings that is close to zero suggests that the subject believes the administration to hold a view towards the policy that would personally benefit the subject.

The four policies rated were Wealth ("Protecting banks and businesses from failure"), Well-Being ("Increasing the availability of low-cost medical treatment"), Respect ("Equalizing opportunities for minorities and women"), and Rectitude ("Legalizing prayer in school").

We might assume the well-being policy to be the most widely beneficial of the four, so we choose to ask if the subject-administration rating difference is not more negative for this policy than for the others. This suggests a statistic, based on rating-differences for the individual subjects, of this form:

$$s = (1/5)\sum_{i=1}^{5}\left[(1/3)\sum_{m=1,3,4} D_{im} - D_{i2}\right]$$

where D_{im} is the rating-difference score for the i-th subject on the m-th policy, where $m = 2$ denotes the well-being policy. Large, positive values of s support our hypothesis.

Our null treatment-effect hypothesis is that there is no difference in the degree or direction of subject-administration incongruity among the policy issues. Given our within-subjects design—the randomization of the order of presentation of the topics for each subject—this null hypothesis asserts that any apparent differences in scores from policy to policy reflect only the chance alignment of policies with rating position for the individual subject. To assess the magnitude of s, we need to compare it with the distribution of values of s^H obtained from rerandomizations of the policy orders for the five subjects.

We'll use the SC procedure shufwith(p,m,M,r,TT) to carry out the rerandomizations and the tabulation of the reference distribution for us. As indicated, shufwith() is called with five arguments:

> p   A procedure called by shufwith() to compute one or more statistics from the original randomization of treatment orders and from each subsequent rerandomization. This procedure itself expects to be called by shufwith() with three arguments: p(X,XTR,v) where X is the name of a matrix containing subjects as rows and treatments as columns (the transpose of the form of the matrix M supplied to

shufwith), XTR is the name of a work matrix with dimensions the same as M, and v is the name of a vector in which p() will store its computed statistics.

m The number of statistics computed by the procedure p(X,XTR,v).

M A matrix with subjects as columns and treatments as rows.

r The number of s^Hs to be added to s to complete the approximation to the reference distribution, a value like 999, 1499, 2499, and so forth.

TT The name to be given by shufwith() to the $(r+1) \times m$ matrix it will create. The first row will contain the s statistics, and each successive row will contain the s^H statistics computed from the next successive rerandomization of treatment orders.

For each rerandomization, shufwith() independently shuffles the scores for each subject.

The requisite procedure p() is purpose-written for a particular analysis. The top of Figure 21.3 shows the text of such a procedure, ppptry(), and the bottom shows the results of using it together with shufwith() in the rating-difference or incongruence analysis.

The procedure ppptry() computes our s from a 5 × 4 matrix organized in the same way as Table 21.2, with the subjects as rows and the policies as columns and with the well-being rating-differences in column 2. This second column is first extracted from the XX matrix as the vector vv and then deleted from that matrix; the command delcols(XX,1,2) is interpreted as "beginning with column 2 delete 1 column from the matrix XX." In the last command, the expression mean(XX) returns the mean of each row of XX. These row means are the means for each of the 5 subjects of their rating-differences over the policies other than well-being. As a result, mean(XX) – vv is a vector of differences between the mean rating-difference for the other policies and the rating-difference for well-being. Finally, mean(mean(XX) – vv) gives the mean of that vector, our test statistic,

$$s = (1/5)\sum_{i=1}^{5}\left[(1/3)\sum_{m=1,3,4} D_{im} - D_{i2}\right]$$

which is stored as the first (and, only) element of a vector that is then available to shufwith().

```
proc ppptry(){
  local(XX,vv)
  XX=$1$
  vector(vv,rows(XX))
  getcol(XX,vv,2)
  delcols(XX,1,2)
  $3$[1]= mean(mean(XX)-vv)
}

> load(polrat)
 [poldif ]
> clprint(poldif)
   1    4    0   -1
   2   -8    2    1
   2   -7    1    2
   0   -5   -7    1
  -2   -8    0   -4

> proc ppptry(){}
> pp= poldif'
> shufwith(ppptry,1,pp,2499,TT)
     Full Sample Statistics (theta):    4.6666667
P-values for Statistic Number:   1
    Percent LE theta[ 1 ]:   98.84
    Percent GE theta[ 1 ]:   1.36
    Percent LE abs(theta[ 1 ]):   98.84
    Percent GE abs(theta[ 1 ]):   1.36
```

Figure 21.3: *Assessing the Cause of the Well-Being Incongruence (SC)*

In the second half of Figure 21.3, the rating-differences are retrieved as the matrix `poldif`, the procedure `ppptry()` is made ready for use, and then the two are used by `shufwith()`.

The value of our comparison statistic is $s = 4\frac{2}{3}$, and values of s^H that large or larger occurred for only 1.36% of the 2,500 rerandomizations. This provides reasonably strong evidence, given the small size of this study, that the greater incongruity in the ratings of the well-being policy was not the result of a chance alignment of subjects with the policy-rating order.

Nonrandom Repeated Measures

Our next illustration involves a population. The FTSE100, like the Dow-Jones Industrial Average in the United States, provides an index of stock market activity in the United Kingdom (U.K.). The FTSE100 tracks the average selling price on the London Stock Exchange of shares in the 100 largest companies in the U.K. In this example, we are interested in short-term changes in the price of shares in these 100 companies.

These 100 largest U.K. companies form a population. The data are the selling prices of their shares at the close of trading on three successive Fridays, as reported in the *Independent on Sunday* for 11, 18, and 25 February 1996. The data are too numerous to report in full, but Table 21.3 reports the share prices (in pence, 1/100 Pound Sterling) for the ten companies at the top of the FTSE100 on 11 February.

Table 21.3: *Share Prices for the Ten Largest U.K. Companies*

	Share Prices (pence)		
Company	**11 Feb**	**18 Feb**	**25 Feb**
Glaxo-Wellcome	934.00	943.00	923.00
British Petroleum	529.50	538.00	527.50
Shell Transport	861.00	861.00	851.00
HSBC Holdings	1083.00	1111.00	1072.00
BT	361.50	371.50	369.00
SmithKline Beecham	731.00	739.50	719.00
BAT Industries	567.00	572.00	575.00
Lloyds TSB	320.00	341.50	322.00
Barclays	781.00	810.00	794.00
BTR	324.00	326.00	321.00

What does the change in share value over one week say about the change for the week following? We'll look at the correlation between the proportional change in value from 11 to 18 February and that from 18 to 25 February.

In particular, our statistic is the product-moment correlation between

$$(\text{Change Week 1})_i = \frac{(18\ \text{Feb Price})_i - (11\ \text{Feb Price})_i}{(11\ \text{Feb Price})_i}$$

and

$$(\text{Change Week 2})_i = \frac{(25\ \text{Feb Price})_i - (18\ \text{Feb Price})_i}{(18\ \text{Feb Price})_i}$$

where the change scores are computed for the $i = 1, \ldots, 100$ companies.

Our data set, a population, is unstructured and reasonably large. As a result, random half-samples can be used to assess the stability of this correlation as a population description.

```
 1: READ FILE "C:\DATASETS\FTNUMER.DAT" F11 F18 F25
 2: SUBTRACT F11 F18 WK1
 3: DIVIDE WK1 F11 CWK1
 4: SUBTRACT F18 F25 WK2
 5: DIVIDE WK2 F18 CWK2
 6: CORR CWK1 CWK2 FULLCORR
 7: PRINT FULLCORR
 8: COPY 1,100 FTSE100
 9: REPEAT 100
10:   SHUFFLE FTSE100 FTSE100
11:   TAKE FTSE100 1,50 PICK
12:   TAKE F11 PICK H11
13:   TAKE F18 PICK H18
14:   TAKE F25 PICK H25
15:   SUBTRACT H11 H18 HW1
16:   DIVIDE HW1 H11 HCW1
17:   SUBTRACT H18 H25 HW2
18:   DIVIDE HW2 H18 HCW2
19:   CORR HCW1 HCW2 HCORR
20:   SCORE HCORR STABDIST
21: END
22: MEDIAN STABDIST MEDN
23: PERCENTILE STABDIST (25 75) IQR
24: PERCENTILE STABDIST (10 90) MID80
25: PRINT MEDN IQR MID80


FULLCORR =    -0.34721
MEDN     =    -0.3536
IQR      =    -0.41064    -0.28125
MID80    =    -0.46722    -0.17868
```

Figure 21.4: *Random Half-Sample Stability Analysis (Resampling Stats)*

Figure 21.4 shows, at the top, a Resampling Stats script to carry out the half-sampling and, at the bottom, the result of its execution.

Lines 10 and 11 in the script identify the randomly chosen 50 FTSE100 firms to be included in a particular half-sample. The 11, 18, and 25 February share prices for these firms are used in the half-sample correlation calculation.

The correlation between the two weekly relative changes,

$$r_{(\text{Change Week 1})(\text{Change Week 2})} = -0.347$$

was negative. How are we to interpret this? The relative change in price over the first week for the typical FTSE100 share tended to be reversed, in amount if not in direction, in the second week of the period studied.

The full data set correlation is quite near the median of the half-sample correlations, and both the middlemost 50% and 80% of the half-sample correlations are negative as well.

We have little reason to suspect the stability of the full data set correlation as a description of the linear relation between the two sets of weekly changes.

Exercises

1. Actually, Ramsey and Schaefer (1997) report that the 12 cases in the IL-1 study were not a random sample. Devise, carry out, and report a more appropriate analysis of the results reported, for these available cases, in Table 21.1. Remember that the order of the three phases was the same for all cases.

2. One class of within-cases design calls for repeated observations under each of two or more treatments. Thus, we could randomize both the order in which treatments are administered, and the number of observations to make for each treatment. This exercise provides an example of the simplest, though most widely used, design of this type. Here the order of two treatments is fixed, but we randomly choose the number of observations to be made under the two treatments. We want to see the impact of the change of treatment. Our random choice of a point in the sequence of observations at which we make the change

provides a mechanism for evaluating whether the observed change is substantial enough to be real or simply a chance result. In some substantive areas, the second treatment is regarded as an intervention and we can speak of this design as one in which the point of intervention is randomly chosen.

Consider this example. A laboratory animal is fed a standard diet each day for two weeks and then, on a randomly chosen day in the third week, the standard diet is replaced with an experimental diet. The experimental diet is continued through the fifth week of the study. On each day of the study, the animal's activity-level is assessed during a one-hour period beginning 30 minutes after feeding. We want to know whether the animal's activity-level is higher under the experimental than under the standard diet. We propose to measure the change in activity-level in the following way. First, let D_m designate the m-th day in the study and let A_m designate the activity-level on the m-th day. Then, let D_I be the randomly chosen day of intervention, the first day on which the experimental diet is presented. We can then describe our change measure as

$$ s = (1/4)\sum_{i=1}^{4} A_{I+1+i} - (1/4)\sum_{i=1}^{4} A_{I-6+i} $$

In words, we subtract the average activity-level for the four days ending two days before intervention from the average activity-level for the four days starting two days after the intervention. We average over four days to obtain more reliable measures of activity-level, and we leave a four-day gap between the two measurements to ensure that the two measurements are solidly within the standard-diet and experimental-diet phases of the experiment.

The statistic s should be positive. Is the magnitude of s simply a chance result, a function of our random choice of the point of intervention? In the conduct of the study, the value of I is determined by chance. It is restricted, by design, to one of the integers between 15 (the first day of the third week) and 21 (the final day of the third week). Now, in principle, we can check whether the value of s could be a chance result or not by comparing its value against a reference distribution of s^Hs determined, as for other designs, by the random element in our design and a null treatment-effect hypothesis.

The null treatment-effect hypothesis is that the size of s reflects only the chance choice of I and not the effect of the intervention, change of diet. And, by our design, I can be any of the integers between 15 and 21 inclusive. So, our reference distribution here would contain just seven values, those that result from replacing I with each of the integers from 15 through 21 in the formula for s. Each of these is equally likely under the null hypothesis.

This tiny experiment would not provide very convincing results. At best, s might be the largest of the seven values in the reference set. But this would let us give it a p-value of $1/7 = 0.14$. It can be no smaller because the random choices were so few.

The experiment would be improved if we were to run it with, say, five animals rather than just one. If we randomly choose an intervention point, I_j, in the third week for the j-th animal, $j = 1, ..., 5$, then we could take as our test statistic,

$$s = (1/5)\sum_{j=1}^{5}\left[(1/4)\sum_{i=1}^{4}A_{I_j+1+i} - (1/4)\sum_{i=1}^{4}A_{I_j-6+i}\right]$$

averaging the change in activity-level over the five animals.

The reference distribution for this statistic would have $7^5 = 16,807$ elements if we allow the value of each of the five I_js to range independently from 15 through 21. Or, we could approximate this reference distribution by computing s^H some large number of times, each time letting each I_j be randomly chosen from among the integers 15 through 21.

Table 21.4 shows the results of just such an experiment. Activity-levels are given for days 10 through 26 reading across the rows for each of five animals. These days span the range from ($I_j - 6 + 1$) for $I_j = 15$, to ($I_j + 1 + 4$) for $I_j = 21$, those being the earliest and latest days that could be involved in the computation of s^H. An intervention point was chosen for each animal by randomly choosing one of the integers between 15 and 21, inclusive. These turned out to be 18 for S_1, 16 for S_2, 15 for S_3, 21 for S_4, and 17 for S_5. Note that it is only by chance that the same day was not chosen more than once, each selection was made independently of the others. Carry out an appropriate analysis and report the results.

3. Many designs that call for collecting several observations on the same case do not allow us to randomize the order in which the observations are collected. One obvious example is the trend study in which the same performance or characteristic of a case is to be measured at several points in time.

Table 21.4:   *Daily Activity-Levels for Five Laboratory Animals*

> S1[10:26]

85	102	89	112	91
83	86	99	101	100
106	115	107	118	99
106	91			

> S2[10:26]

116	118	113	105	105
121	121	111	114	110
124	104	101	115	127
108	119			

> S3[10:26]

76	85	76	89	83
78	66	82	80	81
77	88	82	81	88
90	98			

> S4[10:26]

78	81	87	86	79
92	81	69	84	85
68	88	116	87	87
90	103			

> S5[10:26]

109	100	97	112	120
109	113	115	118	116
114	131	102	111	115
118	132			

The data in Table 21.5 are taken from Mead, Curnow, and Hasted (1993) and describe a study of the influence of two dietary additives on the growth of rats. The design is a completely randomized (CR) one. Twenty-seven rats are randomized, 10 to a control condition, 10 to a Thiouracil condition, and 7 to a Thyroxin condition. The latter two groups have chemicals added to their drinking water. Each rat is weighed at the start of the study and at weekly intervals thereafter for four consecutive weeks. At interest is whether either chemical affects the rate of growth.

a. Looking only at the Week 0 data, satisfy yourself with an omnibus comparison that the randomization of rats to treatment groups did not lead to a statistically significant imbalance in the starting weights of the animals.

Table 21.5: *Growth in Weight of Rats under Three Conditions*

Condition	Rat	Week				
		0	1	2	3	4
Control	1	57	86	114	139	172
	2	60	93	123	146	177
	3	52	77	111	144	185
	4	49	67	100	129	164
	5	56	81	104	121	151
	6	46	70	102	131	153
	7	51	71	94	110	141
	8	63	91	112	130	154
	9	49	67	90	112	140
	10	57	82	110	139	169
Thyroxin	11	59	85	121	156	191
	12	54	71	90	110	138
	13	56	75	108	151	189
	14	59	85	116	148	177
	15	57	72	97	120	144
	16	52	73	97	116	140
	17	52	70	105	138	171
Thiouracil	18	61	86	109	120	129
	19	59	80	101	111	126
	20	53	79	100	106	133
	21	59	88	100	111	122
	22	51	75	101	123	140
	23	51	75	92	100	119
	24	56	78	95	103	108
	25	58	69	93	114	138
	26	46	61	78	90	107
	27	53	72	89	104	122

b. Growth rate can be assessed in a variety of ways. Here is the first of two analyses: Compute a linear growth rate for each animal. This is done most easily by regressing weight against week for each animal. The slope coefficient is the linear growth rate. Using this slope as the response score, assess the difference in growth rates between thyroxin and control animals, and between thiouracil and control animals. In each rerandomization analysis, only two of the three treatments, or treatment groups, are assumed exchangeable.

c. For a second growth analysis, assume you are interested in growth during the first part of the study, and in growth during the second part of the study. Specifically, compute two new growth scores for each animal. The first period growth score might be

$$g_i = \frac{\text{Week 2 (Weight)}_i - \text{Week 0 (Weight)}_i}{\text{Week 0 (Weight)}_i}$$

and the second period growth score,

$$h_i = \frac{\text{Week 4 (Weight)}_i - \text{Week 2 (Weight)}_i}{\text{Week 2 (Weight)}_i}$$

Each of these scores is a proportional rate of growth, considering the weight of the animal at the beginning of the period in question. Using each of these measures in turn, assess thyroxin against control and thiouracil against control.

Do the two growth analyses agree or not?

4. This exercise is based on a study reported by Spence, Williams, and Costello (1987). The description of the goals of the study and of how it was carried out have been modified for purposes of this exercise. At issue is how much familiarity infants develop, *in utero*, with their mothers' voices.

The cases in this study are eight women randomly chosen from a maternity ward population of 50. Before giving birth their voices were recorded under instructions to "soothe an infant." Each woman's infant (a Maternal infant) is matched with a Control infant—an infant whose mother was not sampled for the study, is of the same sex as the maternal infant, and was born within five hours of the maternal infant. At approximately 48 hours of age for each pair of infants, the mother's recorded voice was replayed to the two infants several times over a four-hour period. In random sequence, the replay was either unfiltered or filtered to have those acoustic properties that are available *in utero*.

The data in Table 21.6 are the percentages of voice-replays that were responded to by the maternal and control infants. Each row relates to a different case, a mother whose recorded voice is played back both to her (maternal) infant and to an unrelated (control) infant.

Table 21.6:   *Response of Infants to Filtered and Unfiltered Human Voices*

Maternal Infant		Control Infant	
Filtered	Unfiltered	Filtered	Unfiltered
63	72	37	41
73	55	53	61
67	43	65	63
76	77	63	65
44	51	33	43
51	63	49	59
62	66	52	61
63	56	46	57

Each case received four treatments. The treatments correspond to a 2×2 factorial design. The maternal/control factor for infant is fully crossed with the filtered/unfiltered factor for voice replay.

What within-cases comparisons are of interest here? Develop as much of a scientific rationale as you can for each comparison. Carry out the proposed analyses and summarize your findings.

Applications 22

Linear Models: Measured Response

Researchers frequently use linear models to analyze case-based studies. You might know these models as regression, linear regression, or multiple linear regression (MLR) models. You might also be familiar with the use of linear models to represent analysis of variance (ANOVA) designs, including those in which blocking or covariates play a role.

The defining characteristic of the parametric linear model (parametric-LM)—which embraces MLR and ANOVA models—is that the average value of a response attribute can be expressed as a linear function of the values of one or more explanatory attributes. These explanatory attributes can vary in measurable strength among cases, or they can categorize cases, for example, relative to the blocking or treatment groups to which they belong. A model can contain both measured and categorical explanatory attributes.

In these applications, we first review the parametric-LM with its assumptions about the population distributions of response attributes—that they should have normal cumulative distribution functions (cdfs) and constant variance. Then, as a more flexible alternative, we demonstrate a bootstrap, nonparametric approach to the linear modeling of randomly sampled data. Finally, we look at some related linear model analyses that are appropriate for random assignment studies and for nonrandom data aggregations.

The Parametric Linear Model

The parametric-LM requires one or more random samples of cases. Each case is accompanied by scores for a response attribute and for $k \geq 1$ explanatory attributes. The set of explanatory attribute scores for a case is that case's design point. For each distinct design point, the parametric-LM assumes a case population with a corresponding population distribution of response attribute scores. Each of these population distributions has a normal cdf and a common variance, σ^2. The means of these population distributions are modeled as a linear function of the design point—that is, of the explanatory attribute scores. Thus, the mean of the response-distribution for the population from which the i-th case was sampled is modeled as a linear function of the i-th case's scores on the k explanatory attributes:

$$\mu(y_i) = \beta_0 + \sum_{j=1}^{k} \beta_j x_{ij}$$

Based on the population distribution assumptions, the parametric-LM estimates of the βs or linear model parameters have known sampling distributions. This makes it possible to estimate confidence intervals (CIs) for the parameters and to test simple hypotheses about them.

The case populations can be either natural or prospective populations.

Sampling Natural Populations

For example, we might draw a random sample of 500 40-year-old people residing in King County, State of Washington, and determine for each case:

1. Income for calendar year 1997, the response attribute score

2. Whether male or female, an explanatory attribute score

3. Years of education, a second explanatory attribute score

Our linear model for mean 1997 income could be this one:

$$\mu(y_i) = \beta_1 x_{i1} + \beta_2 x_{i2} + \beta_3 x_{i3}$$

Where x_{i1} is 1 if the i-th case is male and 0 if female, x_{i2} is 1 if the i-th case is female and 0 if male, and x_{i3} is the result of subtracting 12 from the number of years of education for the i-th case. Cases terminating their formal education with a high school diploma will have a score of zero on this attribute, cases with one or more years of college or university study will have positive scores, and those who dropped out of school before graduating from high school will have negative scores.

Finally, $\mu(y_i)$ is the mean 1997 income in the natural population from which the i-th case was sampled. Several such natural populations would be represented in our sample. These might include, among others:

1. The population of 40-year-old King County residents who are male and have completed 12 years of education

2. The population of 40-year-old King County residents who are female and have completed 12 years of education

3. The population of 40-year-old King County residents who are male and have completed 16 years of education

4. The population of 40-year-old King County residents who are female and have completed 18 years of education

Each of these populations would have a 1997 income distribution, and each distribution would have a variance and cdf as well as a mean. The parametric-LM presumes the cdfs to be normal and the variances to be constant.

Sampling Prospective Populations

We now randomly sample 50 rats from a laboratory population of 500 and then divide the sample randomly into two treatment groups of 25 animals each. The treatment groups receive different diets (Control or Enhanced) and we want to know whether the average Activity Level would be higher if the entire population of rats were on the enhanced diet than if they were all on the control diet.

We assess activity level for the sampled animals when they have been on their respective diets for two weeks. Each animal in the study has scores on two attributes:

1. Activity level, the response attribute

2. Diet treatment, control or enhanced, the explanatory attribute

We model the study results in this way:

$$\mu(y_i) = \beta_0 + \beta_1 x_{i1}$$

where x_{i1} is 1 if the i-th case was randomized to the enhanced diet and 0 if randomized to the control one. And $\mu(y_i)$ is the average activity level for one of two prospective populations of 500 rats, depending on whether the i-th animal was randomized to the control or enhanced diet.

The parametric-LM assumes, again, that the two population distributions of activity-level scores have a common variance and that both have normal cdfs.

Parametric-LM: Population Inference

Usually a linear model is constructed so that the most important population parameters from a substantive point of view are the β-coefficients in the linear model. In the Income example, the three parameters have these interpretations:

1. β_1 – The average annual income in the population of 40-year-old males who have completed 12 years of education

2. β_2 – The average annual income in the population of 40-year-old females who have completed 12 years of education

3. β_3 – The amount by which the average annual income increases with an additional year of education

In the Diet example, the two parameters of the linear model are these:

1. β_0 – The average activity level in the laboratory population if all rats were fed the control diet

2. β_1 – The amount by which the average activity level would exceed β_0 if the population of rats were all fed the enhanced diet

Population inference for linear models such as these can take the form of estimating CIs for these parameters or testing hypotheses about their values. The parametric-LM assumptions simplify such tasks because those assumptions provide all that is needed to:

1. Compute maximum likelihood estimates (MLE) of the β-parameters from the sample data

2. Describe the sampling distributions of these MLEs

If we let $\widehat{\beta}_j$ denote the MLE of the population parameter β_j, then the population linear model has its sample analog,

$$\widehat{\mu}(y_i) = \widehat{\beta}_0 + \sum_{j=1}^{k} \widehat{\beta}_j\, x_{ij}$$

where $\widehat{\mu}(y_i)$ is the estimated mean of the distribution of response attribute scores for the population of cases having the same explanatory variable scores as the i-th case.

The Studentized form of an estimated β-parameter,

$$t(\widehat{\beta}_j) = \frac{\widehat{\beta}_j - \beta_j}{\widehat{\mathrm{SE}}(\widehat{\beta}_j)}$$

has a sampling distribution with the same cdf as that of the t-random variable with $(n - k)$ degrees of freedom (df) where n is the sample size, k is the number of β-parameters in the linear model (including any intercept, β_0), and $\widehat{\text{SE}}(\widehat{\beta}_j)$ is an unbiased estimate of the standard error (SE) of $\widehat{\beta}_j$ computed, on the basis of the parametric-LM assumptions, from the sample.

This Studentization of the β-parameter estimate is used to compare the fit of the current model against that of the model in which that β_j has been constrained to be zero—or, equivalently, against the fit of the model from which we have dropped the explanatory attribute associated with that parameter. Large values of $t(\widehat{\beta}_j)$, positive or negative, provide evidence against the constraint and evidence for the inclusion of that explanatory attribute in the model.

Sometimes we want to assess more complicated constraints on the β-parameters. For example, in the Income example, we might be interested in testing the hypothesis that the two parameters β_1 and β_2 are equal, that 40-year-old males with a high school education and 40-year-old females with a high school education have the same average annual income. This hypothesis gives rise to a second, constrained linear model:

$$\mu(y_i) = \beta_0 + \beta_1 x_{i3}$$

where β_0 is the common average income among 40-year-old high school graduates and x_{i3} remains the years of education in excess of 12. Constraining the β-parameters of a linear model always gives rise to a second, constrained linear model with fewer β-parameters than the unconstrained model.

The constraining hypothesis is tested by comparing the fits of the constrained and unconstrained models. The fit of a particular model is assessed by the sum, over the sampled cases, of the squared differences between the response attribute scores and the estimated means of the population distributions from which those scores were sampled. Thus, the fits of the unconstrained and constrained models are given by:

$$\text{SSR}_{\text{U}} = \sum_{i=1}^{n} [y_i - \widehat{\mu}(y_i | \text{Unconstrained Model})]^2$$

and

$$\text{SSR}_\text{C} = \sum_{i=1}^{n} [y_i - \hat{\mu}(y_i|\text{Constrained Model})]^2$$

If the number of β-parameters in the unconstrained and constrained models are k and ℓ (constrained to be fewer than k), then the statistic

$$\text{F} = \frac{[\text{SSR}_\text{C} - \text{SSR}_\text{U}]/(k - \ell)}{\text{SSR}_\text{U}/(n - k)}$$

has a sampling distribution with the same cdf as the F-random variable with $(k - \ell)$ and $(n - k)$ df provided both of the following hold:

1. The assumptions of the parametric-LM are met.

2. The constrained or null-hypothesis model is the correct model.

Large values of the F-ratio statistic provide evidence that the additional parameters of the unconstrained model are needed for a correct model.

Table 22.1:   *River Thames Oxygen Production Data Set*

	Oxy	Chlor	Light
1	2.16	33.8	329.5
2	4.13	47.8	306.8
3	2.84	100.7	374.7
4	4.65	105.5	432.8
5	-0.42	33.4	222.9
6	1.32	27.0	352.1
7	4.04	46.0	390.8
8	1.97	139.5	232.6
9	1.63	27.0	277.7
10	1.16	22.5	258.5
11	0.61	16.5	210.0
12	1.94	71.3	361.8
13	1.70	49.4	300.4
14	0.20	19.3	96.9
15	0.98	71.6	151.8
16	0.06	13.4	126.0
17	-0.19	11.8	67.8

An Illustration of Parametric-LM Inference

Table 22.1 shows data collected as part of an investigation into the factors affecting the productivity of plankton in the River Thames at Reading, England (Mead, Curnow, and Hasted, 1993). On each of 17 days, observations were made of:

1. Oxygen Production (Oxy): Milligrams per cubic meter per day

2. Chlorophyll Abundance (Chlor): Micrograms per cubic meter per day

3. Light Abundance (Light): Calories per square meter per day

Oxy is the response attribute, y, and Chlor and Light the explanatory attributes, x_1 and x_2, in a linear model:

$$\mu(y_i) = \beta_0 + \beta_1 x_{i1} + \beta_2 x_{i2}$$

```
> oxymodel<- lm(Oxy~Chlor+Light,data=thames)
> summary(oxymodel)

Call: lm(formula = Oxy ~ Chlor + Light, data = thames)
Residuals:
    Min      1Q   Median      3Q    Max
  -1.55 -0.3538 -0.03404 0.4287 2.028

Coefficients:
              Value Std. Error  t value Pr(>|t|)
(Intercept) -1.4057    0.5934   -2.3691   0.0327
      Chlor  0.0096    0.0068    1.4158   0.1787
      Light  0.0099    0.0023    4.2948   0.0007

Residual standard error: 0.8943 on 14 degrees of freedom
Multiple R-Squared: 0.6964
F-statistic: 16.05 on 2 and 14 degrees of freedom, the p
-value is 0.0002379

Correlation of Coefficients:
      (Intercept)    Chlor
Chlor -0.1038
Light -0.7837        -0.4423
```

Figure 22.1:   *Modeling Oxygen, Chlorophyll, and Light (S-Plus)*

Figure 22.1 displays an S-Plus analysis of the Thames data. The response and explanatory attribute scores are in a data-frame titled `thames`. For purposes of analysis, we assume the 17 days were chosen at random from a large population of potential dates.

In Figure 22.1, the `lm()` function is used to find estimates of the parameters of the linear model:

$$\widehat{\mu}(y_i) = -1.4057 + 0.0096\, x_{i1} + 0.0099\, x_{i2}$$

Displayed together with the parameter estimates are values of the Studentization of each

$$t(\widehat{\beta}_j) = \frac{\widehat{\beta}_j - \beta_{j0}}{\widehat{\mathrm{SE}}(\widehat{\beta}_j)}$$

where the null-hypothesized values of the parameters are taken to be zero, $\beta_{j0} = 0$, $j = 1, 2, 3$. The p-value associated with the t-test of the β-coefficient for Light would lead us to reject that null hypothesis. On the other hand, there is a fair chance (p $= 0.18$) of an estimate of the Chlor β-coefficient taking a value as large as ± 0.0096 when that β-parameter is zero.

The `Residual standard error` reported in Figure 22.1 is, in our notation, $\sqrt{\mathrm{SSR}/(n-k)}$. The `F-statistic: 16.05 on 2 and 14 degrees of freedom` compares the fits of the present model and a null model with only an intercept parameter, the Light and Chlor parameters both constrained to be zero: $(n-k) = 14$ and $q = (k-\ell) = 2$. The small p-value associated with this test argues against the one-parameter, null model. Our linear model cannot exclude both Light and Chlor as explanatory attributes.

Based on the t-test of the Chlor parameter, however, β_1 could well be zero, so we now simplify our linear model by dropping x_1, Chlor. In Figure 22.2, we obtain estimates for the parameters of this more parsimonious linear model:

$$\mu(y_i) = \beta_0 + \beta_2\, x_{i2}$$

Under the parametric-LM assumptions we know enough about the sampling distribution of $\widehat{\beta}_2$, the estimate of the Light-parameter, that we can calculate a $(1 - 2\alpha)100\%$ CI for the parameter. We follow the pattern established in

Applications 5 for finding a CI for the mean of a population distribution with a normal cdf.

```
> oxymodel2<- lm(Oxy~Light,data=thames)
> summary(oxymodel2)

Call: lm(formula = Oxy ~ Light, data = thames)
Residuals:
    Min      1Q  Median     3Q    Max
 -1.641 -0.4642 -0.1109 0.5689 1.953

Coefficients:
             Value Std. Error t value Pr(>|t|)
(Intercept) -1.3185   0.6096  -2.1630   0.0471
      Light  0.0114   0.0021   5.3119   0.0001

Residual standard error: 0.9238 on 15 degrees of freedom
Multiple R-Squared: 0.6529
F-statistic: 28.22 on 1 and 15 degrees of freedom, the p
-value is 0.00008707

Correlation of Coefficients:
       (Intercept)
Light -0.93
```

Figure 22.2:   *Oxygen Production Modeled by Light Abundance (S-Plus)*

The sampling distribution of

$$t(\widehat{\beta}_j) = \frac{\widehat{\beta}_j - \beta_j}{\widehat{SE}(\widehat{\beta}_j)}$$

has as its cdf that of the t-random variable with $(n - k)$ df. The score distribution for the t-random variable is symmetric about zero. As a result, its α and $(1 - \alpha)$ quantiles can be expressed as $F_{t,(n-k)}^{-1}(\alpha)$ and $-F_{t,(n-k)}^{-1}(\alpha)$. We can substitute each of these quantiles for $t(\widehat{\beta}_j)$ in the equation and then solve each of the resulting equations for β_j to give us the upper and lower limits to our CI:

$$\text{UL}(\beta_j) = \hat{\beta}_j - \left[F_{t,(n-k)}^{-1}(\alpha) \times \widehat{\text{SE}}(\hat{\beta}_j) \right]$$

and

$$\text{LL}(\beta_j) = \hat{\beta}_j + \left[F_{t,(n-k)}^{-1}(\alpha) \times \widehat{\text{SE}}(\hat{\beta}_j) \right]$$

Putting these results together, our parametric-LM $(1 - 2\alpha)100\%$ CI for β_j is

$$\hat{\beta}_j \pm \left[F_{t,(n-k)}^{-1}(\alpha) \times \widehat{\text{SE}}(\hat{\beta}_j) \right]$$

From Figure 22.2, we have $\hat{\beta}_2 = 0.0114$ and $\widehat{\text{SE}}(\hat{\beta}_2) = 0.0021$. The linear model of Figure 22.2 has $k = 2$ parameters estimated from a random sample of $n = 17$ cases. The $\alpha = 0.05$ quantile of the t-distribution with 15 df takes the value -1.753. A 90% CI for the Light-parameter, then, is

$$0.0114 \pm (1.753 \times 0.0021) = (0.0077, 0.0151)$$

If the parametric-LM assumptions have been met, we can be 90% confident that this interval overlaps the value of the Light-parameter.

In this section, our parametric-LM examples have featured one or more measured explanatory attributes. All the examples, parametric and otherwise, of Applications 18, 19, and 20 could have been presented in linear model form with treatment and block factor levels represented in the model by indicator variables. These are explanatory attributes that take the value $x_{ij} = 1$ if the i-th case is at factor level j and the value $x_{ij} = 0$ if the i-th case is at a factor level other than the j-th. Some earlier examples were presented that way, and further illustrations of linear models with indicator variables appear in the following sections.

Nonparametric Linear Models

The parametric-LM with its requirement of normal, constant variance population distributions of response attribute scores for each pattern of explanatory variable scores is widely utilized, both in linear regression and in the ANOVA. As noted for earlier applications, however, neither normality nor homogeneity of variance are typical of empirical research data.

In this section, we develop nonparametric alternatives to the parametric-LM. These alternative linear models utilize bootstrap resampling and can provide a more realistic basis for population inference.

One of the guiding principles of our resampling approach to inference is that the statistical analysis of a study should be faithful to the design of that study.

In linear model terminology, each unique pattern of explanatory attribute scores in a study constitutes a design point for that study. The number of design points can be as large as n, that is, every case in the sample has a unique pattern of explanatory attribute scores, and as small as two, that is, the explanatory attributes only indicate to which of two populations a case belongs.

The parametric-LM sampling distributions for the Studentized β-parameter estimates and F-ratios of SSRs depend not only on the normality and variance homogeneity of population distributions of the response attribute but also on a particular sampling model. It is assumed that those hypothetical replications of the study necessary to define these sampling distributions all use the same set of design points as were present in the original study. Only the response attribute scores change from one hypothetical replication to another. Statisticians refer to this as fixed design point sampling.

In some studies, the design points are fixed in advance. For example, a random sample of 40 patients is to be randomized in equal numbers among four dosages of a new drug, 0 mg/50 Kg body weight (a placebo), 10 mg/50 Kg body weight, 20 mg/50 Kg body weight, and 40 mg/50 Kg body weight. The explanatory attribute Drug Dosage would contain, for this study, exactly 10 entries of 0, 10, 20, and 40, respectively.

In other studies, the design points are not known until after the study is completed. The Oxygen Production experiment is a case in point. None of the values of Chlor or Light can be known until observations at the sampled dates are made. The design points are not fixed but, rather, sampled along with the response attribute.

The parametric-LM is used routinely for both fixed and sampled design point studies. Experience with the use of bootstrap resampling for nonparametric linear model inference, however, leads us to choose our bootstrap samples in the same way the original sample was chosen. We distinguish between fixed and sampled design points when creating and sampling estimated population distributions.

Bootstrap LM: Fixed Design Points

In Applications 19, we assessed, from a parametric perspective, the stress on randomly sampled devices of either being immersed in water or dropped from a height. The data and motivating question are reproduced here.

Table 19.1 shows data for a study described in Box, Hunter, and Hunter (1978). A random sample of 48 devices was selected from those in a warehouse. This sample, in turn, was randomly divided into three smaller groups. To the devices in the Control group nothing was done. Each of those in the Water group was immersed in water. Each of those in the Drop group was dropped from a certain height. All 48 devices then were tested, producing the Performance data in Table 19.1. Larger scores indicate poorer performance. Did the water or drop treatments worsen performance?

Table 19.1: *Performance Data for Control and Stressed Devices*

Control	0.38 0.26 0.41 0.33 0.33 0.37 0.54 0.76
	0.51 0.55 0.53 0.41 0.47 0.49 0.42 0.34
Water	0.53 0.35 0.38 0.45 1.09 0.46 0.57 0.47
	0.39 0.74 0.32 0.74 0.48 0.37 0.52 0.44
Drop	0.51 0.63 0.46 0.47 0.42 0.45 0.41 0.39
	0.35 0.41 0.49 0.40 0.58 0.46 0.38 0.48

The random division of the randomly sampled 48 devices produced random samples, each of size 16, from three prospective populations, the warehouse population of these devices unstressed (control population), stressed by immersion (water population), or by dropping (drop population). Each population has a corresponding population distribution of performance scores. Earlier, we estimated two contrasts among the three distribution means,

$$(\widehat{\mu}_{\text{Water}} - \widehat{\mu}_{\text{Control}}) = 0.075 \text{ and } (\widehat{\mu}_{\text{Drop}} - \widehat{\mu}_{\text{Control}}) = 0.012$$

and tested the null hypotheses of mean differences of zero against alternative hypotheses of positive differences, poorer average performances in the stressed populations. The parametric t-tests for the two contrasts yielded p-values of 0.07 and 0.40, respectively.

We'll test these same hypotheses again, subject to the following:

1. We'll use a linear model to describe the three population distribution means.

2. We'll lift the requirements that the three population distributions have the same variances and normal cdfs.

Our linear model takes this form

$$\mu(y_i) = \beta_0 + \beta_1 x_{i1} + \beta_2 x_{i2}$$

where y is the response attribute, performance, and our two explanatory attributes, x_1 and x_2, are indicator variables; x_{i1} is 1 if the i-th case was randomly sampled from the water population and 0 otherwise, whereas x_{i2} is 1 if the i-th case was randomly sampled from the drop population and 0 otherwise. Cases sampled from the control population will have scores of 0 on both indicator variables. The three design points,

$$(x_{i1}, x_{i2}) = \begin{cases} (0,0) \\ (1,0) \\ (0,1) \end{cases}$$

and the number of times each will appear in the sample, 16, are fixed by the design of the study.

Our choice of indicator variables models the three population distribution means as

$$\mu(\text{Performance}|\text{Control}) = \beta_0$$

$$\mu(\text{Performance}|\text{Water}) = \beta_0 + \beta_1$$

$$\mu(\text{Performance}|\text{Drop}) = \beta_0 + \beta_2$$

Or, put another way, the β-parameters have these interpretations,

$$\beta_0 = \mu(\text{Performance}|\text{Control})$$

$$\beta_1 = \mu(\text{Performance}|\text{Water}) - \mu(\text{Performance}|\text{Control})$$

$$\beta_2 = \mu(\text{Performance}|\text{Drop}) - \mu(\text{Performance}|\text{Control})$$

The last two parameters assess the impact, on average, of the two stress treatments.

We can find plug-in estimates of the β-parameters for our linear model using the `lm()` function in S-Plus or a similar ordinary least squares (OLS) regression command in another package. Figure 22.3 shows the interaction in S-Plus.

The plug-in estimates of β_1 and β_2 are 0.75 and 0.012, identical to the Applications 19 mean contrast estimates. Further, the one-tail p-values for the parametric-LM tests of these parameters ($\beta = 0$ versus $\beta > 0$) are $(0.1340/2) = 0.0670$ and $(0.8102/2) = 0.4051$, in agreement with the t-test results of Applications 19. However, we now want the corresponding nonparametric results.

Our approach to finding nonparametric p-values for the two tests will be to estimate bias-corrected and accelerated (BCA) CIs for each parameter that are bounded from below by the null-hypothesized value of zero. In carrying out the bootstrap sampling, we consider that we have three random samples in this study and draw our samples from three estimated populations. That is, we use each of the three sample distributions in Table 19.1 as the nonparametric estimate of a population distribution.

```
> perform<- c(congrp,watgrp,drpgrp)
> zed<- rep(0,times=16)
> uno<- rep(1,times=16)
> watind<- c(zed,uno,zed)
> drpind<- c(zed,zed,uno)

> lmod<- lm(perform~watind+drpind)
> summary(lmod)

Coefficients:
             Value Std. Error t value Pr(>|t|)
(Intercept) 0.4438  0.0348   12.7676  0.0000
    watind  0.0750  0.0492    1.5259  0.1340
    drpint  0.0119  0.0492    0.2416  0.8102
```

Figure 22.3: *Plug-in Estimates of Linear Model β-parameters (S-Plus)*

In Figure 22.4, we continue the S-Plus dialog. Three arguments are prepared for the `bootstrap()` function:

1. The data matrix xmat containing the response attribute in the first column and the two indicator variables in columns 2 and 3

2. A function lm2ev to return the values of β_1^* and β_2^* from each bootstrap sample

3. A vector grps with entries (1, 2, and 3) indicating the rows of the data matrix to be used in forming the three estimated populations for resampling

```
> xmat<- cbind(perform,watind,drpind)

> lm2ev<- function(X)
+ {
+ foo<- lm(X[,1]~X[,2]+X[,3])
+ foo$coef[2:3]
+ }

> lm2ev(xmat)
 X[, 2]    X[, 3]
  0.075 0.011875

> uno<- rep(1,16)
> grps<- c(uno,2*uno,3*uno)

> bfit<- bootstrap(xmat,lm2ev,group=grps)

> limits.bca(bfit)
               2.5%          5%        95%       97.5%
X[, 2] -0.02775604 -0.01246802 0.1740354 0.1978525
X[, 3] -0.06287822 -0.04786320 0.0700000 0.0795992

> limits.bca(bfit,prob=c(0.086,0.335))
                 8.6%            33.5%
X[, 2] -0.00001103019  0.0496865584
X[, 3] -0.03685387231 -0.0001082724

> limits.bca(bfit,prob=c(0.087,0.337))
                 8.7%            33.7%
X[, 2] -2.608025e-017  4.980034e-002
X[, 3] -3.660329e-002 -1.184336e-016
```

Figure 22.4: *BCA CIs for β_1 and β_2 (S-Plus)*

Some convergent guessing at potential quantiles for the lower bounds of the two BCA CIs leads to estimates of the two p-values of 0.09 (0.087) for the water-stress and 0.34 (0.337) for the drop-stress. Only the last two steps in this convergence are shown in the figure.

Bootstrap LM: Sampled Design Points

The Oxygen Production experiment, for which a parametric analysis was presented earlier, involved a single, random sample of cases: 17 randomly chosen days on which observations were made. We associate a distribution of attribute scores with this population. The i-th member of the population is represented in this distribution by an ordered set of three scores:

$$(y_i, \, x_{i1}, \, x_{i2}) = (\text{Oxy, Chlor, Light})$$

The set of three measurements could have been made on any one of the days making up the population. Our researcher, however, made measurements only on the randomly sampled 17 days. The result is a sample distribution, each element of which is, again, an ordered set of three scores. Both the population and sample distributions are multivariate, each element in the distribution is an ordered set of scores, not a single score. Only one of the three scores, Oxy, is treated as a response, so, although we have multivariate observations, we have a univariate response.

Our nonparametric linear model for sampled design points is different from the one for fixed design points. We no longer postulate a population distribution of response attribute scores at each design point. Our new linear model,

$$\widehat{y}_i = \beta_0 + \sum_{j=1}^{k} \beta_j x_{ij}$$

has on its left side an estimated response attribute value, rather than a population mean. This linear model characterizes the dependence of response attribute scores on explanatory attribute scores in a single multivariate population distribution.

The β-parameters in our new model provide the best estimates of the y scores for the N cases in the population as a linear function of their x scores. "Best" can be given different definitions, but a common one is the least-squares definition. The β-parameters minimize the average squared difference between estimated and actual response attribute scores. That is,

$$(1/N)\sum_{i=1}^{N}(y_i - \widehat{y}_i)^2 = (1/N)\sum_{i=1}^{N}\left[y_i - (b + \sum_{j=1}^{k}c_j x_{ij})\right]^2$$

is minimized for $b = \beta_0$, $c_j = \beta_j$, $j = 1,\ldots, k$. No other values that are substituted for b or any of the c_js will produce a smaller average squared error of estimation.

Based on this definition of the β-parameters, we use OLS regression to obtain plug-in estimates from a sample multivariate distribution. We can use bootstrap resampling to estimate CIs for these parameters. Figure 22.5 shows the parameter estimates, and 90% and 95% BCA CIs. The estimates take the same values as we obtained for the parametric, fixed design point model, $\widehat{\beta}_1 = 0.0096$ and $\widehat{\beta}_2 = 0.0099$. Neither 95% CI includes zero, although that for β_1 might come close.

```
> oxyreg<- function(X)
+ {
+ foo<- lm(X[,1]~X[,2]+X[,3])
+ foo$coef[2:3]
+ }

> oxyreg(thames)
      X[, 2]       X[, 3]
 0.009559641 0.009944357

> nonpoxy<- bootstrap(thames,oxyreg)

> limits.bca(nonpoxy)
              2.5%         5%       95%      97.5%
X[, 2] 0.0006082 0.002334 0.01945 0.02313
X[, 3] 0.0060926 0.006801 0.01370 0.01459
```

Figure 22.5: *Bootstrap: Oxy Modeled by Chlor and Light (S-Plus)*

In our parametric, fixed design point linear modeling of these data, we were unable to reject the null hypothesis that $\beta_1 = 0$ ($p = 0.18$). In consequence, we then fit a second linear model using only Light as an explanatory attribute.

The argument for dropping Chlor is not as convincing in this nonparametric, sampled design point modeling. We see at the top of Figure 22.6 that the nondirectional p-value for the hypothesis that $\beta_1 = 0$ is appreciably smaller, about 0.04.

```
> limits.bca(nonpoxy, prob=c(0.01,0.02,0.03))
                  1%               2%              3%
X[, 2] -0.002109499 4.458416e-008 0.001007209
X[, 3]  0.005440773 5.893248e-003 0.006223489

> oxyreg2<- function(X)
+ {
+ foo<- lm(X[,1]~X[,3])
+ foo$coef[2]
+ }

> oxyreg2(thames)
     X[, 3]
 0.01139419

> fit2<- bootstrap(thames,oxyreg2)

> limits.bca(fit2)
            2.5%        5%       95%    97.5%
X[, 3] 0.007485 0.008314 0.01455 0.01532
```

Figure 22.6: *Bootstrap: Oxy Modeled by Light Alone (S-Plus)*

For purposes of comparison, however, we continue in Figure 22.6 to fit a linear model with just Light as an explanatory attribute and to estimate a nonparametric BCA CI for Light. The nonparametric 90% CI (0.0083, 0.0146) is a bit shorter than the parametric 90% CI for the comparable parameter (0.0077, 0.0151). Because of the shift from fixed design point to sampled design point resampling, our hypothesis tests and CI estimates have slightly different interpretations.

Bootstrap LM: Nested Model Comparisons

In Concepts 11, we noted the difficulty in finding nonparametric estimates of null population distributions from which bootstrap samples could be drawn. Almost always, the null hypothesis, that some population parameter, θ, takes a specified value, θ_0, is not sufficient to define a nonparametric null population distribution estimate. Because of this, we have emphasized the use of CI estimation in testing population distribution hypotheses.

Linear Constraints

In the linear model context, a CI can be used to test any hypothesis that can be expressed as a single linear constraint on the β-parameters. If the number of β-parameters is k, then a linear constraint on those parameters takes the form

$$\sum_{j=1}^{k} C_j \beta_j = \Gamma$$

where the C_js are a set of k weights (often, many of them zero) and Γ is a numeric constant (often zero). Thus the hypothesis that $\beta_2 = 0$ is a linear constraint on the k parameters: $\Gamma = 0$, $C_2 = 1$, and all other C_js are zero. The hypothesis that $\beta_2 = \beta_3$ also is a linear constraint on the k β-parameters: $\Gamma = 0$, $C_2 = 1$, $C_3 = -1$, and all other C_js are zero. Other substantive hypotheses can require more complicated linear constraints. Lunneborg (1994) provides a guide to the phrasing of linear constraints.

A null hypothesis phrased as a linear constraint on the β-parameters is tested by choosing an appropriate α and estimating a $(1 - 2\alpha)100\%$ CI for $\sum C_j \beta_j$.

The null hypothesis can be rejected if Γ falls above, below, or outside that interval, depending on the alternative to the null hypothesis.

Placing a linear constraint on the β-parameters of an unconstrained linear model implies a second, constrained linear model. This second model will have $(k - 1)$ rather than k parameters, the result of dropping or combining explanatory attributes. Our CI test of the linear constraint hypothesis can be thought of as a model comparison test. Rejecting the hypothesis implies a choice of the unconstrained model. Failing to reject the hypothesis implies a choice of the constrained model.

Multiple Linear Constraints

Sometimes a substantive hypothesis requires imposing more than one linear constraint on the parameters of a model. In the stressed device performance example, we had as the parameters of our three-treatment model

$\beta_0 = \mu(\text{Performance}|\text{Control})$

$\beta_1 = \mu(\text{Performance}|\text{Water}) - \mu(\text{Performance}|\text{Control})$

$$\beta_2 = \mu(\text{Performance}|\text{Drop}) - \mu(\text{Performance}|\text{Control})$$

A hypothesis that all three treatments, control, water immersion, and dropping, have the same average effect on performance implies two constraints on these parameters, that $\beta_1 = 0$ and that $\beta_2 = 0$. The constrained model would have one, rather than three parameters. How can we compare these two models? What do we have as the bootstrap nonparametric alternative to the parametric F-ratio test?

Parametric F-Ratio Statistic

The F-ratio test is a straightforward extension of the parametric t-test. For a parametric test of the single hypothesis that $\beta_1 = 0$, we would compute the Studentized statistic:

$$t(\widehat{\beta}_1) = \frac{(\widehat{\beta}_1 - 0)}{\widehat{\text{SE}}(\widehat{\beta}_1)}$$

If our alternative hypothesis were nondirectional, we could use the square of the t-statistic,

$$t^2(\widehat{\beta}_1) = \frac{(\widehat{\beta}_1 - 0)^2}{\widehat{\text{Var}}(\widehat{\beta}_1)}$$

and reject the null hypothesis for large values of t^2. Though not computationally efficient, this t^2 statistic can be expressed as

$$t^2(\widehat{\beta}_1) = (1/1)\left[(\widehat{\beta}_1 - 0)\left[\widehat{\text{Var}}(\widehat{\beta}_1)\right]^{-1}(\widehat{\beta}_1 - 0)\right]$$

For a parametric test of the two hypotheses, $\beta_1 = 0$ and $\beta_2 = 0$, we would compute the F-ratio statistic. This latter turns out to be just a two-dimensional version of the t^2 statistic. If we use matrix notation, we can see this clearly. To form the F-ratio we replace the scalar $(\widehat{\beta}_1 - 0)$ with the two element vector

$$(\widehat{\boldsymbol{\beta}} - \mathbf{0}) = \begin{pmatrix} \widehat{\beta}_1 - 0 \\ \widehat{\beta}_2 - 0 \end{pmatrix}$$

and the scalar $\widehat{\text{Var}}(\widehat{\beta}_1)$ with the 2×2 matrix

$$\widehat{V} = \begin{pmatrix} \widehat{\mathrm{Var}}\left(\hat{\beta}_1\right) & \widehat{\mathrm{Cov}}\left(\hat{\beta}_1,\hat{\beta}_2\right) \\ \widehat{\mathrm{Cov}}\left(\hat{\beta}_2,\hat{\beta}_1\right) & \widehat{\mathrm{Var}}\left(\hat{\beta}_2\right) \end{pmatrix}$$

containing both the sampling variances of the two estimators and the covariance between their sampling distributions. These substitutions give the F-ratio in this form:

$$\mathrm{F}\!\left(\hat{\beta}\right) = (1/2)\!\left[\left(\hat{\beta} - \mathbf{0}\right)^T \left[\widehat{V}\right]^{-1}\!\left(\hat{\beta} - \mathbf{0}\right)\right]$$

where $\left(\hat{\beta} - \mathbf{0}\right)^T$ is the row-form of the vector $\left(\hat{\beta} - \mathbf{0}\right)$, $\left[\widehat{V}\right]^{-1}$ is the reciprocal or inverse of the matrix $\widehat{V}$, and the factor $(1/2) = (1/q)$ adjusts for the number of parameter hypotheses, q. Throughout this development, we use bold-face characters to distinguish multi-element vectors and matrices from single-element scalars.

The F-ratio is not limited to two hypotheses about the β-parameters, nor to simple forms for those hypotheses. If each of q hypotheses can be expressed as a linear constraint on the β-parameters

$$\sum_{j=1}^{k} C_{j\ell}\beta_j = \Gamma_\ell, \; \ell = 1, \ldots, q$$

they can be represented compactly in matrix notation as $C^T\beta = \Gamma$ where β is a k-element vector of the β-parameters, C is a $k \times q$ matrix of constraint weights, one column for each of the constraints, and Γ is a q-element hypothesis vector.

In this notation the F-ratio is equal to (Q/q) where

$$Q = \left(C^T\hat{\beta} - \Gamma\right)^T \left[\widehat{S}\right]^{-1}\!\left(C^T\hat{\beta} - \Gamma\right)$$

and $\widehat{S} = C^T\widehat{V}C$ is the $q \times q$ estimated sampling variance-covariance matrix for the q linear constraints,

$$g_\ell = \sum_{j=1}^{K} C_{j\ell}\widehat{\beta}_j \, , \, \ell = 1, \ldots, q$$

Because of the parametric linear model assumptions, the F-ratio usually has a computational form considerably simpler than this matrix equation. More important, perhaps, those assumptions provide a null sampling distribution for the statistic (Q/q), that of the F-random variable with q and $(n-k)$ df.

Nonparametric Q Statistic: Overview

We've noted the dependence of the F-ratio statistic on a core statistic Q. Scaling Q by $(1/q)$ provides a test statistic that has the parametric F-random variable null sampling distribution, but only when the parametric-LM assumptions are met. However, building on the notion that Q is a multidimensional form of Studentization, Davison and Hinkley (1997) show how to use the bootstrap-t ideas of Concepts 9 in a way that allows us to determine whether the hypothesis vector $\mathbf{\Gamma}$ falls within the $(1-2\alpha)100\%$ nonparametric bootstrap confidence region for the q-dimensional parameter $\mathbf{C}^T\boldsymbol{\beta}$. Because our parameter is multidimensional, its possible values span a q-dimensional space rather than a one-dimensional range.

The basis of this nonparametric-Q approach is this: We can build up a Monte Carlo approximation to the null sampling distribution of Q from the bootstrap sample instances of the Q^* statistic,

$$Q_b^* = \left(\mathbf{C}^T\widehat{\boldsymbol{\beta}}_b^* - \mathbf{C}^T\widehat{\boldsymbol{\beta}}\right)^T \left[\widehat{\mathbf{S}}_b^*\right]^{-1} \left(\mathbf{C}^T\widehat{\boldsymbol{\beta}}_b^* - \mathbf{C}^T\widehat{\boldsymbol{\beta}}\right), b = 1, \ldots, B$$

This bootstrap sampling distribution is an appropriate null sampling distribution because $\mathbf{C}^T\widehat{\boldsymbol{\beta}}$ is the true value of the constraint vector in the population from which the bootstrap samples are drawn. We next find the $(1-2\alpha)$ quantile of this bootstrap sampling distribution, $F_{Q^*}^{-1}(1-2\alpha)$.

Then, we compute Q from the linear model sample data,

$$Q = \left(\mathbf{C}^T\widehat{\boldsymbol{\beta}} - \mathbf{\Gamma}\right)^T [\widehat{\mathbf{S}}]^{-1} \left(\mathbf{C}^T\widehat{\boldsymbol{\beta}} - \mathbf{\Gamma}\right)$$

Finally, if $Q \leq F_{Q^*}^{-1}(1-2\alpha)$, then $\mathbf{\Gamma}$ falls within the $(1-2\alpha)100\%$ confidence region for $\mathbf{C}^T\boldsymbol{\beta}$.

We can put this last finding into the context of constraints and nested model comparisons:

1.   If $Q \leq F_{Q^*}^{-1}(1 - 2\alpha)$, all q constraints on the β-parameters are warranted, and we might choose to replace the unconstrained model with the constrained one.

2.   If $Q > F_{Q^*}^{-1}(1 - 2\alpha)$, not all q constraints on the β-parameters are warranted, and we should not replace the unconstrained model with the constrained one.

Nonparametric Q Statistic: Contributing Elements

It is convenient to use linear algebra to outline what is needed for the computation of Q (and Q_b^*).

We begin by defining some matrices and vectors and then go on to describe computations involving these arrays.

1. The matrix X is our design matrix. It has n rows and k columns. Each row corresponds to a case, and each column corresponds to an explanatory attribute. Typically, the first column of X is filled with 1s, to accommodate an intercept, β_0, in the linear model.

2. The n-element response vector y contains the response attribute scores for the n cases.

3. In linear algebra notation, the least-squares estimates of the βs can be obtained from X and y as the k-element vector

$$\widehat{\beta} = (X^T X)^{-1} X^T y,$$

where X^T is the transpose of X and $(X^T X)^{-1}$ is the regular inverse of the minor product moment of X. The latter is a $k \times k$ symmetric matrix.

4. The estimated response scores for our linear model then are given in the n-element vector

$$\widehat{y} = X\widehat{\beta} = X(X^T X)^{-1} X^T y$$

and the errors of estimation make up the n-element vector

$$e = y - \widehat{y}$$

5. The $n \times n$ matrix

$$H = X(X^T X)^{-1} X^T$$

is known in the regression literature as the hat matrix, the name stemming from the above equation in which the pre-multiplication of y by H produces "y hat." The diagonal elements of H are extracted as the n-

element vector of leverages $\boldsymbol{h}$. The element h_i assesses the extent to which the estimate $\widehat{y}_i$ depends upon the observed y_i.

6. The leverages are used to correct the raw residuals to compensate for this dependence. These corrected residuals (Davison and Hinkley, 1997) make up a vector $\boldsymbol{r}$ with elements

$$r_i = \frac{e_i}{\sqrt{(1 - h_i)}}$$

7. Let $\boldsymbol{D_r}$ be a $n \times n$ diagonal matrix with the diagonal elements supplied by the vector $\boldsymbol{r}$. Then, the product
$$\boldsymbol{W} = \boldsymbol{D_r X}$$
weights each row of the design matrix by the corresponding modified residual.

8. These results provide what is needed (Davison and Hinkley, 1997) to estimate the sampling distribution variances and covariances of the $\widehat{\boldsymbol{\beta}}$ elements. These estimates make up the elements of the $k \times k$ matrix
$$\widehat{\boldsymbol{V}} = (\boldsymbol{X}^T\boldsymbol{X})^{-1}(\boldsymbol{W}^T\boldsymbol{W})(\boldsymbol{X}^T\boldsymbol{X})^{-1}$$
$\widehat{V}_{jj}$ estimates the variance of the sampling distribution of $\widehat{\beta}_j$ and $\widehat{V}_{jk}$ estimates the covariance between the sampling distributions of $\widehat{\beta}_j$ and $\widehat{\beta}_k$.

9. Next we introduce the constraint matrix $\boldsymbol{C}^T$ with q rows and k columns, and the hypothesis vector $\boldsymbol{\Gamma}$ with q elements. Together they express the q linear constraints imposed on the β-parameters in our unconstrained linear model,
$$\boldsymbol{C}^T\boldsymbol{\beta} = \boldsymbol{\Gamma}$$
generalizing the definition of a single constraint, $\sum_{j=1}^{k} C_j\beta_j = \Gamma$, to multiple constraints, $\sum_{j=1} C_{j\ell}\beta_j = \Gamma_\ell$, $\ell = 1, \ldots, q$

10. Our plug-in estimates of the linear functions of the β-parameters, of $\boldsymbol{C}^T\boldsymbol{\beta}$, are given by the q elements of the vector
$$\boldsymbol{g} = \boldsymbol{C}^T\widehat{\boldsymbol{\beta}}$$

11. The estimated sampling variances and covariances of these estimates are given by the symmetric $q \times q$ matrix

$$\widehat{\boldsymbol{S}} = \boldsymbol{C}^T\widehat{\boldsymbol{V}}\boldsymbol{C}$$

12. The regular inverse of this variance-covariance matrix, $\left[\widehat{\boldsymbol{S}}\right]^{-1}$, is computed.

13. We now have what we need to calculate our Q statistic:

$$Q = (g - \Gamma)^T [\widehat{S}]^{-1} (g - \Gamma)$$

We have defined, at least symbolically, the required computations. We now describe algorithms incorporating these computations in the development of a bootstrap-t confidence region.

Nonparametric Q *Statistic: Two Algorithms*

Two slightly different algorithms are needed, one for studies involving sampled design points and one for fixed design point studies.

Here is the sampled design point algorithm.

1. From the random sample of cases, provide a matrix of explanatory attribute scores, X, and a vector of response scores, y. The first column of X will contain 1s if the unconstrained linear model includes a constant term or intercept—the β_0 in a model of the form $\widehat{y}_i = \beta_0 + \sum_{j=1}^{k} \beta_j x_{ij}$.

2. From the substantive hypothesis to be tested, create the constraint matrix C with one column for each of q constraints. The column entries are the weights given the β-parameters in that constraint. Complete the constraints by creating the q-element hypothesis vector Γ containing the right sides of the constraints. Often, the elements of this vector will be zeroes.

3. Supply α, defining the needed degree of confidence in the CIs, and B, the number of bootstrap samples to be drawn, for example, $\alpha = 0.025$ and $B = 2000$.

4. From the sample data

   a. Compute estimates of the model parameters, $\widehat{\beta}_j, j = 1, \ldots, k$

   b. Use these to compute estimates of the linear constraints,

   $$g_\ell = \sum_{j=1}^{k} C_{j\ell} \widehat{\beta}_j, \ \ell = 1, \ldots, q$$

   c. Save the $g_\ell, \ell = 1, \ldots, q$, as the vector g.

5. For $b = 1$ to B,

   a. Randomly sample with replacement n cases .

b. Form the design matrix X_b^* and response vector y_b^* for the sampled cases.

c. Find R_b^*, the inverse of the minor product moment of X_b^*.

d. Use these to compute estimates of the model parameters, $\widehat{\beta}_{bj}^*$, $j = 1, \ldots, k$, and of the linear constraints,

$$g_{b\ell}^* = \sum_{j=1}^{k} C_{j\ell} \widehat{\beta}_{bj}^*, \quad \ell = 1, \ldots, q$$

e. Save the $g_{b\ell}^*$, $\ell = 1, \ldots, q$, as the vector g_b^*.

f. Use X_b^*, y_b^*, and the $\widehat{\beta}_{bj}^*$s to find raw residuals, $e_b^* = y_b^* - \widehat{y}_b^*$.

g. From X_b^* and R_b^*, compute the vector of leverages, h_b^*.

h. Use the leverages and raw residuals to compute modified residuals,

$$r_{bi}^* = \frac{e_{bi}^*}{\sqrt{(1 - h_{bi}^*)}}$$

i. Scale the rows of X_b^* by the modified residuals to obtain W_b^*.

j. Use C, R_b^*, and W_b^* to find the estimated variance-covariance matrix for the constraint estimates, $\widehat{S}_b^*$.

k. Compute and save the bootstrap Q-statistic:

$$Q_b^* = (g_b^* - g)^T \left[\widehat{S}_b^* \right]^{-1} (g_b^* - g)$$

6. Compute a $q \times q$ variance-covariance matrix from the B values of g_b^* and invert this matrix to produce $\left[\widehat{S} \right]^{-1}$.

7. Compute the Q-statistic,

$$Q = (g - \Gamma)^T [\widehat{S}]^{-1} (g - \Gamma)$$

8. Find the $(1 - 2\alpha)$ quantile of the distribution of B values of Q_b^*, $F_{Q^*}^{-1}(1 - 2\alpha)$.

Turning now to the fixed design point study, X, R, and h do not change from sample to sample, and they need be computed only once. This streamlines the algorithm somewhat.

1. From the study, provide the design matrix, X, and a vector of response scores, y, for the randomly sampled cases. The first column of X will contain 1s if the unconstrained linear model has an intercept.

2. From the substantive hypothesis to be tested, create the constraint matrix C with one column for each of q constraints. The column entries are the weights given the β-parameters in that constraint. Complete the constraints by creating the vector Γ containing the right sides of the constraints. Often all the elements of this vector are zeroes.

3. Supply α, defining the needed degree of confidence in the CIs, and B, the number of bootstrap samples to be drawn.

4. From the sample data,

    a. Compute R, the inverse of the product moment of X.

    b. From X and R, calculate the hat matrix and extract its diagonal as the vector of leverages, h.

    c. Compute estimates of the model parameters, $\widehat{\beta}_j$, $j = 1, \ldots, k$

    d. Use C and $\widehat{\beta}$ to compute estimates of the linear constraints,

$$g_\ell = \sum_{j=1}^{k} C_{j\ell}\widehat{\beta}_j, \ \ell = 1, \ldots, q$$

    e. Save the g_ℓ, $\ell = 1, \ldots, q$, as the vector g.

5. For $b = 1$ to B

    a. Randomly sample with replacement and according to the study design, n cases.

    b. Form the response vector y_b^* for the sampled cases.

    c. Use X, R, and y_b^* to compute estimates of the model parameters, $\widehat{\beta}_{bj}^*$, $j = 1, \ldots, k$, and of the linear constraints,

$$g_{b\ell}^* = \sum_{j=1}^{k} C_{j\ell}\widehat{\beta}_{bj}^*, \ \ell = 1, \ldots, q$$

    d. Save the $g_{b\ell}^*$, $\ell = 1, \ldots, q$, as the vector g_b^*.

e. Use X, y_b^*, and the $\widehat{\beta}_{bj}^*$s to find raw residuals, $e_b^* = y_b^* - \widehat{y}_b^*$.

f. Use the leverages, h, and raw residuals to compute modified residuals,

$$r_{bi}^* = \frac{e_{bi}^*}{\sqrt{(1 - h_i)}}$$

g. Scale the rows of X by the modified residuals to obtain W_b^*.

h. Use C, R, and W_b^* to find the estimated variance-covariance matrix for the constraint estimates, $\widehat{S}_b^*$.

i. Compute and save the bootstrap Q-statistic:

$$Q_b^* = (g_b^* - g)^T \left[\widehat{S}_b^*\right]^{-1} (g_b^* - g)$$

6. Compute a $q \times q$ variance-covariance matrix from the B values of g_b^* and invert this matrix to produce $\left[\widehat{S}\right]^{-1}$.

7. Compute the Q-statistic,

$$Q = (g - \Gamma)^T \left[\widehat{S}\right]^{-1} (g - \Gamma)$$

8. Find the $(1 - 2\alpha)$ quantile of the distribution of B values of Q_b^*, $F_{Q^*}^{-1}(1 - 2\alpha)$.

Nonparametric Q Statistic: Illustration

The nonparametric Q-statistic has been developed in the most general form, although, in practice, the number of constraints might be small and they might take simple forms. We illustrate by returning to our fixed design point example, stressed device performance.

Figure 22.7 shows a sequence of S-Plus commands that carry out the first four steps of the fixed design points algorithm. The vector `perform` contains the response scores for the 48 sampled devices arranged so that the first 16 are control devices, the next 16 water immersed devices, and the final 16 dropped devices.

In addition to computing y, X, C, Γ, R, h, $\widehat{\beta}$, and g, a vector is prepared on the final line containing values of 1, 2, and 3 to identify the three populations from which our cases were sampled. This latter is needed by the `bootstrap` function to control the generation of bootstrap samples.

```
> yvec<- perform
> intc<- rep(1,48)
> zed<- rep(0,16)
> uno<- rep(1,16)
> watind<- c(zed,uno,zed)
> drpind<- c(zed,zed,uno)
> xmat<- cbind(intc,watind,drpind)
> cmat<- matrix(c(0,1,0,0,0,1),ncol=2)
> gam<- c(0,0)
> rmat<- solve(t(xmat)%*%xmat)
> hvec<- diag(xmat%*%rmat%*%t(xmat))
> beta<- rmat%*%t(xmat)%*%yvec
> gvec<- t(cmat)%*%beta
> grps<- c(uno,2*uno,3*uno)
```

Figure 22.7:   *Preliminary S-Plus Steps for Q-Function, Fixed Design*

The design matrix has three columns, so the population β-vector has three elements. Our two constraints, $\beta_1 = 0$ and $\beta_2 = 0$, are expressed as

$$\begin{pmatrix} 0 & 1 & 0 \\ 0 & 0 & 1 \end{pmatrix} \begin{pmatrix} \beta_0 \\ \beta_1 \\ \beta_2 \end{pmatrix} = \begin{pmatrix} 0 \\ 0 \end{pmatrix}$$

$$\boldsymbol{C}^T \qquad \boldsymbol{\beta} \quad = \quad \boldsymbol{\Gamma}$$

With these objects in the computing environment, the bootstrap resampling phase of the algorithm can be carried out.

At the top of Figure 22.8 is the text of a user-written function `qboot` that carries out the computations outlined in step 5 of the fixed design points algorithm.

From each of the bootstrap samples, this function calculates and returns the g_b^* vector and the Q_b^* statistic. The `bootstrap` function, in turn, accumulates these results over the $B = 1000$ bootstrap samples and stores them as the three columns of `qout$rep`.

```
> qboot
function(y)
{
    bstar <- rmat %*% t(xmat) %*% y
    gstar <- t(cmat) %*% bstar
    estar <- y - (xmat %*% bstar)
    rstar <- estar/sqrt(1 - hvec)
    wstar <- rstar[, 1] * xmat
    vstar <- rmat %*% (t(wstar) %*% wstar) %*% rmat
    sstar <- t(cmat) %*% vstar %*% cmat
    ssinv <- solve(sstar)
    gap <- (gstar - gvec)
    qstar <- t(gap) %*% ssinv %*% gap
    c(gstar, qstar)
}

> qout<- bootstrap(yvec,qboot,group=grps)

> shat<- var(qout$rep[,1:2])
> shati<- solve(shat)
> Q<- t(gvec-gam)%*%shati%*%(gvec-gam)
> Q
          [,1]
[1,]  2.733935

> limits.emp(qout)
                 2.5%           5%         95%         97.5%
qboot1 -0.09312500  -0.0793750   0.08375    0.10065625
qboot2 -0.08940625  -0.0793750   0.08250    0.09564063
qboot3  0.18698310   0.3970347  17.76368   21.25307770

> vv<- qout$rep[,3]
> length(vv[vv>=Q])/length(vv)
[1] 0.643
>
```

Figure 22.8: *Testing Multiple Constraints: S-Plus Q-Function*

We compute $\left[\widehat{\boldsymbol{S}}\right]^{-1}$ (shati) from the multivariate bootstrap sampling distribution for $\boldsymbol{g}^*$ and use this in the computation of our Q-statistic.

The value of Q (2.73) is quite small, compared to the bootstrap sampling distribution of Q^* for which we find $F_{Q^*}^{-1}(0.95) = 17.76$ and $F_{Q^*}^{-1}(0.975) = 21.25$. We estimate the p-value of our Q at 0.64, the proportion of the bootstrap sampling distribution of $Q^* \geq Q$.

No evidence here supports rejecting the twin null hypotheses, $\beta_1 = 0$ and $\beta_2 = 0$. We conclude that neither stress condition (water-immersion or dropping) leads to a deterioration in average performance of devices.

Figures 22.9 and 22.10 show a similar analysis for the joint hypotheses $\beta_1 = 0$ and $\beta_2 = 0$ in the Oxy production model. The null hypotheses now imply that neither Chlor nor Light influences average Oxy.

The Oxy production design samples design points rather than fixing them, and the flow of S-Plus commands in the two figures have been changed from Figures 22.7 and 22.8 to reflect the different design.

Figure 22.9 describes the computations carried out on the sample distribution of Oxy, Chlor, and Light scores. The $\widehat{\beta}_1 = 0.0096$ and $\widehat{\beta}_2 = 0.0099$ are as reported for the OLS regression model in Figure 22.5.

```
> yt<- as.vector(thames[,1])
> xt<- as.matrix(thames[,2:3])
> it<- rep(1,length(yt))
> xt<- cbind(it,xt)
> ct<- matrix(c(0,1,0,0,0,1),ncol=2)
> gamt<- c(0,0)
> bett<- solve(t(xt)%*%xt)%*%t(xt)%*%yt
> gt<- t(ct)%*%bett
> gt
             [,1]
[1,] 0.009559641
[2,] 0.009944357
```

Figure 22.9: *Preliminary S-Plus Steps for Q-Function, Sampled Design*

The top of Figure 22.10 shows the contents of the function qboot2 to be applied to each bootstrap sample score distribution. This function returns, as did the qboot of Figure 22.8, a g_b^* vector as well as the Q statistic computed from the b-th bootstrap sample, Q_b^*.

In this second example, our Q statistic takes a larger value, 41.88. This time the test statistic takes a value that is greater than $F_{Q^*}^{-1}(0.95) = 18.46$ though not quite as large as $F_{Q^*}^{-1}(0.99) = 49.97$. We have $0.01 < p(Q) < 0.05$.

We might have sufficient evidence, then, to reject the two hypotheses and conclude that either Chlor or Light or both have a statistically significant influence on the average Oxy level of the Thames at Reading. Note: Our use of the S-Plus default value of 1,000 for B makes the estimation of quantiles beyond 0.99 very problematic. If it is important to do so, values of B between 2,000 and 5,000 are required.

```
> qboot2
function(Z)
{
    ystar <- Z[, 1]
    xstar <- cbind(rep(1, length(ystar)), Z[, 2:3])
    istar <- solve(t(xstar) %*% xstar)
    hstar <- diag(xstar %*% istar %*% t(xstar))
    bstar <- istar %*% t(xstar) %*% ystar
    gstar <- t(ct) %*% bstar
    estar <- ystar - (xstar %*% bstar)
    rstar <- estar/sqrt(1 - hstar)
    wstar <- rstar[, 1] * xstar
    vstar <- istar %*% (t(wstar) %*% wstar) %*% istar
    sstar <- t(ct) %*% vstar %*% ct
    ssinv <- solve(sstar)
    gap <- (gstar - gt)
    qstar <- t(gap) %*% ssinv %*% gap
    c(gstar, qstar)
}

> Z<- as.matrix(thames)
> qout2<- bootstrap(Z,qboot2)

> shat2<- var(qout2$rep[,1:2])
> Q2<- t(gt-gamt)%*%solve(shat2)%*%(gt-gamt)
> Q2
            [,1]
[1,] 41.87955

> limits.emp(qout2,prob=c(0.95,0.99))
                95%         99%
qboot21  0.02131930  0.02692994
qboot22  0.01331824  0.01501808
qboot23 18.46482701 49.96767372
```

Figure 22.10: *Testing Multiple Constraints: S-Plus Q-Function*

Figure 22.11 describes the cdfs for the F-random variable with 2 and 14 df (dotted line) and for the bootstrap sampling distribution of Q^* obtained from the Oxy problem (solid line). We've first divided each entry in the latter distribution by $q = 2$ to give them the same scale as that for the F-random variable. The F-cdf is the one we would have used to test our two β-parameter hypotheses ($\beta_{\text{Chlor}} = 0$ and $\beta_{\text{Light}} = 0$) if we had reason to make the parametric-LM assumptions (see Figure 22.1), whereas the Q^*-cdf is the one we used for our nonparametric test of those two hypotheses.

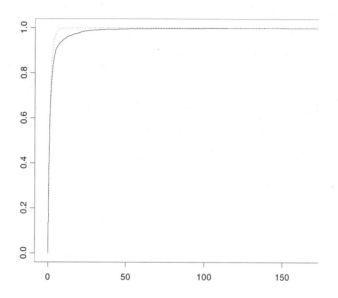

Figure 22.11: *Comparison of the cdfs for* $\mathsf{F}_{2,14}$ *and* Q^*/q

The two cdfs diverge in the quantile range (0.90 to 0.99) where hypothesis testing decisions are made. Although the Q^*-cdf is specific to the Oxy problem data and hypotheses, it would appear from the figure that, for a given level of significance, the nonparametric test statistic, Q/q, must take a larger value than would be required of the parametric F-ratio.

For a given data set and hypotheses, the two test statistics, F and Q/q, probably take different values. This is because they estimate V, the variance covariance matrix for the β-estimates, differently. The parametric F-test statistic employs an estimate,

$$\widehat{V} = \widehat{\sigma}^2 \left(X^T X \right)^{-1} = \left[\left(X^T X \right)^{-1} X^T \right] D_{\widehat{\sigma}^2} \left[X \left(X^T X \right)^{-1} \right]$$

based on the requirement that each response attribute score, y_i, is sampled from a population distribution with a common variance, σ^2. The nonparametric Q-statistic, however, does not require variance-homogeneity and uses a more robust variance-covariance estimate (Davison and Hinkley, 1997),

$$\hat{V} = (X^T X)^{-1} (W^T W)(X^T X)^{-1} = \left[(X^T X)^{-1} X^T \right] D_{r_i^2} \left[X(X^T X)^{-1} \right]$$

To facilitate comparison, the two estimates have been expressed, on the far right sides of both equations, as a certain algebraic function of the unconstrained design matrix, X, and an $(n \times n)$ diagonal matrix. While each of the n entries along the main diagonal of $D_{\hat{\sigma}^2}$ takes the same value, $\hat{\sigma}^2$, those along the main diagonal of $D_{r_i^2}$ can assume different values, the squares of the corrected residuals from the unconstrained model, $r_i^2 = e_i^2/(1 - h_i)$.

For the Oxy problem data and constraints, the F-ratio statistic takes the value 16.05. The proportion of the sampling distribution of the F-random variable with 2 and 14 dfs as large or larger than 16.05 is only p $= 0.0002$. The value of $Q/2$, on the other hand, is 20.94, and the proportion of our bootstrap sampling distribution of $Q^*/2$ taking values this large or larger is, as we have seen, somewhere between 0.01 and 0.05.

Multiparameter Confidence Region Hypothesis Tests

Within the linear model context, we have outlined and illustrated a bootstrap-t approach to the simultaneous testing of hypotheses about more than one β-parameter. Because such a wide variety of between-cases designs can be represented in linear model form, the approach is easily adapted to multiple treatment comparison problems.

The S-Plus functions qboot and qboot2 are reusable as they are given here. Each requires that the sample data first be processed as in Figure 22.7, for fixed design point samples, or as in Figure 22.9, for sampled design point samples.

There does not appear to be an equivalent BCA CI approach to determining a confidence bound for multiple parameters.

Prediction Accuracy

Once a linear model has been chosen and its parameters estimated, that fitted model is often used to predict the response scores for new cases, based on the explanatory attribute scores for those cases. How accurate are those predictions?

As we saw earlier, the sum of squared residuals,

$$\text{SSR(Model } K) = \sum_{i=1}^{n}(y_i - \widehat{y}_i)^2 = \sum_{i=1}^{n}\left(y_i - \sum_{j=1}^{k}\widehat{\beta}_j x_{ij}\right)^2$$

provides a measure of the fit of the k-parameter linear model to the sample response scores. Dividing the SSR by n gives us a mean squared residual (MSR),

$$\text{MSR(Model } K) = \frac{1}{n}\sum_{i=1}^{n}(y_i - \widehat{y}_i)^2$$

which can be interpreted as an average of n squared errors of prediction.

But the MSR assesses prediction accuracy only for the cases used in fitting the prediction model. Because the β-parameter estimates were chosen so as to minimize the MSR, this measure of prediction accuracy provides an overly-optimistic estimate of the accuracy of future predictions made with the fitted model.

Efron and Tibshirani (1993) describe a bootstrap approach to correcting the MSR or apparent error rate for this optimism. The idea is to estimate the optimism for a series of k-parameter models fit to bootstrap samples. An aggregate optimism estimate is obtained by averaging and then used to correct the MSR.

1. From the multivariate score distribution for the random sample, $x = (X, y)$, compute the MSR for the k-parameter linear model. Note that X denotes the design matrix part of the sample and y denotes the vector of response scores.

2. For $b = 1, \ldots, B$

    a. From the estimated population score distribution, $\widehat{X}$, draw a random bootstrap sample, $x_b^* = (X_b^*, y_b^*)$ of size n.

    b. Fit the k-parameter linear model to x_b^*, obtaining the parameter estimates $\widehat{\boldsymbol{\beta}}_b^*$.

    c. Use $\widehat{\boldsymbol{\beta}}_b^*$ and X_b^* to obtain response score predictions, $\widehat{y}_b^*$.

d. Compute $a_b^* = (1/n)\sum_{i=1}^{n}(y_{bi}^* - \widehat{y}_{bi}^*)^2$.

e. Use $\widehat{\boldsymbol{\beta}}_b^*$ and X, the real world design matrix, to obtain response score predictions, $\widehat{\mathbf{y}}_b$.

f. Compute $t_b^* = (1/n)\sum_{i=1}^{n}(y_i - \widehat{y}_{bi})^2$.

g. Compute and save $o_b^* = (t_b^* - a_b^*)$.

3. Compute $\widehat{O} = (1/B)\sum_{b=1}^{B} o_b^*$, the estimated optimism of the MSR.

4. Compute the estimated true prediction error rate, $\widehat{\text{TPE}} = \text{MSR} + \widehat{O}$.

The bootstrap sample estimate of optimism, o_b^*, is the difference between the average squared error for future case predictions, for the cases contributing scores to x, and the average squared error for present case predictions, for the cases contributing scores to x_b^*. This difference is the bootstrap analog to the optimism we need to estimate, the difference between true and apparent error rates.

The number of bootstrap samples employed in the optimism estimate can be small. Efron and Tibshirani (1993) demonstrate the technique employing only $B = 10$ bootstrap samples. Implicit in the logic of the correction is that the new cases to which we want the correction to apply are to be drawn from the same population as the random sample used to fit the model.

We illustrate optimism correction with data from Exercise 1 in Applications 17. Table 17.3, taken from Kleinbaum, Kupper, and Muller (1988), gives the systolic blood pressure, SBP, and the Quetelet Index of body size, QUET, for a random sample of $n = 32$ employed male adults.

The Quetlet Index is computed as

$$\text{QUET} = 100 \times \left[\frac{\text{Weight in Pounds}}{(\text{Height in Inches})^2} \right]$$

We use these sample data to estimate the intercept and slope of a linear prediction equation:

$$\text{Pred}(\text{SBP}_i) = \beta_0 + \beta_1(\text{QUET}_i)$$

Table 17.3: *Blood Pressure and Body Size Data*

SBP	135	122	130	148	146	129	162	160
QUET	2.876	3.251	3.100	3.768	2.979	2.790	3.668	3.612
SBP	144	180	166	138	152	138	140	134
QUET	2.368	4.637	3.877	4.032	4.116	3.673	3.562	2.998
SBP	145	142	135	142	150	144	137	132
QUET	3.360	3.024	3.171	3.401	3.628	3.751	3.296	3.210
SBP	149	132	120	126	161	170	152	164
QUET	3.301	3.017	2.789	2.956	3.800	4.132	3.962	4.010

The estimated values for the intercept and slope are $\widehat{\beta}_0 = 70.576$ and $\widehat{\beta}_1 = 21.492$. How accurately will SBP be estimated for future cases using the following prediction equation?

$$\text{Pred}(\text{SBP}_i) = 70.576 + 21.492\,(\text{QUET}_i)$$

Figure 22.12 shows the computation of an optimism estimate in S-Plus. The MSR for our model is 90.25 mm^2 Hg. The square root of the MSR is 9.5 mm Hg. We can interpret this latter as the amount by which, on average, we miss SBP when predicting it from our model—in the original sample of cases. We can expect to be off by about 9.5 mm Hg. That is an underestimate of what would be true for predictions of SBP made for new cases.

For each of $B = 25$ bootstrap samples, the user-written function `optim` returns a bootstrap estimate of the optimism of MSR. The average of the 25 optimism estimates is 11.27 mm^2 Hg. After increasing our MSR by this amount, we see that we can expect to mispredict the SBP for a new case by about 10.1 mm Hg.

Linear Models for Randomized Cases

The data from studies in which available cases are randomized among treatments can be studied using linear models. The key is to include in the linear model one or more indicator variables with scores indicating the treatment groups to which cases were randomized.

```
> predmodl<- lm(sbp~quet)
> msr<- mean(predmodl$res^2)
> msr
[1] 90.25072
> sqrt(msr)
[1] 9.500038

> xmat<- cbind(rep(1,32),quet)
> yvec<- sbp
> zmat<- cbind(yvec,xmat)

> optim
function(Z)
{
    yy <- Z[, 1]
    xx <- Z[, 2:3]
    bet <- solve(t(xx) %*% xx) %*% t(xx) %*% yy
    e1 <- yy - (xx %*% bet)
    e2 <- yvec - (xmat %*% bet)
    ape <- mean(e1^2)
    tpe <- mean(e2^2)
    c(ape, tpe, (tpe - ape))
}

> bootopt<- bootstrap(zmat,optim,B=25)
> mn<- apply(bootopt$rep,2,mean)
> mn
  optim1    optim2    optim3
 84.9174 96.18461 11.26722

> sqrt(msr+mn[3])
   optim3
 10.07561
```

Figure 22.12: *Correcting for the Optimism of the MSR (S-Plus)*

We saw an example in Applications 20 for a study in which the time taken by children to form a pattern from colored blocks was assessed. The children had been randomly allocated to one of two instructions groups. Half the children were instructed to begin by forming a row of the design, and half were instructed to start with a corner of the design. The researcher was concerned that the time taken by a child to complete the task could be influenced as well by the child's field independence and so a measure of this, the score on an embedded figures test (EFT), was included in the analysis.

This gave rise to a linear model

$$\widehat{y}_i = \beta_1 x_{i1} + \beta_2 x_{i2} + \beta_3 z_i$$

where

$$x_{i1} = \begin{cases} 1, \text{ if the } i\text{-th child was assigned to the row instructions group} \\ 0, \text{ otherwise} \end{cases}$$

$$x_{i2} = \begin{cases} 1, \text{ if the } i\text{-th child was assigned to the corner instructions group} \\ 0, \text{ otherwise} \end{cases}$$

and z_i is the deviation of the i-th child's EFT score from the median EFT score.

These attributes were chosen for the model so that, when the model is fit to the study data, the coefficients β_1 and β_2 have these interpretations:

1. β_1: Model estimate of the time score for a child with a median EFT score given row instructions for the block design task

2. β_2: Model estimate of the time score for a child with a median EFT score given corner instructions for the block design task

The difference $s = (\beta_1 - \beta_2)$ assesses the difference in treatment effects, adjusted for the EFT covariate.

The null treatment-effect hypothesis, that a child would have the same time score, as well as the same EFT score, whichever instructions group that child was randomized to, allows us to develop a null reference distribution for s. We can rerandomize cases to treatment groups, carrying with them their response and covariate scores, and recompute the statistic s^H for each such rerandomization.

Linear Models for Nonrandom Studies

In the absence of a random sample of cases or the random allocation of available cases among treatments, the linear model, as other statistical techniques, cannot support population or causal inference. However, the linear model and its parameters or, more correctly, its fitted constants, can provide a useful description when they are developed in the context of a nonrandom study design.

Table 22.2:   *Birth Weight, Head Circumference, and Gestation*

Birthwt	Headcir	Gest	Birthwt	Headcir	Gest
3680	35.5	42	3240	35.5	40
3180	33.0	38	3530	35.0	41
3100	33.5	39	3690	37.0	39
3550	34.5	41	4160	39.5	41
3410	33.5	38	3600	36.0	39
3010	35.0	37	3250	34.5	41
2820	33.0	35	3390	34.0	38
3870	36.0	37	3560	34.0	40
3890	37.0	40	3020	34.0	37
3400	35.5	37	2860	34.0	34
3400	35.5	39	3130	34.5	39
3300	36.0	38	3370	36.5	36
4080	35.0	42			

Here is an example. Lovie (1991) reports data on three attributes of 25 newborns, part of a larger study on methods of assisted delivery. The 25 cases contributing to Table 22.2 were selected by Lovie to illustrate particular points about regression and cannot be regarded as a random sample from any well-defined population. The three at-birth attributes are gestation in weeks (Gest), birth weight in grams (Birthwt), and the baby's head circumference in centimeters (Headcir).

Lovie looked at birth weight as a function of head circumference and gestation. How stable are those relations in this data set?

Figures 22.13 and 22.14 describe a Resampling Stats analysis of the regression weights obtained from a series of randomly chosen half-samples of the original data set. The script appears as Figure 22.13, and the output of one run makes up Figure 22.14.

The fitted equation for the full data set takes the form

$$\text{Pred(Birthwt)} = (130.12 \times \text{Headcir}) + (80.86 \times \text{Gest}) - 4278.6$$

```
'the Table 22.2 data form the vectors bwt, hdc and ges
size bwt nn
copy 1,nn cid
divide nn 2 xx
round xx n2
if n2 > xx
  subtract n2 1 n2
end
regress noprint bwt hdc ges bwts
print bwts
repeat 100
  shuffle cid scid
  take scid 1,n2 hcid
  take bwt hcid bwth
  take hdc hcid hdch
  take ges hcid gesh
  regress noprint bwth hdch gesh bwth
  take bwth 1 xx
  score xx head
  take bwth 2 xx
  score xx gest
  take bwth 3 xx
  score xx intc
end
median head mdnhead
median gest mdngest
median intc mdnintc
percentile head (25 75) iqrhead
percentile gest (25 75) iqrgest
percentile intc (25 75) iqrintc
print mdnhead iqrhead
print mdngest iqrgest
print mdnintc iqrintc
```

Figure 22.13: *Half-Sample Stability, Resampling Stats Script for Regression*

```
BWTS     =    130.12       80.86      -4278.6
MDNHEAD  =    127.21
IQRHEAD  =    109.28      147.1
MDNGEST  =    78.473
IQRGEST  =    62.251      98.767
MDNINTC  =    -4030.4
IQRINTC  =    -4588.7     -3544.9
```

Figure 22.14: *Half-Sample Stability of Regression Weights*

The 100 half-sample weights for head circumference had a median value of 127 with 50% of the values clustered between 109 and 148. The median half-sample weight for gestation was 78, and the middle half of the weights were between 62 and 99. There would appear to be nothing in these results to cast doubt on the stability, over this data set, of the description provided by the linear model for the full data set.

Exercises

1.   A body mass index (BMI, weight in Kg divided by the square of height in meters) was assessed for 20 anorexic patients on their admission to a hospital for treatment and again at follow-up, after discharge from the hospital. A BMI was also calculated, based on each patient's stated preferred weight at time of admission. Data for Table 22.3 are taken from Hand et al. (1994, Data Set 376).

a. Model the follow-up BMI as a function of admission and preferred BMIs. Does the preferred BMI help explain the outcome?

b. Carry out and interpret stability analyses.

Table 22.3:   *Change in Body Mass Index for Hospitalized Anorexics*

Admission BMI	16.84	16.26	14.33	14.30	11.98	13.59	15.03
Preferred BMI	18.01	19.22	18.65	17.54	12.05	17.38	19.16
Follow-up BMI	24.03	18.50	16.61	16.57	18.99	15.62	25.96
Admission BMI	13.95	14.02	12.22	12.00	16.05	15.86	11.55
Preferred BMI	12.31	17.96	15.63	16.84	18.08	19.22	15.06
Follow-up BMI	13.82	21.09	21.38	12.34	13.17	17.32	16.07
Admission BMI	15.76	14.12	14.44	18.12	12.60	15.52	
Preferred BMI	21.10	17.46	18.15	18.54	18.01	15.89	
Follow-up BMI	17.22	13.06	14.72	21.16	18.92	18.01	

2. You can demonstrate the process of estimating the prediction error with these random sample data reported in Hand et al. (1994, Data Set 48). The scores in Table 22.4 are from 34 randomly selected batches of peanuts in storage. For each batch, the percentage of noncontaminated peanuts (% Non) was determined as was the level of aflatoxin in parts per billion (Aflatox).

How accurately can the percentage of noncontaminated peanuts be predicted from the aflatoxin measurement?

Table 22.4:   *Percentage Noncontaminated Peanuts and Aflatoxin Concentration*

% Non	Aflatox	% Non	Aflatox	% Non	Aflatox
99.971	3	99.942	18.8	99.863	46.8
99.979	4.7	99.932	19.9	99.811	46.8
99.982	8.3	99.908	21.7	99.877	58.1
99.971	9.3	99.97	21.9	99.798	62.3
99.957	9.9	99.985	22.8	99.855	70.6
99.961	11	99.933	24.2	99.788	71.1
99.956	12.3	99.858	25.8	99.821	71.3
99.972	12.5	99.987	30.6	99.83	83.2
99.889	12.6	99.958	36.2	99.718	83.6
99.961	15.9	99.909	39.8	99.642	99.5
99.982	16.7	99.859	44.3	99.658	111.2
99.975	18.8				

3. The data in Table 22.5 are taken from Hand et al. (1994, Data Set 301) and relate to a calibration experiment. The experimental setup is this: For each of five blood samples of known lactic acid concentration (1, 3, 5, 10, and 15 ppm), from 3 to 5 concentration measurements are made with a new instrument.

The resulting data are to be used to calibrate the instrument, that is, to develop an equation by which lactic acid concentration (True Conc) can be predicted from the instrument reading (Instr Rdg). Use these data to develop the equation. You may want to test whether including the square of instrument reading helps to predict true concentration.

Once you have chosen a model, estimate the error, taking optimism into account. Remember, true concentration is the response and instrument reading is the explanatory attribute. The resampling, however, must mimic the original sampling design. We'll take that to mean that four instrument readings were randomly sampled at a true concentration of 1 ppm, five instrument readings were randomly sampled at a true concentration of 3 ppm, and so forth. You will need to make provision, then, for drawing samples of the appropriate sizes from each of five different estimated populations.

Table 22.5:   *True Blood Lactic Acid Concentration and Instrument Readings*

True Conc	Instr Rdg	True Conc	Instr Rdg
1	1.1	5	8.2
1	0.7	5	6.2
1	1.8	10	12.0
1	0.4	10	13.1
3	3.0	10	12.6
3	1.4	10	13.2
3	4.9	15	18.7
3	4.4	15	19.7
3	4.5	15	17.4
5	7.3	15	17.1

4. The data in Table 22.6 are adapted from Hand et al. (1994, Data Set 70) and were collected from an experiment to investigate silver iodide use in cloud seeding to increase rainfall. The data are from 24 days in the Summer of 1975 when researchers judged conditions were suitable for seeding. On each of these days the decision to seed or not was made randomly (A = 1 indicates seeding, A = 0 indicates no seeding).

The response attribute Y shows the amount of rainfall falling in a target area during a six-hour period on that day. Rainfall can also be influenced by other factors. Two of these are included here as potential covariates: C, the percentage cloud cover on the day, and P, the pre-wetness or amount of precipitation in the target area one hour in advance of when seeding occurred or would have occurred.

The study involves randomly allocating available cases, 24 suitable days, among two treatments.

Formulate a linear model for rainfall, Y, including an indicator variable for seeding, A. Fit the model. Test the significance of the weight for the indicator variable by rerandomizing days to seeded/unseeded treatments. Does seeding have an effect?

Table 22.6:   *Data from a Cloud Seeding Experiment*

A	0	1	1	0	1	0	0	0
C	13.40	37.90	3.90	5.30	7.10	6.90	4.60	4.90
P	0.274	1.267	0.198	0.526	0.250	0.018	0.307	0.194
Y	12.85	5.52	6.29	6.11	2.45	3.61	0.47	4.56

A	0	1	1	1	0	1	1	0
C	12.10	5.20	4.10	2.80	6.80	3.00	7.00	11.30
P	0.751	0.084	0.236	0.214	0.796	0.124	0.144	0.398
Y	6.35	5.06	2.76	4.05	5.74	4.84	11.86	4.45

A	0	1	0	1	1	0	1	0
C	4.20	3.30	2.20	6.50	3.10	2.60	8.30	7.40
P	0.237	0.960	0.230	0.142	0.073	0.136	0.123	0.168
Y	3.66	4.22	1.16	5.45	2.02	0.82	1.09	0.28

5. The data in Table 22.7 are taken from Dusoir (1997) and show measurements made on a random sample of 31 black cherry trees. The first column lists the volume in cubic feet, the second column lists the diameter in inches at 4.5 feet above ground, and the third column shows the height in feet.

Develop an equation for predicting tree volume as a function of height and diameter. You might want to transform either explanatory attribute or include powers of either or both to arrive at a best model.

When you have chosen a model, work out, adjusting for optimism, the accuracy of volume predictions.

Table 22.7: *Black Cherry Trees: Volume, Diameter, and Height*

Volume	Diameter	Height
10.3	8.3	70
10.3	8.6	65
10.2	8.8	63
16.4	10.5	72
18.8	10.7	81
19.7	10.8	83
15.6	11	66
18.2	11	75
22.6	11.1	80
19.9	11.2	75
24.2	11.3	79
21	11.4	76
21.4	11.4	76
21.3	11.7	69
19.1	12	75
22.2	12.9	74
33.8	12.9	85
27.4	13.3	86
25.7	13.7	71
24.9	13.8	64
34.5	14	78
31.7	14.2	80
36.3	14.5	74
38.3	16	72
42.6	16.3	77
55.4	17.3	81
55.7	17.5	82
58.3	17.9	80
51.5	18	80
51	18	80
77	20.6	87

6. The data in Table 22.8 for a random sample of 32 cases are taken from Kleinbaum, Kupper, and Muller (1988). In this exercise, begin by fitting a regression model of the form

$$\widehat{\text{SBP}}_i = \beta_0 + \beta_1 \text{QUET}_i + \beta_2 \text{AGE}_i + \beta_3 \text{SMK}_i$$

which relates a case's systolic blood pressure, SBP, to body-size, age in years, and smoking status. The Quetelet Index, QUET, is a ratio of weight to the square of height and SMK is 1 for a smoker, 0 for a nonsmoker.

Begin to explore the simplification of the model by using the Q statistic to test if both $\beta_2 = 0$ and $\beta_3 = 0$.

Table 22.8: *Systolic Blood Pressure, Body Size, Age, and Smoking Status*

SBP	135	122	130	148	146	129	162	160
QUET	2.876	3.251	3.100	3.768	2.979	2.790	3.668	3.612
AGE	45	41	49	52	54	47	60	48
SMK	0	0	0	0	1	1	1	1

SBP	144	180	166	138	152	138	140	134
QUET	2.368	4.637	3.877	4.032	4.116	3.673	3.562	2.998
AGE	44	64	59	51	64	56	54	50
	1	1	1	1	0	0	1	1

SBP	145	142	135	142	150	144	137	132
QUET	3.360	3.024	3.171	3.401	3.628	3.751	3.296	3.210
AGE	49	46	57	56	56	58	53	50
SMK	1	1	0	0	1	0	0	0

SBP	149	132	120	126	161	170	152	164
QUET	3.301	3.017	2.789	2.956	3.800	4.132	3.962	4.010
AGE	54	48	43	43	63	63	62	65
SMK	1	1	0	1	0	1	0	0

Applications 23

Categorical Response Attributes

Most of our examples and exercises have featured measured or quantitative response attributes. In this final Applications, we look specifically at a series of experimental designs and research questions that focus on one or more categorical or qualitative attributes of cases, particularly where such a categorical attribute is treated as a response to some treatment.

Cross-Classification of Cases

We consider the following data structure. Each of n cases is assigned to a cell in a table with I rows and J columns. The rows and columns of the table correspond to the levels of two categorical attributes. A case is assigned to the cell at the intersection of the i-th row and j-th column if that case has the i-th level of the row attribute and the j-th level of the column attribute. We let n_{ij} indicate the number of cases assigned to the cell at the intersection of the i-th row and j-th column, $n_{i.}$ indicate the number of cases assigned to the i-th row of the table, and $n_{.j}$ indicate the number of cases assigned to the j-th column of the table.

Independent Row and Column Classification

You might be familiar with a statistical test for such a table. If the n cases are sampled independently of one another, then we can test the null hypothesis of row-by-column independence—that the two ways of classifying the cases operate independently one of another.

Actually, the cases can be randomly sampled in any of three different ways:

1. All n cases can be randomly sampled from one population and then categorized relative to the levels of both row and column attributes.

2. Independent random samples can be drawn from each of I populations, corresponding to the levels of the row attribute of the table, for example, male and female. The samples are of sizes $n_{i.}$, $\sum n_{i.} = n$. The sampled cases are then categorized relative to the levels of their column attributes and assigned to cells within the table.

3. Independent random samples can be drawn from each of J populations, corresponding to the levels of the column attribute of the table, for example, Republican, Democrat, or Independent political party affiliation. The samples are of sizes $n_{.j}$, $\sum n_{.j} = n$. The sampled cases are then categorized relative to the levels of their row attributes and assigned to cells within the table.

Under any of these sampling scenarios, each sample distribution is drawn from a multinomial population distribution. A multinomial population distribution does not consist of N numeric values, a univariate distribution, or numeric vectors, a multivariate distribution. Rather, each of the N elements belongs to one of a known and limited number of categories.

1. In the single sample design, the population distribution spans $I \times J$ categories. We let π_{ij} indicate the proportion of the N population cases that fall into the ij-th category. Thus,

$$\pi_{i.} = \sum_{j=1}^{J} \pi_{ij}$$

is the proportion falling in the i-th row category,

$$\pi_{.j} = \sum_{i=1}^{I} \pi_{ij}$$

is the proportion falling in the j-th column category, and

$$\sum_{i=1}^{I} \sum_{j=1}^{J} \pi_{ij} = 1.0$$

2. When we draw samples from each of I row populations, then each population distribution spans J column categories. Here, we let $\pi_{j|i}$ indicate the proportion of cases in the i-th row population falling in the j-th column category, $\sum_{j=1}^{J} \pi_{j|i} = 1.0$.

3. Similarly, when we sample each of J column populations, then each population distribution spans I row categories. Now we let $\pi_{i|j}$ indicate

the proportion of cases in the j-th column population falling in the i-th row category, $\sum_{i=1}^{I} \pi_{i|j} = 1.0$.

For these population category proportions, the independence hypothesis takes these different forms:

1. For the single multinomial sample, the independence hypothesis is that

$$\pi_{ij} = \pi_{i.} \times \pi_{.j}, \text{ for } i = 1, \ldots, I \text{ and } j = 1, \ldots, J$$

Plug-in estimates of these hypothetical proportions can be computed from the tabled sample distribution as

$$\widehat{\pi}_{ij} = \widehat{\pi}_{i.} \times \widehat{\pi}_{.j}$$

where $\widehat{\pi}_{i.} = (n_{i.}/n)$ and $\widehat{\pi}_{.j} = (n_{.j}/n)$.

2. For the I independent multinomial samples from row populations, the independence hypothesis is that the proportion of cases in the j-th category is the same across all I populations,

$$\pi_{j|1} = \pi_{j|2} = \ldots = \pi_{j|I} = \pi_j, \text{ for } j = 1, \ldots, J$$

Plug-in estimates of these common category proportions are obtained from the tabled sample distributions as

$$\widehat{\pi}_j = \frac{\sum_{i=1}^{I} n_{ij}}{\sum_{i=1}^{I} n_{i.}} = \frac{n_{.j}}{n}$$

3. And for J independent multinomial samples from column populations, the independence hypothesis is that the proportion of cases in the i-th category is the same across all J populations,

$$\pi_{i|1} = \pi_{i|2} = \ldots = \pi_{i|J} = \pi_i, \text{ for } i = 1, \ldots, I$$

Plug-in estimates of these common category proportions are obtained from the tabled sample distributions as

$$\widehat{\pi}_i = \frac{\sum\limits_{j=1}^{J} n_{ij}}{\sum\limits_{j=1}^{J} n_{.j}} = \frac{n_{i.}}{n}$$

Chi-Squared Test of Independence

The classical chi-squared test of independence uses the test statistic

$$X^2 = \sum_{i=1}^{I}\sum_{j=1}^{J} \frac{\left(n_{ij} - m_{ij}\right)^2}{m_{ij}}$$

where n_{ij} is the number of cases falling in the ij-th cell of the table, and m_{ij} is our sample estimate of the number expected in that cell under the independence hypothesis or model. These modeled frequencies are estimated from the observed frequencies as

$$m_{ij} = \widehat{\pi}_{ij}\, n = \left(\frac{n_{i.}}{n}\right)\left(\frac{n_{.j}}{n}\right) n = \frac{n_{i.}\, n_{.j}}{n}$$

for the single sample,

$$m_{ij} = \widehat{\pi}_j\, n_{i.} = \left(\frac{n_{.j}}{n}\right) n_{i.} = \frac{n_{i.}\, n_{.j}}{n}$$

for the row samples, and

$$m_{ij} = \widehat{\pi}_i\, n_{.j} = \left(\frac{n_{i.}}{n}\right) n_{.j} = \frac{n_{i.}\, n_{.j}}{n}$$

for the column samples. The three sampling schemes lead to the same cell estimates.

Intuitively, large values of the X^2 statistic provide support for rejecting the hypothesis of independence or of constant row or column category proportions.

Asymptotically, that is, as the sample size or sizes become large, the cumulative distribution function (cdf) for the null sampling distribution of the X^2 statistic approaches that of the chi-squared random variable with $(I-1) \times (J-1)$ degrees of freedom (df). This is true for all three sampling schemes.

Thus the classic test of independence refers the X^2 statistic to the $\chi^2_{(I-1)(J-1)}$ distribution and assigns as the p-value for the statistic the proportion of that distribution that is as large or larger than X^2:

$$p(X^2) = 1 - F^{-1}_{\chi^2_{(I-1)(J-1)}}(X^2)$$

The X^2 statistic also is known as the Pearson chi-squared statistic to distinguish it from a second statistic,

$$G^2 = 2 \sum_{i=1}^{I} \sum_{j=1}^{J} \left[n_{ij} \times \log\left(\frac{n_{ij}}{m_{ij}}\right) \right]$$

the likelihood-ratio chi-squared statistic, which can be used to test the same null hypotheses. Asymptotically, G^2 also has as its null sampling distribution that of the chi-squared random variable with $(I-1) \times (J-1)$ df. The two statistics take similar values.

Exact Test of Independence

For small sample sizes or for tables with many near-empty cells, the large-sample requirements of the X^2 or G^2 statistics might not be satisfied. As a result, the p-values derived from the cdf of the $\chi^2_{(I-1)(J-1)}$ random variable can be quite inaccurate.

However, for small tables, the null reference distribution can be completely specified. Using that distribution to assign a p-value to the X^2 or G^2 statistic provides an exact, rather than asymptotic, test of the null hypothesis.

Table 23.1 provides an example. A small sample of marbles ($n = 7$) is drawn from an urn at random and without replacement. The marbles are then classified by color and size.

Table 23.1: *A Sample of Marbles Classified by Size and Color*

	Blue	Red	White	Total
Large	2	0	1	3
Small	0	3	1	4
Total	2	3	2	7

Under the hypothesis of independence, the cell frequencies can be modeled from the marginal frequencies. For example, the number of large blue marbles expected under independence would be

$$m_{11} = \frac{n_{1.} \times n_{.1}}{n} = \frac{3 \times 2}{7} = 0.85714$$

and similar computations could be carried out for each of the cells of the table.

The value of the Pearson X^2 statistic for this table is 4.958. The statistics package SC reports an asymptotic, χ^2cdf-based p-value of 0.083. Because of the small sample sizes overall and in the table cells, this asymptotic result might be inaccurate.

SC has an exact test function for the two-way table of frequencies, ecta(). This function uses a sophisticated network algorithm to track through the alternative tables that would be permissible under the null hypothesis and to determine the probability of obtaining a table at least as removed from independence as the observed table. The exact p-value for this table is 0.143. The asymptotic result, in this case, overstates the case against the null hypothesis.

Rerandomization Test of Population Independence

The network algorithm is more sophisticated than we can explain here. However, we can demonstrate a Monte Carlo approximation to the exact test. The exact test we shall approximate, the one carried out by ecta(), is the most widely used. This test is referred to as a conditional test because the tables considered permissible under the null hypothesis are restricted to those that have the same row and column totals as the original table. This restriction is mildly controversial, but it makes sense in many experimental designs.

In the final section of Concepts 12, we established that the rerandomization approach to hypothesis testing could be extended to test population hypotheses. What is required for the extension is that we be able to take as our null hypothesis that we have random samples from two or more identical population distributions. We have eschewed that null hypothesis heretofore as too broad. When you sample from binomial or multinomial distributions, however, identicalness often is an appropriate null hypothesis.

We'll develop our rerandomization test as if we had drawn independent random samples from I multinomial distributions, one for each row category. That is not a formal requirement. Our rerandomization test works just as well under either of the other two sampling schemes.

```
'Chi-Sq (Monte Carlo) Exact p-value
'
copy (2 0 1) row1              'Input section reads in a table
copy (0 3 1) row2             'of frequencies by rows,
copy 2000 monte              'defines the reference dist'n size,
size row1 ncols
copy 2 nrows                  'saves the number of cols and rows,
sum row1 ndot1
sum row2 ndot2
concat ndot1 ndot2 rowtots    'computes row and column totals,
add row1 row2 coltots         'and reforms the frequencies as a
concat row1 row2 nvec         'vector, row following row
'
```

Figure 23.1: *Input Phase of Exact Chi-Squared Test (Resampling Stats)*

Figures 23.1 through 23.5 show a Resampling Stats script for approximating the exact p-value for a X^2 statistic computed from a two-way table of frequencies. With the exception of the input phase (Figure 23.1), the script can be reused with any size table. The balance of the script assumes only that the input phase has prepared the following

monte	The number of rerandomized tables to be evaluated under the null hypothesis
nrows	The number of rows in the table
ncols	The number of columns in the table

rowtots   A vector of row totals from the table

coltots   A vector of column totals from the table

nvec   A vector of observed cell frequencies, the first row of the
           table followed by the second row, and so on

```
sum coltots tabtot              'Total sample size, n
divide coltots tabtot scal      'J-element vector, elements n(.j)/n
copy 0 i
repeat nrows
  add i 1 i
  take rowtots i x
  multiply x scal y             'm(ij) for the ith row of table
  if i=1
    copy y mvec
  end
  if i>1
    concat mvec y mvec
  end
end
subtract nvec mvec x            'vector: n(ij)-m(ij)
square x x                      'differences squared and
divide x mvec x                 'divided by m(ij),then
sum x chisq                     'summed to produce X^2
print chisq
'
```

Figure 23.2: *The Modeled Frequencies and X^2 Statistic (Resampling Stats)*

The second portion of the script, shown in Figure 23.2, first computes mvec, a second vector of $(I \times J)$ cell frequencies, this time as modeled under the null or independence hypothesis. The entries in this vector are organized in the same order as the observed cell frequencies are organized in nvec.

The script uses these two vectors to compute the X^2 statistic, which is then printed to the output screen.

Under the general independence hypothesis, cases can be rerandomized subject only to the restriction that the row and column totals for a rerandomized table are the same as for the original table. We can create such tables in a variety of ways. Some are more complicated to implement than others, but they are all equivalent in the sense that building a long random sequence of tables will give the same reference distribution for our X^2 statistic whatever the method of table construction. Thus assured, we choose one of the simpler approaches to null table construction.

Our chosen algorithm appears to keep each case in its original column, while randomly reallocating cases among the rows. This is only an appearance because the algorithm does not work at the case level, but does ensure that row and column totals are preserved during the rerandomization.

```
copy 0 j
repeat ncols
  add j 1 j
  take coltots j x          'x holds n(.j)
  set x j y                 'y holds n(.j) copies of j
  if j=1
    copy y indvec
  end
  if j>1
    concat indvec y indvec  'at end, indvec holds n(.1) 1s,
  end                       'n(.2) 2s, etc. through n(.j) js
end
'
```

Figure 23.3:   *Preparing the Column Indicator for Resampling Stats*

The portion of the Resampling Stats script in Figure 23.3 creates a column-indicator vector. This is an n-element vector, each element of which contains the name of a column. For our marble sample, the contents of indvec would be (1 1 2 2 2 3 3). That is, two cases are to be rerandomized to column 1, three to column 2, and two to column 3, in agreement with the observed table of frequencies.

Before beginning our resampling we have one last bit of housekeeping to deal with. We've created several n-element vectors. We need to know how to create the successive rows of a table from that vector. In Figure 23.4, our script creates two vectors top and bot, each with I elements. The $n_{i.}$ cases for the i-th row will be taken from vector positions top(i) through bot(i).

The final portion of the script, Figure 23.5, describes the construction of a table of frequencies, randomly cross-classifying the n cases while maintaining the row and column totals. Randomness is induced by shuffling the indvec vector. The first $n_{1.}$ cases are then allocated to the first row of the table, carrying their column assignments with them. The next n_2 cases from the shuffled vector are allocated to the second row, and so forth.

```
copy 1 top                    'top(1) contains 1
take rowtots 1 bot            'bot(1) contains n(1.)
subtract nrows 1 rm1
copy 1 i
repeat rm1
  copy i im1
  add i 1 i
  take top im1 told
  take bot im1 bold
  take rowtots i rsiz         'for i>1
  add bold 1 tnew             'top(i) = bot(i-1)+ 1
  add bold rsiz bnew          'bot(i) = bot(i-1)+ n(i.)
  concat top tnew top
  concat bot bnew bot
end
'
```

Figure 23.4:   *Row Divisions for the Case Vector (Resampling Stats)*

The vector `ivec` contains the cell frequencies for the randomized table. They are arranged in the same sequence as was used in creating `nvec` and `mvec`. Because the row and column totals for the randomized table are the same as for the observed table, the vector `mvec` still contains the modeled—under the independence hypothesis—cell frequencies.

The vectors `ivec` and `mvec` provide what is needed to compute `stat`, the X^2 statistic for the randomized table. The script keeps track of the proportion of X^2 statistics computed from null-randomized tables that are as large or larger than the one computed for the observed table of frequencies.

Earlier, we learned that the exact p-value for the test of row-by-column independence for the table of frequencies given in Table 23.1 was 0.143. Four successive runs of the Resampling Stats script produced Monte Carlo approximations to this p-value of 0.136, 0.152, 0.142, and 0.149. Each was based on a random sequence of 5,000 null-randomized tables. If your statistical package provides an exact test for the two-way table, use it. If not, you can approximate the associated p-value using the algorithm just demonstrated.

```
copy 1 tail
subtract monte 1 mm1
repeat mm1                        'for each of (R-1) resamples
  shuffle indvec indvec          'randomize case assignments
  copy 0 i
  repeat nrows
    add i 1 i
    take top i ti
    take bot i bi
    take indvec ti,bi x          'identify cases for the ith row
    copy 0 j
    repeat ncols
      add j 1 j
      count x=j cell             'determine cell counts in ith row
      score cell ivec            'and build up IxJ table vector
    end
  end
  subtract ivec mvec x           'compute X^2 statistic in the
  multiply x x x                 'independence hypothesis resample
  divide x mvec x
  sum x stat
  if stat >= chisq               'if at least as large as observed
    add tail 1 tail              'statistic, increase tail count
  end
  clear ivec                     'ready for next resample
end
divide tail monte pval           'compute and report p-value
print pval                       'for the chi-squared statistic
```

Figure 23.5: *Resampling Stats, An Exact p-Value on Null Hypothesis Frequencies*

Randomized Row Assignment

We have described the chi-squared test of row-by-column independence in the context of having either a single random sample of cases or separate random samples for each row category or separate random samples for each column category.

However, the algorithm we used to approximate the exact test p-value carries with it the suggestion that the test for independence might also be valid when a set of available cases is randomly allocated among row categories. As noted earlier, we can interpret the algorithm as one that, under the null hypothesis, rerandomizes cases among rows, the cases carrying with them their column categorizations. This scenario fits a design in which:

1. The row categories correspond to alternative treatments.

2. The column categories correspond to treatment responses.

3. We fix the number of cases to be randomly assigned to each treatment, the row totals.

4. Our null treatment hypothesis is that the individual case would fall in the same response category whatever treatment it was randomized to.

Under this null treatment-effect hypothesis, we can rerandomize cases among row categories or treatments. And, as cases carry their responses with them on rerandomization, the column totals are fixed under rerandomization to be the same as in the original experiment.

When we randomize available cases among treatments, rather than draw random samples, our hypotheses must change. We no longer entertain hypotheses about populations of potential respondents but, rather, about the responses of individual cases making up the available set.

Table 23.2: *Post-Surgical Incidence of Dumping Syndrome*

Surgical Procedure	Dumping Syndrome		
	None	Slight	Moderate
A	18	6	1
B	18	6	2
C	13	13	2
D	9	15	2

Consider the data in Table 23.2 taken from research reported in Forthofer and Lehnen (1983). At one of four hospitals participating in the study, candidates for duodenal ulcer surgery were randomized (we assume) among four surgical procedures ranging in radicalness from least (A) to most (D): A, drainage and vagotomy; B, 25% resection and vagotomy; C, 50% resection and vagotomy; and D, 75% resection. Note that a 25% resection excises 25% of the duodenum.

One undesirable outcome of the surgery is the occurrence of the Dumping syndrome. Table 23.2 reports the classification of each patient as exhibiting none, a slight amount, or a moderate amount of the syndrome.

```
> chisq(dsynd)
   Chisquare=   10.921349    df=  6   / 0.090837881 /
> ecta(dsynd)
   0.060988632
```

Figure 23.6: *Test of Common Dumping Syndrome Outcomes (SC)*

For these patients, do the four procedures differ in dumping syndrome outcome? Figure 23.6 shows an extract from a SC session; dsynd is the name of the 4×3 matrix of frequencies from Table 23.2. The X^2 statistic takes the value 10.92. The asymptotic p-value, based on the cdf for the χ_6^2 random variable, is 0.091, but the exact p-value is 0.061. This latter might not be quite small enough to overturn the null treatment hypothesis—that a patient would exhibit the same degree of dumping syndrome following any of the four surgical treatments.

The chi-squared statistic treats the row and column categories as purely nominal, any ordering of categories is ignored. We could change the order of columns or rows or both in the table and the same X^2 statistic would result. In the dumping syndrome example, however, the row categories are ordered, by radicalness of surgery, and the column categories are ordered, by extent of dumping syndrome response. The chi-squared statistic would not be the best choice if our substantive or alternative hypothesis took these orderings into account. In fact, the researchers expected that the severity of dumping syndrome would increase with the radicalness of the surgical procedure.

One alternative statistic that does account for the ordering of row and column categories is the Goodman-Kruskal gamma. Each pair of cases is classified as either concordant, discordant, or tied. A pair of cases is classified as concordant if one member of the pair ranks higher than the other on both row and column attributes. The pair is classified as discordant if one case ranks higher than the other case on the row attribute, while the second case ranks higher than the first on the column attribute. The pair is tied if both members have the same rank on at least one of the two attributes. The Goodman-Kruskal γ is the ratio

$$\gamma = \frac{P_C - P_D}{P_C + P_D}$$

where P_C and P_D are, respectively, the number of concordant and discordant pairs (Agresti, 1990). Like the product moment correlation coefficient, γ takes values between -1.0 and $+1.0$, depending on whether the association is negative or positive.

```
> gkg(sgy,dmp)
    gamma=  0.38547969
    t=  3.1227587    / 0.99837994 / 0.0016200597 /


> shuffl_k(pergkg,1,sgy,nt,ns,999,RR)

    Full Sample Statistics (theta):    0.38547969

P-values for Statistic Number:  1
    Percent LE theta[ 1 ]:   99.9
    Percent GE theta[ 1 ]:    0.2
    Percent LE abs(theta[ 1 ]):   99.4
    Percent GE abs(theta[ 1 ]):    0.7
```

Figure 23.7: *Test for Positive Association with γ Statistic (SC)*

In Figure 23.7, we first use the SC function gkg() to find $\gamma = 0.385$ for the surgery/dumping table data. The two vectors, sgy and dmp, contain the ordered scores from Table 23.2 for each of the 105 patients on surgical radicalness (from 1 to 4) and on dumping severity (from 1 to 3). We next use the rerandomization shuffl_k() to create a null reference distribution for γ. The user-written procedure pergkg computed γ for each of 999 rerandomizations of the sgy assignments against the dmp outcomes. Only 2 of the 1,000 elements of the resulting null reference distribution (p = 0.002) were as large or larger than 0.385. The null hypothesis of no association can be rejected in favor of the alternative hypothesis of a positive association.

For γ, as for X^2, there is an asymptotic sampling distribution, and the SC function gkg reports a large sample p-value of 0.0016, very close to our Monte Carlo approximation of the appropriate exact p-value.

Nonrandom Cross-Classifications

X^2 and G^2 are both test statistics. We cannot tell from their size alone whether they are large or not. We need a null sampling or reference distribution to gauge their size. As a result, the two statistics are ill-suited simply to describe the extent of the relatedness of column and row categories in a two-way table.

Table 23.3 is taken from a study of 30 U.S. and U.K. men who have chosen not to have children (Lunneborg, 1999). The 30 men are not a random sample. In the table, they are cross-classified by two attributes:

1. The row attribute indicates whether they regard their dads as good, disinterested, or abusive.

2. The column attribute indicates whether they report that Dad did or did not influence their decision not to have children.

Table 23.3:   *Fathers and Their Influence on Fathering*

	Influenced Decision		
	Yes	No	Total
Good Dads	2	11	13
Disinterested Dads	6	4	10
Abusive Dads	7	0	7
Total	15	15	30

For a small table such as this, we might not need to summarize any relation with a statistic; the table might say it all. It should be reasonably clear that the direction of the tabled frequencies cannot be overly influenced by any particular case.

There are association statistics that can be computed for a table like this. The Goodman and Kruskal tau, for example, has an interpretation similar to R^2 in multiple linear regression; it assesses the proportional reduction in uncertainty about the column classification of cases when one considers the row classification rather than just the column totals (Agresti, 1990). Our statistical package SC reports $\tau = 0.45$ for Table 23.3, a 45% reduction in uncertainty.

The stability of this description can be investigated using the subsampling approach of Concepts 16. Some thought should be given to subsampling small data sets in contingency table analyses. Half-samples can be quite small and, because the association statistic is sensitive to the distribution of small frequencies over several table cells, considerable variability might be expected if the subsamples are reduced to half-sample size.

```
> subsamp(substau,1,XX',25,100,ST)
   Full Sample Statistics:   0.45435897

   Description of random selection of subsamples:   100
   range=        0.331909   ( 0.347578  to  0.679487 )
   mean=         0.456171
   median=       0.458181
   sd=           0.0670809
   iqr=          0.107774   ( 0.394672  to  0.502446 )
   MAD=          0.0450262

> subsamp(substau,1,XX',15,100,ST)
   Full Sample Statistics:   0.45435897

   Description of random selection of subsamples:   100
   range=        0.833333   ( 0.166667  to  1 )
   mean=         0.498982
   median=       0.459821
   sd=           0.172477
   iqr=          0.208615   ( 0.383221  to  0.591837 )
   MAD=          0.109694
```

Figure 23.8:   *Subsample Stability of τ for Fathers Influence (SC)*

In Figure 23.8, we show the results of two subsample stability analyses. We first ask SC randomly to omit 5 cases each time. The observed value of τ is quite near the median of the 100 subsample τ values. The full range of the 100 was from 0.35 to 0.68. However, when we ask SC randomly to omit one-half of the cases, reducing the subsample sizes to 15, distributed over five of the six table cells, the range increases considerably, from 0.17 to 1.0, although the median of the 100 half-sample statistics remains very near the full data set description. Either analysis should provide sufficient evidence that the τ of 0.45 is not materially the result of one or a few cases.

The 2 × 2 Table

The 2×2 table is a special form of the $I \times J$ table of cross-classified frequencies. With each attribute at only two levels, the Goodman-Kruskal gamma measure of association simplifies to a statistic earlier christened Yule's Q,

$$Q = \frac{n_{11}n_{22} - n_{12}n_{21}}{n_{11}n_{22} + n_{12}n_{21}}$$

Table 23.4 is taken from Balding and Shelley (1993) and summarizes the responses of an opportunity-sample of 1991–1992 English primary school children in Year 4 classes (8 to 9 years of age) to the question "What was the time when you went to bed last night?"

Table 23.4:   *Bedtimes for English School Children*

| | Bedtime | | |
	Before Nine	After Nine	Total
Girls	488	191	679
Boys	371	285	656
Total	859	476	1335

For this set of nonrandomized, available cases, Yule's Q takes the value

$$Q = \frac{(488 \times 285) - (191 \times 371)}{(488 \times 285) + (191 \times 371)} = 0.3249$$

How stable is this description? The children answering the bedtime question were pupils in schools that had volunteered to administer the Primary Health Related Behaviour Questionnaire. The most appropriate subsampling analysis of stability would recompute Q, leaving out one school at a time, thus assessing whether a particular school had a strong influence on the overall description. Given only the data in Table 23.4, we cannot do that.

Where the rows of the 2×2 table are treatments, diagnostic categories, or other explanatory attributes and the columns correspond to alternate outcomes or responses, a useful summary statistic for the table is the odds ratio.

We saw an example of this earlier, in Applications 13. There Table 13.2, summarized the outcome of a study comparing radiation therapy with surgery in the treatment of cancer of the larynx (Agresti, 1990).

Table 13.2: *Outcomes of a Randomized Treatment Study*

| | Cancer | |
Treatment	Controlled	Not Controlled
Surgery	21	2
Radiation	15	3

The obtained odds of success for surgery, the ratio of controlled to not controlled cancers, were $(21/2) = 10.5$. The obtained odds of success for radiation were $(15/3) = 5$. The odds ratio in favor of surgery is the ratio of these two odds, that for surgery in the numerator,

$$s = (10.5/5) = 2.1$$

The odds of success following surgery were about twice those following radiation.

In our analysis of these data, we assumed the patients available for the study were randomized independently to either surgery or radiation. This assumption told us how to rerandomize patients under the null treatment hypothesis that each patient's response to treatment would have been the same under either treatment. From a long random sequence of such rerandomizations, we then developed a null reference distribution we could use to assess the significance of our odds ratio.

We cannot tell from the data alone whether our assumptions about how the study was carried out were correct and, hence, whether our analysis was appropriate. To carry out an appropriate analysis we would need to know

1. If the patients constituted a random sample or simply a group of available cases

2. If the patients were assigned to the alternate therapies randomly or on a nonrandom basis

3. What random assignment mechanism was used, if therapy assignment was random

Can we estimate a CI for the odds ratio? We'll need randomly sampled cases for that. Can we compute a p-value for the odds ratio? If the cases aren't randomly sampled, they must be randomly assigned to alternate treatments if we are to do so.

Logistic Regression

Linear regression is widely used as a way of modeling the dependence of a measured response attribute on one or more explanatory attributes. In Applications 22, we outlined resampling approaches to many linear model data analysis questions.

When the response attribute is binary rather than measured, that is, when the response must take one of two values or fall into one of two categories, the dependence model of choice is logistic regression. We develop it first in its parametric, multipopulation version.

Parametric Logistic Regression

Let z be an explanatory attribute that takes ℓ (≥ 2) distinct values, $\{z_{[1]}, z_{[2]}, \ldots, z_{[\ell]}\}$. For each of these values, there is a population distribution of response attribute scores. There are only two distinct values in each of these population distributions. We'll name the two values Success and Failure. The proportion of successes in the population response distribution associated with explanatory attribute score $z_{[i]}$ is $\pi_{[i]}$. The proportion of failures in that population distribution, then, must be $\left(1 - \pi_{[i]}\right)$.

The logistic regression model links the proportion of successes to the value of the explanatory attribute in a certain way. This linkage can be expressed by the equation

$$\log\left(\frac{\pi_{[i]}}{1 - \pi_{[i]}}\right) = \beta_0 + \beta_1 z_{[i]}$$

The ratio of the proportion of successes to the proportion of failures is the odds of success. Logistic regression models the log of the odds of success as a linear function of the explanatory attribute score.

Equivalently, if we exponentiate on both sides of this equation, we can write the same model as

$$\left(\frac{\pi_{[i]}}{1 - \pi_{[i]}} \right) = e^{\beta_0 + \beta_1 z_{[i]}} = e^{\beta_0} e^{\beta_1 z_{[i]}} = \gamma_0 \, \gamma_1^{z_{[i]}} = \gamma_0 \, \gamma_1^{(z_{[i]} - 1)} \, \gamma_1$$

where $\gamma_0 = e^{\beta_0}$ and $\gamma_1 = e^{\beta_1}$. In this multiplicative form of the logistic regression model, the odds of success appear on the left side of the equation, and γ_1 is known as an odds multiplier. An increase of one unit in the value of z multiplies the odds by γ_1. The βs of the linear log-odds model and the γs of the multiplicative-odds model are logically related:

1. If $\beta_1 = 0$, then $\gamma_1 = 1$

2. If $\beta_1 > 0$, then $\gamma_1 > 1$

3. If $\beta_1 < 0$, then $0 < \gamma_1 < 1$

The linear log-odds form of the logistic regression model can also be rewritten to express $\pi_{[i]}$ itself as a function of $z_{[i]}$:

$$\pi_{[i]} = \frac{e^{(\beta_0 + \beta_1 z_{[i]})}}{1 + e^{(\beta_0 + \beta_1 z_{[i]})}}$$

The πs, βs, and γs are all unknown population parameters. Based on a random sample of cases at two or more design points, these parameters can be estimated.

Typically, the βs of the linear log-odds model are estimated. Under the logistic regression model assumptions, maximum likelihood estimates can be found for these parameters. These $\widehat{\beta}$s can then be plugged into either of the last two equations to estimate the γs or πs.

Table 23.5 is taken from Lunneborg (1994) and gives systolic blood pressures (SBP) for a random sample of clinic patients who are also characterized as free of coronary heart disease (CHD) or not. In this population, how do the odds of coronary heart disease change with an increase in blood pressure?

Table 23.5:   *Systolic Blood Pressures of CHD-Free and CHD Patients*

Systolic Blood Pressures	
CHD Free:	135 122 130 148 146 129 162 160 144 166
	138 152 138 140 134
CHD Present:	145 142 135 149 180 150 161 170 152 164

Figure 23.9 describes an S-Plus session in which we use the sample distribution of scores in Table 23.5 to fit the logistic regression model

$$\log(\text{Odds of CHD})_i = \beta_0 + \beta_1 \, \text{SBP}_i$$

The log-odds of CHD in the i-th patient is modeled as a linear function of that patient's SBP.

By specifying the distribution of our response attribute to be binomial, we can use the S-Plus generalized linear function glm() to fit a logistic regression. The estimates of the β-parameters give this fitted model:

$$\log\left(\frac{\widehat{\pi}_i}{1 - \widehat{\pi}_i}\right) = -10.790 + 0.070 \, \text{SBP}_i$$

Asymptotically, the $\widehat{\beta}$s have sampling distributions with normal cdfs. S-Plus reports a large-sample standard error (SE) estimate for $\widehat{\beta}_1$ of 0.036. If the asymptotic sampling distribution properties held true when $n = 25$, then, under the null hypothesis that $\beta_1 = 0$, the Studentized form,

$$t(\widehat{\beta}_1) = \frac{\widehat{\beta}_1 - \beta_{1(0)}}{\widehat{\text{SE}}(\widehat{\beta}_1)} = \frac{0.06988}{0.03614} = 1.93$$

is an observation from the t-random variable with $(n - 2) = 23$ df.

The t-random variable results are untrustworthy for small samples, even when model assumptions are correct. A more reliable test for practical sample sizes is given by a likelihood-ratio test (Agresti, 1996). This test utilizes a statistic called the deviance difference.

```
> sbp<- c(135,122,130,148,146,129,162,160,144,166,138)
> sbp<- c(sbp,152,138,140,134,145,142,135,149,180,150)
> sbp<- c(sbp,161,170,152,164)
> chd<- c(rep(0,15),rep(1,10))
> logr1<- glm(chd~sbp,family=binomial)
> summary(logr1)

Call: glm(formula = chd ~ sbp, family = binomial)
Deviance Residuals:
      Min        1Q     Median        3Q       Max
 -1.534934 -0.8879167 -0.5777831 0.9890732 1.780715
Coefficients:
                  Value Std. Error    t value
(Intercept) -10.78990521 5.40796279 -1.995189
        sbp   0.06987895 0.03614137  1.933489
(Dispersion Parameter for Binomial family taken as 1)
    Null Deviance: 33.65058 on 24 degrees of freedom
Residual Deviance: 29.0325 on 23 degrees of freedom
Number of Fisher Scoring Iterations: 3
Correlation of Coefficients:
    (Intercept)
sbp -0.9965492

> 1-pchisq((33.65058-29.0325),df=1)
[1] 0.03163663

> exp(0.06987895)
[1] 1.072378

> (exp(0.06987895))^10
[1] 2.011317
```

Figure 23.9: *Fit of the SBP-CHD Logistic Regression Model (S-Plus)*

The deviance provides an assessment of the goodness of fit of a logistic regression model and is computed as

$$\text{Dev(SBP Model)} = G^2(\text{SBP Model}) = 2 \sum_{i=1}^{n} y_i \log\left(\frac{y_i}{\widehat{\pi}_i}\right)$$

where y_i is either 1 if the i-th patient in the sample had CHD, or 0 if the i-th patient was free of CHD, and $\widehat{\pi}_i$ is the model estimate of the proportion of CHDs in the population of patients with the same SBP as the i-th patient. The likelihood ratio chi-square statistic is applied at the case rather than at the cell level.

The deviance difference compares the goodness of fit of an unconstrained model with that of a model in which the β-parameters have been constrained by a null hypothesis. In this instance, the null hypothesis constrains β_1 to be zero. This gives the constrained logistic regression model

$$\log\left(\frac{\widehat{\pi}_i}{1 - \widehat{\pi}_i}\right) = \beta_0$$

a model known as the null model because there are no longer any explanatory attributes in the model.

The deviance difference test statistic for the hypothesis that $\beta_1 = 0$, then, is

$$\text{Dev(Null Model)} - \text{Dev(SBP Model)} = 33.65058 - 29.0325 = 4.618$$

based on the values reported by S-Plus. If the null hypothesis is true and the logistic regression model assumptions are met, this deviance difference is a value in the distribution of the chi-squared random variable with df equal to the number of constraints imposed by the null hypothesis. Here, there is just one constraint and S-Plus computes the proportion of the χ_1^2 distribution with values of 4.618 or larger at 0.0316.

That p-value might be small enough for us to conclude that, in the population sampled, the influence of an increase in SBP on CHD incidence is real. How big is the influence? The estimated odds multiplier for SBP is computed from the model fit as

$$\widehat{\gamma}_1 = e^{\widehat{\beta}_1} = e^{0.0699} = \exp(0.0699) = 1.07$$

Increasing SBP by one unit, 1 mm Hg, increases the odds of CHD by 7%. Increasing SBP by ten units—say, from 155 mm Hg to 165 mm Hg—doubles the odds of CHD: $1.07^{10} = 2.01$.

We have introduced logistic regression with a single explanatory attribute. As in linear regression, the models for log-odds can include a mix of measured and categorical attributes on the explanatory side. Also as in linear regression, the logistic regression study design can call for fixed or sampled design points.

Nonparametric Logistic Regression

Statistical inference for parametric logistic regression models rests on the assumption that either the current model or an associated null model is correct. This means that, in the real world, the proportion of successes actually is a linear function of the explanatory attributes we've chosen to measure.

In reality, our regression models are almost always wrong. They are very likely incomplete; we can't have measured everything we should have. Nonetheless, these models can provide a useful basis for summarizing relations, testing certain scientific hypotheses, and predicting the responses of future cases.

In describing the parametric logistic regression, we offered no explanation for the apparently arbitrary way in which the model formulates the dependence of the success proportions in the sampled populations on the values of k (≥ 1) explanatory attributes:

$$\log\left(\frac{\pi_i}{1 - \pi_i}\right) = \beta_1 + \sum_{j=2}^{k+1}\beta_j z_{ij}$$

Why is the linear model for the log-odds of success, rather than for the odds of success or for the proportion of successes? In the parametric context, this choice facilitates the mathematics of finding good estimates, that is, maximum likelihood estimates, of the linear model parameters, the βs.

When we take a nonparametric view, we no longer need this aid to estimation. There are, however, good reasons for maintaining the log-odds link to the binary response in our linear model:

1. Both $\pi_{[i]}$ and $\left[\pi_{[i]}/(1 - \pi_{[i]})\right]$ are limited in the values they can assume. The first can take values only between 0 and 1. The second can take only positive values. The log-odds, however, is unbounded. It can take any value, positive or negative. This means that we are free to find the best estimates of the β-parameters without worrying that they will produce impossible proportion or odds estimates from acceptable values of z (Agresti, 1996).

2. If successes and failures are the result of dichotomizing an underlying measured response attribute with a normal cdf, then the log-odds transformation of $\pi_{[i]}$ restores the linear relation that was present before dichotomization (Lunneborg, 1994).

3. We might have substantive reasons for believing that an increase in the value of an explanatory attribute will be accompanied by a multiplicative increase in the odds of success rather than by an additive increase. By fitting the log-odds model, we can estimate, by exponentiation, the requisite odds multipliers (Forthofer and Lehnen, 1983).

In Applications 22, we estimated—or computed, when we did not have random samples of cases—the linear model parameters using least squares. That is, we chose values $\widehat{\beta}_0$ and $\widehat{\beta}_1$ to minimize, over our n cases, the average squared difference between the observed response attribute score for a case, y_i, and the score predicted for that case by the fitted linear model, $\widehat{y}_i$. This gave us (a) maximum likelihood estimates for the parametric-LM, (b) plug-in estimates of relevant population parameters in the bootstrap nonparametric linear model, and (c) good-fitting models for available case data.

The technique commonly used to find maximum likelihood estimates of the parametric log-odds linear model parameters is a variation on least squares termed iteratively reweighted least squares (IRLS). Again, the overall goal is to minimize, in some sense, the squared differences between observed responses, 1s for successes and 0s for failures, and model estimates of the probability of a success, $\widehat{\pi}$s taking values between 0 and 1.

The parametric-LM assumed that the variance in each response distribution would be the same, though the means would differ. This variance homogeneity cannot be true for binary response populations. The variance of the binary 0/1 distribution decreases as the mean of that distribution, π, moves away from 0.50. It is sensible to weight more heavily cases sampled from distributions with smaller variances. Thus, we should use weighted least squares rather than ordinary least squares in finding parameter estimates. But the case weights, dependent on the π_is, are as unknown as the βs. Thus the need is for iteratively reweighting. In effect, IRLS alternates between estimating the βs based on a current set of case weights and revising the case weights based on those updated β estimates. The alternation continues until both parameter estimates and case weights are unchanging.

We can use the IRLS algorithm outside the bounds of the parametric logistic regression model in the same way that we extended the use of ordinary least squares for fitting linear models in Applications 22. We begin by using IRLS to obtain plug-in estimates of population parameters, facilitating bootstrap population inference.

Bootstrap BCA Confidence Bounds

Classical population inference for logistic regression model parameters is hampered by its large sample requirements. Asymptotically, $\widehat{\beta}_j$ may have a normal cdf sampling distribution with estimable SE, but that does not help us estimate confidence bounds for β_j when our sample sizes are moderate rather than large. Our bootstrap BCA estimates do not rely on asymptotic assumptions about sampling distribution cdfs and can be of use here. As well, we can estimate CIs for population characteristics for which there are not even classic asymptotic results.

In Figure 23.10, we return to the SBP model for CHD. We use the `bootstrap()` command in S-Plus to find a lower confidence bound for γ_1, the odds mulitplier for SBP in the logistic regression model. Our estimate of that parameter (Figure 23.9) was 1.07. How confident can we be that γ_1 is greater than 1.0?

We also estimate CIs for two population proportions, $\pi_{[160]}$, the proportion of those cases with SBP of 160 mm Hg who have CHD, and $\pi_{[180]}$, the proportion of those cases with SBP of 180 mm Hg who have CHD, as well as a lower confidence bound for the difference, $\pi_{[180]} - \pi_{[160]}$.

The one-time function `app23ci()` extracts $\widehat{\beta}_1$ and computes from the $\widehat{\beta}$s the estimated proportions, $\widehat{\pi}_{[160]}$ and $\widehat{\pi}_{[180]}$, and the estimated difference between those proportions, $\widehat{\pi}_{[180]} - \widehat{\pi}_{[160]}$.

Applying this function to the original sample data of Figure 23.9 yields:

$$\widehat{\beta} = 1.072, \ \widehat{\pi}_{[160]} = 0.596, \ \widehat{\pi}_{[180]} = 0.857, \text{ and } \left[\widehat{\pi}_{[180]} - \widehat{\pi}_{[160]}\right] = 0.260$$

Based on a random sequence of 2,000 bootstrap samples from our nonparametric estimate of the bivariate population distribution of CHD and SBP, we can be 95% confident that the odds multiplier for SBP is greater than 1.008 and 97.5% confident that the difference in proportions, $\pi_{[180]} - \pi_{[160]}$, exceeds 0.127. We can also be 95% confident that $\pi_{[160]}$ falls within the interval [0.282, 0.879] and that $\pi_{[180]}$ falls within the interval [0.336, 0.993].

Model-Based Binary Population Estimates

In Concepts 7, we looked briefly at alternatives to the nonparametric estimation of population distributions. One such was the model-based estimate.

Model-based estimation is an alternative when the response distribution is binary. Consider this example.

```
> app23ci
function(X)
{
    tmp <- glm(X[, 1] ~ X[, 2], family = binomial)
    aa <- exp(tmp$coef[2])
    lo160 <- tmp$coef[1] + (tmp$coef[2] * 160)
    lo180 <- tmp$coef[1] + (tmp$coef[2] * 180)
    bb <- exp(lo160)/(1 + exp(lo160))
    cc <- exp(lo180)/(1 + exp(lo180))
    dd <- cc - bb
    out <- c(aa, bb, cc, dd)
    out
}
> chdsbp<- cbind(chd,sbp)
> app23ci(chdsbp)
   X[, 2] (Intercept) (Intercept) (Intercept)
 1.072378    0.5964576   0.8567194    0.2602618
> logregci<- bootstrap(chdsbp,app23ci,B=2000)
> limits.bca(logregci)
                    2.5%        5%        95%       97.5%
     X[, 2] 0.9973459 1.0088128 1.1412844 1.1652008
(Intercept) 0.2827135 0.3385413 0.8350301 0.8783332
(Intercept) 0.3369416 0.4550157 0.9844018 0.9922932
(Intercept) 0.1273630 0.1635136 0.4674200 0.5296237
```

Figure 23.10: *S-Plus BCA CIs:* γ_1, $\pi_{[160]}$, $\pi_{[180]}$, *and* $\left[\pi_{[180]} - \pi_{[160]}\right]$

We have a random sample of n cases and for each we have a trivariate score: (y_i, x_{i1}, x_{i2}), $i = 1, \ldots, n$. y_i is the i-th case's response attribute score and takes a value of 0 or 1, and x_{i1} and x_{i2} are scores for that case on explanatory attributes x_1 and x_2.

Our nonparametric estimate of the underlying trivariate population distribution is a collection of copies of these n trivariate scores. When we sample from this estimated distribution and draw the i-th case, we bring the trivariate score for this case, (y_i, x_{i1}, x_{i2}), into our bootstrap sample. Each time the i-th case is selected, the response score for that case is fixed at 0 or 1, whatever response score accompanied the i-th case in the real-world sample.

Now let's say that we are certain that the binary population distributions from which we have sampled can be described by the logistic regression model,

$$\log\left(\frac{\pi_i}{1 - \pi_i}\right) = \beta_0 + \beta_1 x_{i1} + \beta_2 x_{i2}$$

We can use that knowledge to generate a model-based estimate of the trivariate population distributions.

1. Use the real-world sample data to fit the logistic regression model, providing estimates of the β-parameters: $\widehat{\beta}_0$, $\widehat{\beta}_1$, and $\widehat{\beta}_2$.

2. For each of the n cases in the real world sample, compute an estimated proportion of successes using the formula

$$\widehat{\pi}_i = \frac{e^{\widehat{\beta}_0 + \widehat{\beta}_1 x_{i1} + \widehat{\beta}_2 x_{i2}}}{1 + e^{\widehat{\beta}_0 + \widehat{\beta}_1 x_{i1} + \widehat{\beta}_2 x_{i2}}}$$

3. Compose the model-based population distribution, $\widehat{X}$, as a collection of copies of n trivariate scores, each of the form $(\widehat{\pi}_i, x_{i1}, x_{i2})$.

4. When resampling from $\widehat{X}$, replace $(\widehat{\pi}_i, x_{i1}, x_{i2})$ either with $(1, x_{i1}, x_{i2})$ or with $(0, x_{i1}, x_{i2})$ depending on the following: Generate a random number from the uniform distribution over the interval $(0,1)$. If the random number is less than or equal to $\widehat{\pi}_1$, choose $(1, x_{i1}, x_{i2})$, otherwise choose $(0, x_{i1}, x_{i2})$.

This model-based approach acknowledges the uncertainty of the response score for a case with explanatory attribute scores (x_{i1}, x_{i2}). Some cases in the population with these scores will have a response of 1, others a response of 0.

Logistic Regression in Randomized Studies

We consider now studies in which we no longer have one or more random samples of cases. Rather, we have a set of available cases randomly distributed over K treatment groups. For example, in the study summarized in Table 23.6 and taken from Hand et al. (1994, Data Set 19), 960 plum cuttings made in the autumn were the available cases. The cuttings were randomly divided into two groups of 480. One group was left intact (long) and one group was trimmed in length (short). The long group was further subdivided, at random, into two groups of 240, one to be planted immediately, and one to be planted in the following Spring. The same random division into planting time groups was made for the short cuttings. Survival status of the cuttings subsequently was recorded, and serves as the response for our analysis.

Table 23.6: *Survival of Plum Cuttings*

		Survival Status	
Length	Planting Time	Dead	Alive
Long	Immediately	84	156
	Spring	156	84
Short	Immediately	133	107
	Spring	209	31

We want to model the outcome of the study using the log-odds linear model

$$\log\left(\frac{\pi_{jk}}{1 - \pi_{jk}}\right) = \beta_0 + \beta_1\, x_j + \beta_2\, x_k$$

where π_{jk} is the modeled proportion of alive cuttings among those planted at length j (0: short or 1: long) and time k (0: Spring or 1: immediately). Correspondingly, x_j is an explanatory attribute that takes the values 0 for short and 1 for long and x_k is an explanatory attribute that takes the values 0 for Spring and 1 for immediately.

In the multiplicative version of this model

$$\left(\frac{\pi_{jk}}{1 - \pi_{jk}}\right) = \gamma_0\, \gamma_1^{x_j}\, \gamma_2^{x_k}$$

$\gamma_1 = e^{\beta_1}$ is the odds multiplier for leaving the cutting long, and $\gamma_2 = e^{\beta_2}$ is the odds multiplier for planting the cutting immediately.

We want to determine whether either of these odds multipliers is greater than 1.0 by more than a chance amount.

As a start, we use the IRLS algorithm to compute the βs and γs from the study data.

```
> surv<- c(rep(0,84),rep(1,156),rep(0,156),rep(1,84))
> surv<- c(surv, rep(0,133),rep(1,107),rep(0,209))
> surv<- c(surv,rep(1,31))
> long<- c(rep(1,480),rep(0,480))
> immed<- c(rep(1,240),rep(0,240),rep(1,240),rep(0,240))

> logcut<- glm(surv~long+immed,family=binomial)
> logcut$coef
 (Intercept)      long      immed
   -1.731407 1.017656  1.427501
> exp(logcut$coef)
 (Intercept)      long      immed
    0.1770352 2.766702  4.168271
> g<-exp(logcut$coef)
```

Figure 23.11: *Odds Multipliers for Long and Immediately Planted Cuttings (S-Plus)*

The S-Plus interaction in Figure 23.11 determines the odds multipliers to be $\gamma_1 = 2.77$ for long versus short cuttings and $\gamma_2 = 4.17$ for planting immediately rather than delaying until Spring.

The rerandomization of cuttings to assess the chances of obtaining multipliers as large or larger than these, solely as a result of random assignment, is described in Figure 23.12.

```
> shuf1<- function(x)
+ {
+ tmp<- glm(surv~long+x,family=binomial)
+ exp(tmp$coef[3])
+ }
> shgp1<- long+1
> pgam2<-
+bootstrap(immed,shuf1,sampler=samp.permute,group=shgp1)
> length(pgam2$rep[pgam2$rep>=g[3]])
[1] 0
> max(pgam2$rep)
[1] 1.773211
> 1/1001
[1] 0.000999001
```

Figure 23.12: *Rerandomization Causal Inference for Odds Multipliers (S-Plus)*

The one-time function shufl() saves γ_2^H from a rerandomization of the time-of-planting. The vector shgpl is used to control the rerandomization, ensuring that it takes place separately for the two levels of length of cutting. None of the 1,000 values of γ_2^H is as large as γ_2, and we estimate the p-value at 0.001. Similar results are found when cases are rerandomized between lengths but separately for the two times of planting. The two planting factors are, statistically speaking, very significant contributors to survival.

Odds Multipliers in Nonrandom Studies

Cialdini and Schroeder (1976) report the results of a study in which donations to the American Cancer Society were solicited from passersby by student associates of the researchers. Potential donors were invited to contribute in one of four ways:

1. Control: Standard request

2. Even a Penny: Standard request followed by "Even a penny will help."

3. Even a Dollar: Standard request followed by "Even a dollar will help."

4. Social Legitimization: Standard request modified to "We've already received some contributions, ranging from a penny up, and I wonder if you would be willing to help?"

Table 23.7:   *Compliance with Solicitations to Contribute*

Treatment	Giving	Not Giving	Total
Control	10	21	31
Even a Penny	14	16	30
Even a Dollar	20	11	31
Social Legitimization	18	13	31

There is no evidence in the report that potential donors were selected at random or randomized among the approaches. We'll treat the study as a nonrandom one. Data of interest from the study are given in Table 23.7.

```
> giv<- c(rep(1,10),rep(0,21),rep(1,14),rep(0,16))
> giv<- c(giv,rep(1,20),rep(0,11),rep(1,18),rep(0,13))
> eap<- c(rep(0,31),rep(1,30),rep(0,62))
> ead<- c(rep(0,61),rep(1,31),rep(0,31))
> socl<- c(rep(0,92),rep(1,31))

> comply<- glm(giv~eap+ead+socl,family=binomial)
> comply$coef
 (Intercept)        eap        ead      socl
  -0.7417583 0.6082269 1.339553 1.06718
> exp(comply$coef)
 (Intercept)        eap        ead      socl
   0.4762758 1.837171 3.817338 2.90717
```

Figure 23.13: Odds Multipliers for Contribution Requests (S-Plus)

The odds multipliers for the three non-control requests are worked out by S-Plus in Figure 23.13. They are:

1. $\gamma_1 = 1.84$, the odds of giving are nearly doubled by the Even a Penny request, compared with the control request.

2. $\gamma_2 = 3.82$, the odds of giving are nearly quadrupled by the Even a Dollar request, compared with the control request.

3. $\gamma_3 = 2.91$, the odds of giving are nearly tripled by the Social Legitimization request, compared with the control request.

In fact, these are coefficients for a saturated model, one that fits the odds of giving perfectly.

We could have worked out the odds multipliers directly, without the aid of IRLS estimation. For example, γ_1 is the multiplier that converts the Control odds of giving into the Even a Penny odds:

$$\left(\frac{10}{21}\right) \times \gamma_1 = \left(\frac{14}{16}\right)$$

or

$$\gamma_1 = \left(\frac{14}{16}\right) \times \left(\frac{21}{10}\right) = 1.837$$

How stable are these odds multipliers? We can recompute the γs for a series of randomly chosen half-samples and look at the resulting distributions to help us decide.

How should the half-samples be selected? The data suggest that the study was designed to include equal numbers of cases at the four treatment levels. In the language of Applications 16, the cases are design-divided and we should carry out our subsampling within those divisions, rather than overall.

```
> subs23<- cbind(giv,eap,ead,socl)
> subsfunc
function(X)
{
    ii <- 0
    tmp <- matrix(nrow = 100, ncol = 3)
    while(ii < 100) {
        ii <- ii + 1
        aa <- sample(1:31, size = 15)
        bb <- sample(32:61, size = 15)
        cc <- sample(62:92, size = 15)
        dd <- sample(93:123, size = 15)
        pick <- c(aa, bb, cc, dd)
        YY <- X[pick, ]
    xxx <-
glm(YY[,1]~YY[,2]+YY[,3]+YY[,4],family=binomial)
        tmp[ii,  ] <- exp(xxx$coef[2:4])
    }
    tmp
}

>xout<- subsfunc(subs23)
> apply(xout,2,median)
[1] 1.750000 3.999090 2.999825
> apply(xout,2,quantile,0.25)
[1] 1.312496 2.999650 2.249997
> apply(xout,2,quantile,0.75)
[1] 2.471336 5.499998 4.571309
```

Figure 23.14: Stability Analysis of Odds Multipliers (S-Plus)

The special-use S-Plus function subsfunc() listed in Figure 23.14 draws random subsamples of 15 cases from each of the four treatment levels and computes the three odds multipliers for each of these half-samples. This is repeated for 100 half-samples.

The medians of the three sets of half-sample odds multipliers are all quite near the full data set descriptors:

$\gamma_1 = 1.837$, median of half-sample estimates: 1.750

$\gamma_2 = 3.817$, median of half-sample estimates: 3.999

$\gamma_3 = 2.907$, median of half-sample estimates: 3.000

And the ranges of the middlemost 50% of the half-sample odds multipliers are entirely consistent with the full data set γs.

The descriptions would appear to be stable.

Exercises

1. Table 23.8 is taken from Hand et al. (1994, Data Set 409) and summarizes for a recent year examination results of students taking honors degrees in mathematics at English and Scottish universities.

Table 23.8:   *Mathematics Honors Degrees, England and Scotland*

Country	Sex	Degree Class				
		1	2-1	2-2	3	Total
England	Male	782	1390	1346	825	4343
	Female	339	637	785	382	2143
Scotland	Male	100	104	112	52	368
	Female	47	76	89	49	261

The four Degree Classes order the degrees by quality. A Class 1 degree is superior to a Class 2.1, which in turn is superior to a Class 2.2. And, the Class 2.2 degree is superior to a Class 3. Our interest in these data is in how the women fare relative to the men, and whether this is the same at Scottish universities as at English universities. How might the contents of Table 23.8, the population of English/Scottish honors degrees in mathematics, be described relative to these two questions?

The challenge to be met in making the description is to take account of the ordering of the classes. A technique that has been worked out for use with ordered response categories is an extension of logistic regression known as proportional odds modeling (see Lunneborg, 1994, for a fuller description). The technique is usually employed as a way of drawing population inferences from random samples, but we can use some of the modeling results for sample (or, as here, population) description. Because of the large numbers of cases in each cell of the table, the results will be quite stable—as a few half-sample replications of the analysis will attest.

Take the English university male as the control condition and include indicator variables for Scottish university and for female student.

If you do not have access to a statistical package that will fit proportional odds models, perform a series of three logistic regressions. For the first, let the success category be degree class 1, and let the failures be those in the other three classes. For the second analysis, let the success category include both degree classes 1 and 2-1. For the third analysis, let success include all but degree class 3. How consistent across these analyses are the odds multipliers for female? For Scottish universities?

Perform at least one of the analyses on a sequence of, say, three randomly chosen half-samples.

2. Table 23.9 taken from Upton (1986) summarizes the results of a study of helpfulness among pedestrians in an English shopping precinct. The design produced random samples from 8 prospective populations. Each of these populations is charactrerized by sex of the pedestrian, sex of requester, and ethnicity of requester. The samples ranged in size from 23 to 28. Pedestrians were approached by a requester who asked "Excuse me, do you have change for $5p$?" The study was intended to investigate how the helpfulness of the response was influenced by sex of pedestrian, sex of requester, and race of requester. Use logistic regression to model the log-odds of giving help as a function of these three explanatory attributes.

Use bootstrap resampling to estimate CIs for each of the three odds multipliers. In arranging your resampling, draw from eight populations each containing varying frequencies of just two values, Gave Help, and Refused Help.

Table 23.9: *Helpfulness of English Pedestrians*

	Pedestrians			
	Male		Female	
	Gave Help	Refused	Gave Help	Refused
Requesters				
English Male	20	4	21	5
Female	23	0	24	3
Asian Male	9	15	21	5
Female	25	2	17	11

Table 23.10: *Heart Disease and Snoring in a Random Sample of Men*

	Snoring History			
	Nonsnorer	Occasional	Most Nights	Every Night
Heart Disease	24	35	21	30
No Heart Disease	1355	603	192	224

3. The data in Table 23.10 are from Hand et al. (1994, Data Set 24) and come from a large, randomly sampled survey. The data extracted here were chosen to investigate whether snoring was related to heart disease. Define the

population parameter of interest, θ, as the population version of the Goodman-Kruskal gamma. Estimate a CI for θ based on the plug-in estimator. How strong is the evidence that snoring and heart disease are related in the population?

Postscript

Generality, Causality, and Stability

An article in a recent issue of The American Statistician (Ludbrook and Dudley, 1998) reports that a survey of "252 prospective, comparative studies reported in five, frequently cited biomedical journals revealed that experimental groups were constructed by randomization in 96% of the cases and by random sampling in only 4%." However, Ludbrook and Dudley report that in 84% of the randomization studies, treatment group comparisons were carried out using t- or F-tests, valid only for random samples.

The authors of the article, medical researchers, note that these results draw "attention to a serious misunderstanding between statisticians (especially teachers and consultants) and biomedical scientists who employ statistical analyses. ... Statisticians appear to believe that biomedical researchers do most experiments by taking random samples, and therefore recommend statistical procedures that are valid under the population model of inference. In fact, randomization of a nonrandom sample, not random sampling is more usual ... and the statistical procedures best adapted to this model are those based on permutation."

The inappropriate overreliance on population inference certainly is not limited to the analysis of biomedical data. It is common in the behavioral sciences as well. The cause might not be so much a misunderstanding between statistician and researcher as a historic practical consideration. Until quite recently, the computational resources necessary for easy, routine use of rerandomization and other resampling approaches have not been available to researchers. As a result, statistician and researcher have had little choice but to push on with the most familiar tools at hand, those of parametric population inference.

By now, you will have learned that we need no longer be hindered in carrying out more appropriate and realistic analyses of our data. The thesis of this text is that the analysis of experimental data should be faithful to the design of the research study. We now have the tools that allow us to pursue that goal.

Study Design and Resampling

In Concepts 7 through 16, we developed three resampling approaches to statistical inference. Our goal was to overcome the limitations of classical parametric inference and provide for more realistic data analyses.

Each approach was directed at a particular class of research design. Realistic data analysis requires that the statistical inference methodology used in an analysis be appropriate to the research design from which the data were obtained.

Figure 6.1 is reproduced here to remind us that the major considerations in selecting the appropriate mode of statistical inference—and, hence, a resampling approach to statistical inference—are whether cases have been randomly sampled and, if not, whether the available cases were randomly assigned to different levels of treatment.

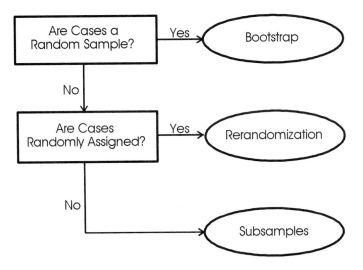

Figure 6.1: *Resampling Strategies and Random Designs*

Random Samples: Bootstrap Inference

Where we have a random sample of cases from a population, our goal in analysis is to generalize from the sample, to draw inferences about the population sampled.

1. We use the idea of the sampling distribution for an estimator, t, of some population characteristic or parameter, θ, to assess, through standard error (SE) and confidence interval (CI) estimates, how accurately we have estimated the parameter.

2. We can use either our CI for θ or the null sampling distribution for the estimator t to test a statistical hypothesis about the value of θ, that is, that $\theta = \theta_0$, or to compute a p-value assessing the strength of support in the sample data for that hypothesis.

We cannot draw additional random samples, x, from the real-world population distribution X as needed to form the sampling distribution of t. Instead, we must create a numeric estimate, $\widehat{X}$, of the population distribution from which we can resample. We focus in this text on nonparametric population distribution estimates, built from copies of the sample distribution. Resamples from the estimated population distribution, x^*, are called bootstrap samples, and the distribution of the estimator computed over these resamples, t^*, is called the bootstrap sampling distribution.

1. The standard deviation of the bootstrap sampling distribution of t^* is a useful estimate of the SE of t.

2. The first of two improved CI estimates for θ based on a bootstrap sampling distribution is the bootstrap-t CI. This is our method of choice when θ is a location parameter (such as a mean, median, or quantile), and we know how to compute both t^* and a SE estimate for t^* from each bootstrap sample.

3. The second of our improved CI estimates, the bias-corrected and accelerated (BCA) CI, is quite generally applicable, and we use it whenever we cannot estimate bootstrap-t confidence bounds.

Traditionally, statistical population hypothesis testing led either to a rejection of the null hypothesis or to a failure to reject. It is now more common to report a p-value, the proportion of the null sampling distribution of the estimator that is at least as far removed from θ_0 in the direction of the alternative hypothesis as the observed t.

We have introduced two ways of using a bootstrap sampling distribution in population hypothesis testing.

1. If we can compute both the estimator and an estimate of its SE from each bootstrap sample, then we can use the notion of an approximately pivotal transformation to test the hypothesis that $\theta = \theta_0$. We refer the value of the Studentization of our plug-in estimator t,

$$t(t) = \frac{t - \theta_0}{\widehat{SE}(t)}$$

to the bootstrap sampling distribution of

$$t(t_b^*) = \frac{t_b^* - t}{\widehat{SE}(t_b^*)}$$

We have $p = F_{t(t^*)}[t(t)]$ or, $p = 1 - F_{t(t^*)}[t(t)]$, depending on the direction of the alternative hypothesis, as the p-value for the test, and we can reject the null hypothesis if this p is small enough.

2.   Where we do not know an approximately pivotal form for our estimator, we obtain the BCA CI for the appropriate level of confidence and reject the null hypothesis if θ_0 lies below or, depending on the direction of the alternative hypothesis, above the interval. Alternatively, we can assign a p-value to our test by finding the confidence level of the BCA interval that is bounded by θ_0.

The bootstrap approach to statistical population inference allows us to use sampling distributions that are faithful to the following:

1. The size and nature of our sample

2. The size of our population

3. Our knowledge about that population

4. Our choice of estimator or test statistic

Random Assignment: Rerandomization

Many experiments call for dividing a set of cases among two or more treatment groups. There are advantages, from the standpoint of scientific and statistical inference, to making this division on a random basis. Scientifically, the randomization of cases permits us to conclude that differences in response to treatment are caused by differences in treatment, rather than by other, confounding differences among cases. Statistically, the randomization provides a probabilistic basis for determining whether observed treatment differences are large enough to have a treatment rather than a chance explanation.

1. If the cases to be allocated among treatments are a random sample from a natural case population, then their random division among K treatments yields random samples from each of K prospective populations. We can then use bootstrap methods for both population and causal inferences.

2. If the cases constitute only an available set, the random assignment of cases to treatments provides the statistical basis for local causal inference.

By studying the results of the rerandomization of the same cases, under an appropriate null hypothesis, we can assess the chances of accounting for an observed difference in response to treatment solely on the basis of the randomization of cases. Where a result is highly unlikely to be due to case randomization, we can argue that the treatments themselves are responsible for or are causative of the treatment response differences.

In Concepts 12 through 15, we developed rerandomization analyses for three classes of studies:

1. Randomization of cases to treatment groups. Each case receives one of the treatments to be compared.

2. Randomization of cases to treatment-sequence groups. Cases receive all treatments to be compared, but in a randomly determined sequence.

3. Randomization to points of intervention. At the randomized point of intervention, the case is changed from one treatment to another.

Nonrandom Designs: Subsampling

When the cases in a study are neither a random sample nor randomly allocated among treatments, any statistical inferences to be drawn can be neither generalizable to a population of cases nor can they have causal implications. Each of these inferences depends on a random mechanism in the study design.

We still have an interest in summarizing or describing the results of such studies. Such descriptions can be regarded as, in a limited sense, inferences. They are inferences about the available cases, so they are local.

For the local description to be useful to us scientifically, we need assurance that it truly characterizes the available cases. In particular, we ask about the stability of the description. If the cases are homogeneous, the description should be a stable one, not changing markedly with the withdrawal of some cases. Lack of homogeneity casts doubt on the stability of the description.

Our approach to assessing stability is to recompute the description on a sequence of subsamples. Each subsample contains some but not all the available cases. In Concepts 16, we described two approaches to subsampling.

1.   Source-structured data sets: If there are G sources, we create G subsamples by eliminating each source in turn. Does eliminating one source have a large effect on the description? If so, the overall description is unstable, it depends too heavily on the one source.

2. Unstructured data sets: If data sets are unstructured we create subsamples by randomly eliminating half of the cases. The descriptions of randomly chosen half-samples should fluctuate about, but be consistent with the full data set description.

Realistic Data Analysis through Resampling

The goal of resampling inference is to facilitate realistic data analyses. We can let the analysis follow not only the design of the study, but be responsive to the quantity and precision of the data. We are freed from making unwarranted and, frequently, quite unrealistic assumptions about the source of our data. We can choose as our statistics, parameters, and data set descriptors quantities that are of substantive interest to us, rather than those known to have nice mathematical properties.

Resampling Tools

In the applications, we have used three statistical computing packages, Resampling Stats, S-Plus, and SC. The first, as its name implies, is resampling-oriented, but has limited statistical capabilities. The second and third are general statistical packages with wide statistical repertoires linked to their resampling capabilities through powerful programming languages. With only a little practice, any resampling analysis can be carried out using one of these computing packages.

These are not the only packages that support resampling inference. Bootstrap resampling can be carried out by major statistical packages such as MINITAB, SPSS, SAS, and STATA, and subsampling and Monte Carlo approximations to rerandomization or permutation tests can be created using the programming languages in those programs. Specialist resampling programs such as StatXact (Mehta and Patel, 1995), BOJA (Dalgleish, 1995), RT (Manly, 1996), and SIMSTAT (Péladeau, 1994) are available as well. The list of computing tools will expand as resampling grows more popular.

References

Agresti, A. (1990) *Categorical Data Analysis.* New York: Wiley.

Agresti, A. (1996) *Introduction to Categorical Data Analysis.* New York: Wiley.

Albert, A., & E. K. Harris (1987) *Multivariate Interpretation of Clinical Laboratory Data.* New York: Marcel Dekker.

Balding, J., & C. Shelley (1993) *Very Young People in 1991-2.* Exeter, UK: Schools Health Education Unit, University of Exeter.

Basak, I., W. R. Balch, & P. Basak (1992) Skewness: Asymptotic critical values for a test related to Pearson's measure. *Journal of Applied Statistics,* **19**, 479–487.

Booth, J. G., R. W. Butler, & P. Hall (1994) Bootstrap methods for finite populations. *Journal of the American Statistical Association,* **89**, 1282–1289.

Box, G. E. P., W. G. Hunter, & J. S. Hunter (1978) *Statistics for Experimenters: An Introduction to Design, Data Analysis, and Model Building.* New York: Wiley.

Brown, S. R., & L. E. Melamed (1993) Experimental design and analysis. In Lewis-Beck, M. S. (ed.) *Experimental Design and Methods.* Thousand Oaks, CA: Sage.

Bruce, P., J. Simon, & T. Oswald (1995) *Resampling Stats: User's Guide.* Arlington, VA: Resampling Stats, Inc.

Cialdini, R. R., & D. A. Schroeder (1976) Increasing compliance by legitimizing paltry contributions: When even a penny helps. *Journal of Personality and Social Psychology,* **34**, 599–604.

Dalgliesh, L. I. (1995) Software review: Bootstrapping and jackknifing with BOJA. *Statistics and Computing,* **5**, 165–174.

Daly, F., D. J. Hand, M. C. Jones, A. D. Lunn, & K. J. McConway (1995) *Elements of Statistics.* Wokingham, UK: Addison-Wesley.

Davison, A. C., & D. V. Hinkley (1997) *Bootstrap Methods and Their Application.* Cambridge: Cambridge University Press.

Draper, N. R., & H. Smith (1981) *Applied Regression Analysis.* (2d. Ed.) New York: Wiley.

Dusoir, A. E. (1997) *SC: Statistical Calculator, v1.56.* Belfast, NI: Mole Software.

Edgington, E. S. (1995) *Randomization Tests.* (3d. Ed.) New York: Marcel Dekker.

Efron, B. (1979) Bootstrap methods: Another look at the jackknife. *Annals of Statistics,* **7,** 1–26.

Efron, B., & R. J. Tibshirani (1993) *An Introduction to the Bootstrap.* New York: Chapman & Hall.

Forthofer, R. N., & R. G. Lehnen (1983) *Public Program Analysis: A New Categorical Data Approach.* Belmont, CA: Lifetime Learning.

Good, P. (1994) *Permutation Tests.* New York: Springer-Verlag.

Hand, D. J., F. Daly, A. D. Lunn, K. J. McConway, & E. Ostrowski (1994) *A Handbook of Small Data Sets.* London: Chapman & Hall.

Kleinbaum, D. G., L. L. Kupper, & K. E. Muller (1988) *Applied Regression Analysis and Other Multivariable Methods.* Boston: PWS-Kent.

Koopmans, L. H. (1987) *Introduction to Contemporary Statistical Methods.* (2d. Ed.) Boston: Duxbury.

Leach, C. (1991) Nonparametric methods for complex data sets. In Lovie, P., & A. D. Lovie (eds.) (1991) *New Developments in Statistics for Psychology and the Social Sciences.* London: British Psychological Society.

Loftus, G. R., & E. F. Loftus (1988) *Essence of Statistics.* New York: Knopf.

Lovie, P. (1991) Regression diagnostics: A rough guide to safer regression. In Lovie, P., & A. D. Lovie (eds.) (1991) *New Developments in Statistics for Psychology and the Social Sciences.* London: British Psychological Society.

Ludbrook, J., & H. Dudley (1998) Why permutation tests are superior to *t* and *F* tests in medical research. *American Statistician,* **52,** 127–132.

Lunneborg, C. E. (1994) *Modeling Experimental and Observational Data.* Belmont, CA: Duxbury.

Lunneborg, P. W. (1999) *The Chosen Lives of Childfree Men.* Westport, CT: Greenwood.

Manly, B. F. J. (1991) *Randomization and Monte Carlo Methods in Biology.* London: Chapman & Hall.

Manly, B. F. J. (1996) *RT, a Program for Randomization Testing.* (Vers 2.0) Dunedin, NZ: Centre for Applications of Statistics and Mathematics, University of Otago.

Manly, B. F. J. (1997) *Randomization, Bootstrap and Monte Carlo Methods in Biology.* (2d. Ed.) London: Chapman & Hall.

MathSoft (1997) *S-Plus User's Guide, Version 4.0.* Seattle: Data Analysis Products Division, MathSoft, Inc.

Mead, R. (1988) *The Design of Experiments: Statistical Principles for Practical Applications.* Cambridge: Cambridge University Press.

Mead, R., R. N. Curnow, & A. M. Hasted (1993) *Statistical Methods in Agriculture and Experimental Biology.* (2d. Ed.) London: Chapman & Hall.

Mehta, C. R., & N. R. Patel (1995) *StatXact 3 for Windows.* Cambridge, MA: Cytel Software Corporation.

Noreen, E. W. (1989) *Computer Intensive Methods for Testing Hypotheses: An Introduction.* New York: Wiley.

Péladeau, N. (1994) *SIMSTAT User's Guide.* Montreal: Provalis Research.

Ramsey, F. L., & D. W. Schaefer (1997) *The Statistical Sleuth.* Belmont, CA: Duxbury.

Regan, P., M. Williams, & S. Sparling (1977) Voluntary expiation of guilt: A field experiment. *Journal of Personality and Social Psychology,* **24**, 42–45.

Risebrough, R. W. (1972) Effects of environmental pollutants upon animals other than man. *Proceedings of the 6th Berkeley Symposium on Mathematics and Statistics.* p. 443–463. Berkeley: University of California Press.

Robertson, C. C. (1991) Computationally intensive statistics. In Lovie, P., & A. D. Lovie (eds.) (1991) *New Developments in Statistics for Psychology and the Social Sciences.* London: British Psychological Society.

Senn, S. (1993) *Cross-Over Trials in Clinical Research.* Chichester, UK: Wiley.

Shao, J., & D. Tu (1995) *The Jackknife and Bootstrap.* New York: Springer-Verlag.

Spence, A. D., G. J. Williams, & R. Costello (1987) Prenatal experience with low-frequency maternal-voice sounds influences neonatal perception of maternal voice samples. *Infant Behavior and Development*, **10**, 133–142.

Sprent, P. (1998) *Data Driven Statistical Methods*. London: Chapman & Hall.

Tibshirani, R. (1988) Variance stabilization and the bootstrap. *Biometrika*, **75**, 433–444.

Upton, G. J. G. (1986) Cross-classified data. In Lovie, A. D. (ed.) (1986) *New Developments in Statistics for Psychology and the Social Sciences*. London: British Psychological Society.

Westfall, P. H., & S. S. Young (1993) *Resampling-Based Multiple Testing*. New York: Wiley.

Wilcox, R. R. (1997) *Introduction to Robust Estimation and Hypothesis Testing*. San Diego: Academic.

Woodward, J. A., D. G. Bonet, & M-L. Brecht (1990) *Introduction to Linear Models and Experimental Design*. San Diego: Academic.

Index